Port Management and Operations

Port Management and Operations

Maria G. Burns

CRC Press
Taylor & Francis Group
Boca Raton   London   New York

CRC Press is an imprint of the
Taylor & Francis Group, an **informa** business

CRC Press
Taylor & Francis Group
6000 Broken Sound Parkway NW, Suite 300
Boca Raton, FL 33487-2742

© 2015 by Taylor & Francis Group, LLC
CRC Press is an imprint of Taylor & Francis Group, an Informa business

No claim to original U.S. Government works

Printed on acid-free paper by CPI Group (UK) Ltd, Croydon, CR0 4YY
Version Date: 20140808

International Standard Book Number-13: 978-1-4822-0675-3 (Hardback)

Library of Congress Cataloging-in-Publication Data

Burns, Maria G., author.
　　Port management and operations / Maria G. Burns.
　　　　pages cm
　　Includes bibliographical references and index.
　　ISBN 978-1-4822-0675-3 (hardcover : alk. paper) 1. Harbors--Management. 2. Harbors--Economic aspects. 3. Harbors--Finance. I. Title.

HE551.B87 2014
387.1068--dc23　　　　　　　　　　　　　　　　　　　　　　　　　　　　　　　　2014016455

Visit the Taylor & Francis Web site at
http://www.taylorandfrancis.com

and the CRC Press Web site at
http://www.crcpress.com

To my beloved husband, Leonard

Contents

Preface

This book was written with the purpose of redefining the strategic role of global seaports in the present "Post-New Economy Era."

Ports are these remarkable human constructions that over centuries reflect the epitome of global evolution, economic growth, and innovation. As 70.8% of the global surface is covered by water, seaports reflect all sovereign nations' political superiority and financial prosperity. Ports are the pillars of global economy, trade, and transport: 80% of global commodities are carried by water; over 9000 seaports, harbors and inland waterways, and multiple terminals per port facilitate world trade by serving over 50,000 oceangoing ships while generating over 30% of the global GDP on an annual basis.

Historically, the rise and fall of empires has been associated with seaports, either through naval battles at times of war or through sea trade and transport at times of peace. In fact, superpowers and robust economies show their long-standing strength and dominance through seaports.

The shipping industry has phenomenal depth, perspective, and structure, and it comprises a plethora of sciences and arts: for maritime professionals to survive in this highly competitive, rapidly changing environment, they need to possess both practical and theoretical knowledge of as many disciplines as possible, including strategic thinking; global economics; political science; laws and regulations on safety, security, the environment, and so on; trade agreements among countries; contracts; naval architecture; novel ship designs; emerging technologies; engineering; navigation; marketing; risk management; emergency response; incident investigation and root cause analysis; oceanography and weather studies; operations; bunkering; the energy markets; major global commodities; logistics; and so much more.

Today, the role of modern ports and sea trade is more crucial than ever. The power of global key players has never before shifted in such an unpredicted manner, and the necessity for innovation, energy efficiency, and economy efficiency probably has never before been more compelling.

This is a critical era of wealth distribution among nations: global economies still struggle to overcome the 2008 global financial meltdown, while sovereign nations are now classified into "budget deficit nations" versus "budget surplus nations." The years to come will be characterized by intense global competition among the developed and emerging markets whose effects will affect the Eastern and Western Hemispheres. The Western world has wrongly assumed that the Asian economy will either prevail through Westernization or collapse. In the following years, the Western Hemisphere will observe Asia's progress without necessarily assuming a Western cultural or philosophical stance. For example, China's 12th five-year plan (2011–2015) will see Hong Kong as a leading financial, stock-exchange, and trade center, with significant impacts in the Western financial and commodity markets. Most important, a severe currency crisis may seek to redefine the global currency standard.

As the global sea trade will multiply in volume, not all commodity prices will increase. This new era may generate a new necessity of government protectionism, port specialization, and redistribution of power. The industry will be governed by stricter regulations in terms of security, safety, and the environment, with significant commercial and financial consequences to ports and ships alike.

While some global mega-ports will become strategic hub centers to distribute significant volumes of cargoes to the hinterland markets, the majority of seaports will serve as feeder ports. Despite the industry's need for innovation and because of the global system's powerlessness to protect original ideas and discoveries through copyright and patents, novelty may not be encouraged or rewarded financially at a personal or corporate level. On the other hand, powerful economies may be established through cost-efficient factors of production, with the elements of time, safety, security, and product integrity being decisive marketing factors. The new era will signify new trade routes and new strategic ports, determined by outsourcing, insourcing, and global production/consumption distribution patterns.

Economics is a major issue for ports and the shipping industry: nations will have to achieve political stability through overcoming financial obstacles such as (i) monetary deficits, (ii) national and private debt, (iii) interest rates, (iv) inflation, and (v) currency fluctuations and exchange rates. The ongoing currency wars will need to be addressed, as the profitability of any and all trade and transport contracts is determined by the currency stipulations. The rising price of gold and other precious metals, as well as oil price versus natural gas/LNG price, will determine the future commodities markets.

This book examines the ways in which global seaports will be affected by all the changes occurring at a national, regional, and global level. For the sake of good order, this wide spectrum of interrelated port management principles, strategies, and activities is classified in a logical sequence and under four cornerstones: (1) Port Strategy and Structure, (2) Legal and Regulatory Framework, (3) Input: Factors of Production, and (4) Output and Economic Framework. These four pillars are subdivided into the 12 book chapters as illustrated in Figure P.1:

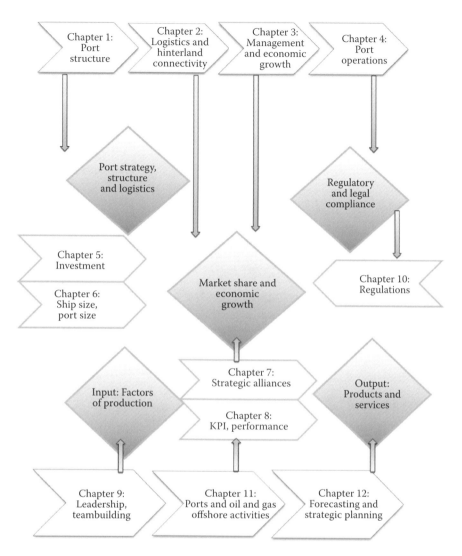

FIGURE P.1 The interrelated port management principles, strategies, and activities as explored in this book.

Maria G. Burns
Managing Director,
Center for Logistics and Transportation Policy,
University of Houston

Foreword

It is with great pleasure that through this Foreword, I introduce the readers to this book authored by Prof. Maria Burns, which maps out the multidisciplinary practices of port management and operations, as well as the distinctive contribution of the shipping industry to the global economy, through trade and development.

As I read the book chapters, I cannot help but reflect on a Greek saying I heard in the Dodecanese: "The sea doesn't separate us, it unites us" («Θάλασσα δεν μας χωρίζει, μας ενώνει»). The reason for saying this is that ships with no ports are useless, but also ports without ships are equally useless, as ships carry almost 90% of world trade. With this in mind, it underlines, in my opinion, the necessity of adequate and correctly educated, certified, motivated, and happy seafarers and port personnel. My late twin brother—a qualified captain—when he was honored by the Greek state some months before dying, couldn't make his prepared speech. He stood up and said only two lines "A ship without a competent crew is nothing, but competent crew without a seaworthy ship is also nothing," thus showing the synergistic correlation between ship and crew.

As I recall the important milestones proclaimed by the modern shipping industry, I am profoundly pleased with modern shipping and its future: Contemporary seaports are involved in myriad ambitious undertakings, all of which require technological innovation, reinventing and reallocating the factors of production, but most importantly a new thought process—all of which are offered in this book. Today, in the aftermath of the 2008 global economic crisis, nations' future will be won or lost in their geography and the efficiency of their ports. Napoleon Bonaparte remarked that "Geography is destiny." Indeed, throughout the history of mankind, nations sought to extend their commercial and political authority by taking advantage of their geographic particularities.

I am a firm believer in Maria's compelling way of writing a maritime book, which combines the most recent port developments and state-of-the art technologies with the traditional maritime practices and seafaring concepts that go back in the millennia. The book is very impressive in terms of analysis, while demonstrating in-depth research, covering a plethora of disciplines. This is no surprise to me, knowing both her Greek lineage and her professional background.

To get into the realm of this book is to experience an aspiring terrain where resourceful ideas and concepts await for you, vibrant with passion for the sea, ports, and ships.

Burns is exhilarating as she explores global ports in a panoramic view, covering the commercial, financial, logistics, operational, technical, regulatory, and legal aspects of port activities, while offering practical advice that heightens the awareness of modern shipping and modern port management.

The outstanding contribution of this book is to reveal the role of seaports as a critical component of modern supply chains, enhanced with guidance into the significance of strategic and tactical port planning, modern maritime professions, working practices, and trends that are frequently only vaguely understood, if not completely overlooked.

The readers will discover in this book a considerable variety of usefulness, encompassing port professionals, ashore executives, and shipboard officers and crew. Burns is a connoisseur of the variety of principles and methods that constitute shipping practice. The book chapters and well-researched material have been organized in a sequence and manner so that the components make up a meaningful entity.

One of the remarkable aspects of Prof. Maria Burns is that she uses her enormous energy in order to develop valuable guidelines for the maritime industry, while combining creative thinking with down-to-earth perception. It is mainly through authorship and research that scientists build their reputation, and only when their views are widely accepted, their contribution might someday become legacy. The supreme ambition of an inventive maritime professional is to perform the type of work that will be both useful and acclaimed by fellow professionals most competent to evaluate its value. In the maritime industry, empirical studies and scientific research are well regarded to the extent that the industry may frequently refer to it to progress and grow.

Burns' book captures the readers' mind, soul, and intellect; it is a well-written expository book of port management and shipping practices, which is truly accessible to everyone.

Nicky Pappadakis
President Emeritus, Intercargo

NICKY (NICHOLAS) PAPPADAKIS
President Emeritus, Intercargo

Chairman of the Malta International Shipping Council (Shipowners' Association under Malta flag), ex-Chairman of the Greek Committee of RINA, immediate past Chairman of INTERCARGO and present Chairman Emeritus, immediate past President of the US Propeller Club International Port of Piraeus, member of the current Board of Governors, Vice President of The Hellenic Chinese Chamber of Commerce, and a former member of the Board of Directors of the Union of Greek Ship-Owners.

He is a member of the Board of Directors of the Maritime Authority of the Cayman Islands, Hellenic Committee of Lloyds Register, Hellenic Committee of Germanischer Lloyd, Hellenic Committee of American Bureau of Shipping, Hellenic Committee of Det Norske Veritas, Mediterranean Committee of China Classification Society, The London Steam-Ship Owners' P&I Club Committee, and HELMEPA & INTERMEPA (Hellenic Maritime Protection Association).

He is a Life Member of NAMEPA and has an ongoing deep concern and commitment for the sea, seafarers, and environment.

Foreword by Panama Canal Authorities

Professor Burns has meticulously authored a book with knowledge and enthusiasm. She efficiently takes the pulse of the supply chain, consisting of canals, ports, shipowners, and the global logistics networks, in an effort to identify the current and future trends.

While reading this book, and its focus on the strategic significance of global seaports, I reflected on the history of the Panama Canal, from the early explorers of the Americas, to October 10, 1913, when the waters of the Atlantic and Pacific oceans first met. This is when US President Woodrow Wilson relayed a telegraph to set off the ignition of 8 tons of dynamite, which created the first version of the Panama Canal. The peoples' ecstatic cheers and cries of 1913 were followed by intelligent strategies, work ethics, and tireless, disciplined work on behalf of the Panama Canal organization for the next 100 years, which has led the Panama Canal and the Panama Flag to take a prominent role as a key player in the twenty-first century shipping industry and exceeded the success of any other manmade Canal the world over.

A new era has commenced for the maritime industry in anticipation of the inauguration of the Panama Canal expansion in 2015. The state-of-the-art "Third Set of Locks Project" is designed to increase Panama Canal's capacity by twofold, via building an additional navigational lane and thus facilitating global trade in this strategic geopolitical region, by significantly increasing economies of scale. The project includes (i) constructing two new locks, on the Atlantic and Pacific sides, respectively, while dredging new channels for each; every lock will consist of three compartments with water-saving basins; (ii) broadening and deepening of the existing channels; (iii) excavating a new Pacific Access Channel with post-Panamax dimensions; and (iv) increasing the existing maximum operating level of Gatun Lake.

Burns has a scintillating rapport with her readers, as she manages to bring together a plethora of disciplines, sciences, and concepts, and explain the reason for the industry's developments over different time periods. Under the principle that history repeats itself, her comprehensive examination of past trends can be used by the readers as a useful tool to speculate potential future developments. This is an eye-opening book, rich in content and quality that should not be missed! I know this work will become a treasure for anyone involved in the maritime field anywhere in the world. Kudos for a great accomplishment!

Yira A. Flores Naylor
Communications and Historic Documentation Section
Panama Canal Authorities

Acknowledgments

I would like to express my immense gratitude to the distinguished individuals and organizations who have offered their invaluable support and contribution.

First and foremost, I would like to thank my beloved husband, Leonard T. Burns, with deepest love, respect, and appreciation. Also, to our parents George and Athanasia, and Lawrence and Frances, and extended family, with all my love.

A very special thanks goes to my publishers, CRC Press/Auerbach/Chapman & Hall/Productivity Press, Taylor & Francis Group, an Informa Business, with their most efficient team:

To Mark Listewnik and Jennifer Abbott, my multitalented Senior Editors, I wish to express my deepest appreciation and gratefulness for the most productive support and feedback throughout this exciting voyage from authorship to publishing. Thank you for everything.

To Jennifer Stair, Project Editor; Stephanie Morkert, Project Coordinator, and Amor Nanas, Project Manager. Thank you for your creativity, diligence, and tireless efforts. Thank you to the entire editorial team, the illustrators, graphic designers, and book cover designers. Your contribution has been tremendous.

Thank you to the University of Houston: Dean William Fitzgibbon, College of Technology; Dr. Ray Cline Jr., Department Chair, Information and Logistics Technology; Prof. Daniel Cassler, Assistant Chair, Information and Logistics Technology; Dr. Mary Ann Ottinger, Vice Chancellor for Research; and all my wonderful colleagues at the UH. Working with you is a privilege.

It is a great honor to host two distinguished individuals in the global maritime industry that generously forworded this book:

Nicky Pappadakis (President Emeritus, Intercargo; Chairman of the Malta International Shipping Council, and one of the wisest—and smartest—maritime leaders of our times.

Yira Flores Naylor, Communication and Historic Documentation Department, Panama Canal Authority. Thank you for generously and zealously sharing with our readers valuable data on the historic timeline and innovative expansion of the Panama Canal.

Thank you for making this book possible with your professionalism, leadership, and integrity.

My lifelong appreciation and gratitude goes to the US Coast Guard for conferring my Honorary Membership to the US Coast Guard Auxiliary. Serving the purposes of the US Coast Guard Auxiliary is a most noble cause. My deepest respect and appreciation also go to the National Maritime Center, Washington DC, and also to the US Coast Guard Auxiliary.

I will forever be thankful to my MENTORS:

- Ceres Hellenic and the Livanos Family: Shipowners George P. Livanos, Fotini Livanos and Peter G. Livanos; Capt. Sotiris Shinas of Ceres/Euronav, Capt. Nicholas and Loukia Tsarouhas; and Dimitri and Stella Tsakos.
- American Bureau of Shipping: my lifelong partner Leonard T. Burns (Manager, Corporate Energy Project Development) and Vangelis Papastathis (Principal Engineer).
- Prime Marine and Shipowners George Kouleris & Stathis Topouzoglou. Thank you—you have set the standards for me!

It is an honor to host in this book the distinguished port authorities, corporations, associations, and their most capable "Corporate Ambassadors" who greatly enhanced this book with primary data, images, and interviews, all of which are duly referenced. I hereby wish to thank each and every professional who generously shared information on their corporate achievements and contributed to this publication.

Thank you to:

1. American Association of Port Authorities: Kurt J. Nagle, President and CEO; Aaron Ellis, Public Affairs Director; and Dr. Rexford B. Sherman, Director of Research and Information Services.
2. A.P. Moller–Maersk Group of Companies: Timothy Simpson, Director of Marketing; and Morten Andersen, Director of Category Management.
3. Baltic and International Maritime Council (BIMCO): Anna Wollin Ellevsen, Legal and Contractual Affairs Officer.
4. British and Irish Legal Information Institute (BAILII).
5. British Financial Conduct Authority: Chris Hamilton, Press Office.
6. Federation of National Associations of Shipbrokers and Agents (FONASBA).
7. The Baltic Exchange: Jonathan C. Williams FICS, General Manager.
8. Gulf Winds: Steve Stewart, Chairman; and Todd Stewart, President.
9. Harvey Gulf: Shane J. Guidry, Chairman and CEO; Chad Verret, Senior Vice President Alaska & LNG Operations; and Michael Caroll, Senior Vice President New Construction and Chief Naval Architect.
10. Louisiana Offshore Oil Port (LOOP), Louisiana, USA.
11. U. S. Department of Transportation, Paul N. Jaenichen, Acting Maritime Administrator (MARAD), Washington, DC.
12. Marine Energy Pembrokeshire, England: David Jones, Project Manager.
13. NYK Group, Nippon Yusen Kabushiki Kaisha, and their most generous, efficient, and capable leaders.
14. Odfjell Group: Capt. George M. Pontikos, Vice President Port Operations; and Mrs. Margrethe Gudbrandsen, Communication Manager.
15. Panama Canal Authorities: Yira A. Flores Naylor, Comunicación y Documentación Histórica, Programa de Ampliación del Canal.
16. Port Anchorage, Alaska.
17. Port Fourchon, Greater Lafourche Port Commission, Louisiana.
18. Port Freeport, Texas, USA: James Nash, Business Development.
19. Port of Antwerp, Belgium.

20. Port of Haifa, Israel: Mendi Zaltzman, CEO; and Zohar Rom, PR Executive and Spokesman.
21. Port of Hong Kong, China.
22. Port of Houston Authorities, Texas, USA.
23. Port of Pembroke, UK.
24. The National Archives, England: Judy Nokes, Information Policy Adviser.
25. Tidal Energy Ltd.: Martin Murphy, Managing Director; and Rebecca Jones, Marketing Manager and PA to MA.
26. Wärtsilä: Capt. Paul Glandt, Director, Ship Power Business Development for North and South America, and USN (Retired).
27. West Gulf Maritime Association, USA. In particular I wish to thank Niels Aalund, Vice President of WGMA, who is a dynamic change-agent in the US Maritime industry, Economy and the Environment.
28. The Houston Maritime Museum and its dynamic leadership: our extraordinary Board Members; Leslie Bowlin, Interim Director; Kristin Josvoll, Director of Operations; Lucia Cerritos, Collections Manager, as well as our Sponsors, Volunteers, Students and Supporters. Your contribution to the maritime, oil & gas industry is invaluable.

Please accept my sincere thanks for bringing life to the theories and practices of Port Management and Operations, with your most impressive innovations and meaningful accomplishments.

Last but not least, thank you to each and every one of my global colleagues, research partners, mentors and students! We are all part of this exhilarating voyage to global growth, innovation and prosperity!!!

Prof. Maria G. Burns

CHAPTER 1

Introduction

1.1 PORT MANAGEMENT AND OPERATIONS: STRATEGY IN THE THROES OF A TRANSITION

In addition to being, historically, the first and primary facilitator of world trade—instigating economic activities and growth—the maritime industry may well proclaim to being incontestably the first industry that is truly global in nature. And yet, if one wishes to understand global economy and take the pulse of regional and national development, production, employment, and growth rates, one simply needs to examine how seaports work.

Napoleon Bonaparte opined that "to know a nation's geography is to know its foreign policy." When it comes to seaports, it can be inferred that to understand a nation's seaports is to perceive its underlying economic fundamentals: to forecast the commodity markets with unfailing accuracy, one simply need visit a port on a regular or seasonal basis and observe the ship types and sizes, while assessing the commodities these ships carry. To acquire an overall picture of the market cycles, one can observe a port's short-term traffic, in conjunction with the port's long-term strategy. This includes partnerships with oil majors, terminal operators, shippers, major liner companies, cruise lines, and so on. Last but not least, forecasting by extrapolation—of the real and projected jobs in different industrial segments of a particular geographic location—is possible through the examination of port authority leasing contracts, concessions, leaseholds, land rents, and so on.

Port services may encompass one or more of the four key business categories of global trade and transport:

1. Landlords, through property ownership, leasing and management.
2. Brokers, by hosting or liaising with bunker brokers commodity brokers, property brokers, ships' agents, ship brokers, stock brokers, and so on.
3. Suppliers, through leasing and handling cargo equipment, and liaising with ship chandlers, and suppliers of spare parts, engines, commodities, tools, and so on.
4. Manufacturers, by hosting or liaising with shipyards, petroleum refineries, and industrial zones. Manufacturing also encompasses broad maritime activities including port design, engineering technology, installation and modification of equipment, manufacturing machinery, and so on.

As fundamental logistic and financial portals, a seaport's efficiency is crucial to ensure the safe, secure, productive, and ecofriendly practices of marine operations. Regardless of their size, location, and specialization, seaports are principally designed to provide shelter to oceangoing or inland ships, while effectively managing numerous dissimilar activities, human force, materials, and financial resources. Port authorities are in charge of harboring and securing ships, while ensuring smooth operations throughout ships' anchorage, pilotage, berthing/unberthing, lightering, mooring/unmooring, loading/unloading operations, and so on. They oversee canal transits and channel passages and supervise cargo movement, transferring of wet, dry, and gaseous cargoes, while handling bulk, containerized, and palletized cargoes.

Based on the above, Port Management may be defined as *the process of organizing, monitoring, and controlling the activities of a seaport in a precarious global industry, in order to accomplish corporate goals, which are in line with its regional and national interests.*

1.1.1 Port Authorities Departments and Activities

Global trade is characterized by high-risk, cutthroat competition and capital intensive activities. Hence, port authorities find it increasingly arduous to adapt to an ever-changing global landscape and antagonize with global ports in an effort to improve their annual performance and achieve sustainable import/export levels. Port authorities liaise with governments, policy makers and law enforcement agencies, shareholders, investors, banks, shipowners, ship managers, cargo forwarders, cargo receivers, classification societies, P&I clubs, underwriters, unions, flags of convenience, commodity brokers, customs brokers, ship brokers, ship agents, ship chandlers, shipyards, repair teams, surveyors, inspectors and auditors, not to mention senior maritime officers, seafarers, and stevedores (longshoremen). Port executives are in charge of purchasing land and facilities; they allocate and maintain warehouses as well as indoor and outdoor storage spaces, while recruiting and training efficient personnel. Figure 1.1 illustrates how port management within a typical supply chain is a nexus of sea trade, multimodal trade, and inland trade.

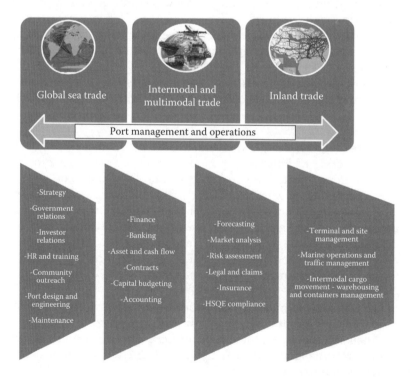

FIGURE 1.1   Port management within a global supply chain. (Courtesy of M.G. Burns.)

While ports' efficiency is typically measured in terms of time, safety, and value for money, the ultimate challenge for modern port managers is the optimum combination and usage of their factors of production, in order to serve the global supply chain. After all, the demand for seaports derives from the demand for commodities and seaborne trade.

1.1.2 Ports' Strategy in the Throes of a Transition

During the contemporary history of shipping, ports empower corporations and consumers to sell and purchase global commodities to an extent, rate, and volume that were previously considered inconceivable. Technology has immensely contributed in the way we do business.

It is imperative for port authorities to cultivate strategists capable of efficiently operating in international market platforms, while taking decisions critical to the port's future employability. Port management functions have been fundamentally reshaped over the past decades, owing to the accelerating change of maritime technology, followed by a major shift of global economic power and trade patterns:

1. *Information Technology* is an umbrella term that covers the satellite systems and software used to facilitate global communication between ports, ships, and supply chain. Ports benefited from improved communication, including the widespread use of satellite communication and Internet-based software onboard ships, enabling cargo handling, loading/unloading operations, and the remote monitoring and controlling of a ship's navigational and engineering performance. Port managers became vital emergency and rescue coordinators and recipients of ships' distress signals, through emergency location beacon devices, such as AIS (Automatic Identification System), that is, an automatic tracking system used for identifying and locating vessels; EPIRBs (Emergency Position Indicating Radio Beacons); PLBs (Personal Location Beacons); SARTs (Search and Rescue Transponders); VHF Radios; and so on.
2. *Maritime Technology* encompasses the novelties of ship design, marine engineering, ship building, and ship operations. Ports have to keep abreast with the new ship types, sizes, and designs that emerged and accommodate their clients' requirements pertaining to safe port operations and efficient cargo handling.
   a. In 1992, the MARPOL 73/78 protocol (IMO's International Convention for the Prevention of Pollution from Ships) enforced its new maritime regulations for double-hull/double-bottom ships, to ensure ship safety and environmental protection.
   b. Containerization was a great breakthrough as it increased the efficiency of high-valued break-bulk cargoes and reduced cargo loading/discharging time by 84% and costs by 35%. The first container ships were operated in 1952 in the United States and Denmark, and since then multiple design, volume, and technological advances have been gestated.

      Port terminals now required container handling facilities and equipment, and a whole new industry was reinvented, that is, container trucks, stevedore operations, minimum port stay requirements, and so on. Subsequently, ports can fully benefit from economies of scale, in terms of higher dock labor productivity per working hour, increased ship size, and reduced traffic. In 2009, over 90% of nonbulk commodities were being transported in containers.

Pursuant to this high demand and the gradual recovery from the 2008's economic crisis, new, larger, and more efficient designs have been launched. As of 2013, the largest and state-of-the-art container type is Maersk's "Triple E Class," which contains three major design advantages: "Economy of scale, Energy efficiency and Environment improvements."

c. Liquefied Natural Gas Carriers (LNG) and Liquefied Petroleum Gas Carriers (LPG).

   i. *LNG carriers* store natural gas that is transformed into a cryogenic liquid (i.e., liquefied through extremely cold temperatures). Typically, the temperature required to condense it ranges between –120°C and –170°C (between –184°F and –274°F). The first LNG carrier sailed from Louisiana, United States, to the United Kingdom in 1959. New gas deposits discovered over the past few years have changed the trade flows and have significantly increased the demand for LNG carriers.

   ii. *LPG carriers* are designed to carry liquefied petroleum gases (e.g., butane, propane, etc.) at a controlled pressure and temperature. They are categorized into three key types: fully pressurized, semipressurized and refrigerated, and fully refrigerated. During the last few years, several larger-capacity fully pressurized vessels have been constructed with spherical tanks.

   Hence, new-designs of LNG/LPG tankers are growing larger and sophisticated to accommodate larger volume of cargoes.

3. *Global Economy and Trade*: The *"New Economy"* concept introduced in the 1980s illustrated a transition from a manufacturing-based economy to a service-based economy. The shift toward a service-oriented, value-added economy led to the geographical reorganization of the supply chain. In 1991, the World Wide Web became broadly available, facilitating business transactions and enabling instant global communication. This breakthrough further promoted the "compulsive outsourcing" concept, in a world where geographical and trade barriers were diminished. In the 1990s, the impact on the maritime sector and global seaports was enormous and far-reaching; a reshuffle of the deck changed the major trade routes with a shift from the West to the East. *Ten years later*, by the dawn of the 2000s, countries like China, India, and Brazil moved from closed, centrally structured systems to capital and export-oriented models. As the rapidly developing economies grew, global ports boomed in an unprecedented growth of global production and seaborne trade.

The global framework of the 1990s signified an era where global diplomatic negotiations increasingly reflected ports' negotiating strength. By 2008, the Great Recession had already cast its shadow on the monetary stability of many countries and their ports, where the external debt crisis was literally perceived by many as "complete evaporation of liquidity." Ports were heavily influenced through their countries' national exports—through diminished production levels—and imports—through diminished customers' purchase power. Again, at a time that well-established ports collectively experienced the market meltdown through dramatically reduced cargo volumes, other ports remained unaffected and frequently enjoyed their precrisis growth rates (Burns-Kokkinaki 2012). According to the latest UNCTAD Global Economic Outlook, developing countries and economies in transition are anticipated to keep on feeding the engine of the global economy, growing by approximately 5.6% in 2012 and 5.9% in 2013, a pace much faster

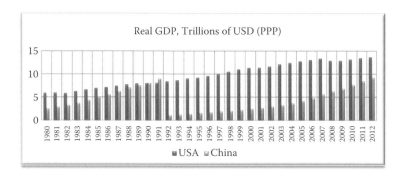

FIGURE 1.2   Real GDP, United States and China (1980–2012). (Courtesy of M.G. Burns, based on data from US Bureau of Economic Analysis [2013] and Statistical Communiqué of the People's Republic of China on the 2012 National Economic and Social Development [2013].)

compared to the advanced economies of the Western Hemisphere (UNCTAD 2012). Figure 1.2 demonstrates China's rapid expansion, as compared to the US economy, which is the world's leader for the past decades.

For purposes of discussion in the book, we will dub the *"Post-New Economy"* era as the aftermath of 2008. This is the era where ports' bargaining power was increasingly relying upon the country's trade agreements, political agenda, and status, whereas the new production map included rapidly developing economies, such as China, India, and Brazil. Ports' leveraging became critical to port strategists, yet led by the managers' persistently explored ways of emerging out of the crisis. Increased control mechanisms in port management were crucial to safeguard their competitive edge toward established partnerships, revenue channels, and competitors.

Since port competition is a by-product of global and regional trade-related competition, it was intensified by the great market boom of the late 1990s and the enormous trade volume boosted to a large extent by the major global exporters of the time, that is, United States, China, Europe, Japan, South Korea, and so on. The paradox of competition in all industries is that it seems to intensify both when the market is booming and during a market crisis. This explains the intense port competition both during the late 1990s, when global ports competed in marketing strategies and modernization to accommodate the large cargo volumes, and during the 2008–2011 global crisis, where ports still competed, in order to ensure optimum occupancy and justify the previous overambitious investments.

What is of utmost significance to modern port managers is to enhance their contract negotiating power despite the global climate of uncertainty and fragile power shifts. The 2008 crisis has encouraged the formation of mega-terminal operators and ship-owning consortiums that have gained serious bargaining power when negotiating with port authorities. While ports operate in a specific geographic area and place their investment capital into a single region, mega-terminal operators spread the risk by global investments, with offshore-type legal and taxation arrangements. As market power is directly analogous to the price elasticity of demand, port authorities at a disadvantage may be forced to lower port tariffs and leasing contract earnings.

Ports' bargaining power is diminished once they enter long-term lease concessions on Public–Private Partnership Projects. Once the lease contract has been finalized, the port will not profit by any possible future market growth. To the contrary, terminal operators will fully benefit from the port's location, regional supply chains, and consumers. When shorter-term lease concessions are agreed, port authorities have more market power. Alternatively, when longer lease arrangements are required, ports can pursue to establish partnership contracts providing for distribution of profits and losses, allocating power between managers, terminals, and shareholders. Flexible leasing agreements can be negotiated with port authorities receiving an annual base fee, plus annual adjustments based on annual performance, profit, and productivity. The terms and conditions could provide port authorities the option of an early termination of the contract, financially benefit from a sublease, and generally exert more of their market power.

Successful ports are made up of visionary leaders and *forward-thinking* professionals who have an ability to recognize the market trends and help the port in achieving its full potential. The sections to follow study the making of an effective port manager and address methods of achieving a port's goals.

At this turning point for ports and economies alike, real, tangible strategies of recovery are needed. Strength in unity is key for ports to work in alliance and be represented by trade associations that can better promote and safeguard their interests. Both global and national organizations must establish policy positions that they represent before the country's political leadership, whereas they actively collaborate with coalitions promoting the interests of the maritime, trade, and transportation industries. At a global level, the International Association of Ports and Harbors (2013, http://www.iaphworldports.org) provides a global alliance of 200 ports in 85 countries. The member ports manage over 60% of the global maritime trade and almost 80% of the global container traffic.

In the United States, the American Association of Port Authorities (AAPA) is one of the largest and the oldest trade associations, representing over 130 port authorities in the Western Hemisphere, including the United States, Canada, Latin America, and the Caribbean (American Association of Port Authorities 2013, http://www.aapa-ports.org). The AAPA's mission, scope, and usefulness in the development of American ports are outlined here.

CASE STUDY: AMERICAN ASSOCIATION OF PORT AUTHORITIES

America's seaports are gateways to the global trade and a vital element in financial prosperity and national sovereignty. Typically, the imports and exports in the ports of the 50 US States exceed $5.5 billion of value, whereas each state depends upon 13 to 15 seaports for the control of its trade.

AAPA—the alliance of major ports in the Western Hemisphere—safeguards and promotes the common pursuits of its diverse members as they are the link between their region and the global transport mechanism. AAPA offers cutting-edge leadership and strategic mentoring on port and connecting infrastructure development, operations, economics, freight transportation, security, environmental programs, and other port-related issues. In addition, AAPA actively reaches out to have conversations with the media, the public at large, policy influencers, and national, state, and local policy makers concerning the critical function of ports at a global level.

Since its inception in 1912, AAPA has established a powerful alliance of seaports in the Western Hemisphere. Some of the benefits it offers are accreditation, advocacy, and promotion of the members' best interests, networking, education, and training. According to Mr. Aaron Ellis, AAPA offers "the common bonds from sharing information and knowing you don't have to re-create the wheel at your own port or organization because someone else has paved the way for you." (*Source:* http://www.cargobusinessnews.com/news/081512/news1.html)

Today, AAPA's corporate members include more than 140 port authorities throughout the Americas, including the United States, Canada, Mexico, Latin America, and the Caribbean. AAPA also has approximately 250 sustaining members, who are individuals and businesses that provide goods and services to seaports and the port industry. AAPA's headquarters are located in Alexandria, Virginia, and directed by Kurt J. Nagle, president and CEO. The association's board of directors elects a new board chairman each year. For the 2013–2014 time frame, AAPA's chairman of the board is Tay Yoshitani, CEO for the Port of Seattle (Washington).

AAPA has specified the three critical challenges faced by modern ports:

- Economic Impact—for contemporary ports to compete in the current global trade and economy, investment is required for modern, navigable seaports with uncongested intermodal freight access.
- Security—safe and secure seaport facilities are fundamental to both protecting our borders and moving goods around the world.
- Environment—seaports working to identify solutions that enhance our coastal resources and reduce environmental impact.

By 2020, a considerable growth of seaborne trade is predicted, as a result of rising world population and disposable income, and the number of travelers moving around US ports will also increase. In order to satisfy these demands, AAPA and its members are dedicated to maintaining our ports navigable, sustainable, and secure.

Here are some important AAPA and port-related events since it was founded in 1912:

1912 December meeting in New York establishes the National Association of Port Authorities with 11 members
1914 Membership extended to ports in the Western Hemisphere, name changed to American Association of Port Authorities
1914 Opening of the Panama Canal revolutionizes shipping routes
1915 First Canadian member joins AAPA
1918 First standing committees formed
1920 First issue of the *Monthly Bulletin* is published
1921 *Monthly Bulletin* renamed *World Ports*
1921 First Latin American member joins AAPA
1930 AAPA formally incorporated in Delaware, giving it permanent legal structure
1930s AAPA efforts result in greater uniformity in port tariffs and practices and a Canon of Ethics for governing public port entities
1936 25th Annual Convention held in San Francisco

1945 *World Ports* resumes full monthly publication after severe cutbacks during WWII

1947 Association's offices established in Washington, DC

1949 Paul Amundsen becomes first full-time AAPA staff person, later becoming executive director

1949 First Caribbean member joins AAPA

1956 Advent of cargo containerization

1961 50th Annual Convention held in Long Beach

1966 Communications Awards Program initiated

1973 Environmental Improvement Awards Program initiated

1974 Richard Schultz becomes AAPA executive director

1979 J. Ron Brinson appointed AAPA executive director

1980s A full-time government relations program was established, research and membership services were greatly expanded, and a full slate of seminars became a regular part of the AAPA activity calendar

1984 AAPA headquarters relocates from Washington, DC, to Alexandria, Virginia

1985 Bylaws amended to allow AAPA delegations broader participation in governance

1986 75th Annual Convention held in Miami

1987 Erik Stromberg named AAPA president and CEO

1991 Separate association delegations created for Latin America and Caribbean ports

1991 First Communications Director hired

1995 Kurt Nagle named AAPA president and CEO

1995 Professional Port Manager (PPM) certification program begins and graduates its first candidate

1998 AAPA web site established

2000 Strategic plan approved, setting four goals of professional development and education, public awareness, relationship building, and representation and advocacy

2001 Terrorist attacks on World Trade Center and Pentagon reshape port and maritime security

2002 Maritime Transportation Security Act enacted

2002 Information Technology Awards program started

2003 Facilities Engineering Awards Program begun

2003 *Seaports Magazine* begins quarterly publication

2003 Quality Partnership Initiative with the US Army Corps of Engineers launched

2004 Port Environmental Management System (EMS) Assistance Program established

2004 "Seaports of the Hemisphere Allied in Relationships for Excellence" (SHARE) Initiative established

2004 Memoranda of Understanding signed with General Secretariat of the Organization of American States and IAPH

2004 Latin American Coordinator position established

2005 Hurricane Katrina disrupts Gulf of Mexico port operations, AAPA Port Employee Emergency Relief Fund established

2005 Seaport Security Manual in both English and Spanish developed by the Security Committee

2006 Emergency Planning and Disaster Recovery Working Group prepares an Emergency Preparedness and Continuity of Operations Planning Manual for Best Practices

2006 Latin American Professional Port Manager (PPM) certification program started

2006 AAPA Cruise Award established

2006 Port Professional Technical Assistance Program established

2007 Memorandum of Understanding signed with the Association of Canadian Port Authorities

2008 "Seaports Deliver Prosperity" awareness initiative inaugurated

2009 Memorandum of Understanding signed with the European Sea Ports Organization

2010 Professional Port Manager (PPM) certification program revised from an individual to a group structure

AAPA'S LEADERSHIP

AAPA is led by its 10-member Executive Committee and 66-member Board of Directors. AAPA has 13 technical and three policy committees, with 350 individual corporate and nearly 200 individual associate members on the committees. AAPA's leadership has established a number of events such as conferences, educational and professional development sessions, and media events, with the purpose of sharing information and connecting its members with the government authorities, policy makers and influencers, maritime professionals, community groups, schools, and so on.

Interview with Aaron Ellis

The Future of US Ports

American seaports have recently displayed a dynamic development with an estimated annual investment exceeding $9 billion, generated by port authorities and private-sector funds. Among the factors to spur this vibrant financial commitment, the most significant factors include the following:

- US economic recovery pursuant to the 2008–2009 recession has led to robust economic growth: the worth of US exports has escalated to 70.1% while that of imports increased to 53%.
- US population growth for the world's third largest population, which by 2050 is anticipated to expand by 16.5% to 362 million.
- US trade deficit narrows as exports rise. A healthier balance of trade and growing exports, primarily toward countries with escalating standards of living, in particular Latin America and Asia. By the 2020s, total US exports are expected to exceed imports for the first time in a generation.

- Near-sourcing, that is, the movement and relocation of industries from overseas, mainly because of increasing labor costs overseas, a thinning labor differential domestically, and complex, time-consuming logistic distributions to market.
- Significant infrastructure projects in the Western Hemisphere.

The Panama Canal expansion is striving to retain its competitive edge in a rapidly growing global trade growth, with growing ship sizes, vibrant trading alliances, and so on. Other premium infrastructure ventures are being initiated in Brazil, Canada, and Mexico, for example, with a wide spectrum of investment portfolios, ranging from mega-ports to power plants.

Because global trade exceeds 25% of US gross domestic product (GDP), generates over 13 million jobs, and earns over $200 million in tax revenue, there is a compelling need for the government to increase the priority for freight movement and rectify the federal underinvestment by funding better accessibility to and interconnectivity with seaports, for example, via road, rail, bridge, tunnel, and navigation infrastructure. America's growing trade volumes will soon surpass the local networks' capacity to handle this transport hyperactivity, to the detriment of growth, time efficiency, and cost efficiency. This is the time for America to boost its global competitiveness, through facilitating global trade, and this will be achieved via port and infrastructure investment.

Interview with Dr. Rexford B. Sherman, AAPA's *Director of Research and Information Services and Latin American Coordinator*

Comparison of the US Ports' Structure with Other Global Port Systems

The US port system is decentralized and far less homogenous than is typically the case for those in other countries. US public ports are created by the states and vary widely in structure—from state to municipal and county entities and semiautonomous special-purpose political subdivisions. In China, public ports are municipal entities. In northern Europe, ports are also municipal entities. In southern Europe, I believe the national governments have a stronger hand.

The US system is extremely competitive—between public ports in the same states, ports in other states, ports in other countries (Mexico, Canada, the Bahamas), and even, in some cases, between public and private ports. The networking is through state (e.g., Florida Ports Council), regional (e.g., Gulf Ports Association), and national/international (e.g., AAPA, IAPH) trade associations and FMC-sanctioning rate-discussion groups. China and Europe also have seaport associations and many also belong to the IAPH. Many US ports have sister port agreements with ports in other countries—some of them made through AAPA. Also, AAPA has cooperative agreements with IAPH and the European Sea Ports Organization and frequently hosts port delegations from overseas—especially China, but others as well. This sort of networking facilitates technical exchanges, personal relationships, and mutual support (particularly in helping AAPA members respond to crisis such as natural and man-made disasters) and in the case of US ports in dealing with federal government issues that affect all ports—trade, security, environment, channel development, and maintenance. I think the impact may be less significant on the logistical chain, because it is more dependent on port relations with carriers and shippers rather than on inter-port relationships, but lobbying through AAPA can facilitate matters by dealing with institutional and political impediments to trade.

Case Study Sources:

AAPA Interviews: **Aaron Ellis,** *Public Affairs Director*
                      Dr. Rexford B. Sherman, *Director of Research and Information*

REFERENCES

AAPA (2013).

2011 Memorandum of Intent signed with the US Department of Commerce to implement the "Partnership with America's Seaports to Further the National Export Initiative."

2011 100th Annual Convention held in Seattle.

1.2 THE HISTORY OF PORTS: ADVANCED THINKING, PLANNING, AND DEVELOPMENT

Whosoever Commands the Sea, Commands Trade; Whosoever Commands the Trade of the World, Commands the Riches of the World, and Consequently the World Itself.

Sir Walter Raleigh (1552–1618)

An effective way of understanding the future is to thoroughly examine the past. In order to tame the contemporary and future corporate challenges associated with ports, it is necessary to reflect upon the historical events that have shaped the modern concepts on successfully designing, planning, and managing seaports. While the following chapters address modern port planning, technologies, and marketing strategies, the aim of this section is to present a concise timeline of the history of ports and comprehend how learning from the past will help modern port decision makers shape a most promising future.

1.2.1 Ports' History and Etymology: A Passage, a Journey, and a Haven

This section provides historic evidence to verify how, since the dawn of mankind, global seaports served as the gates of global trade and facilitators of products exchange. The etymology of the word *port* derives from the ancient Greek *poros* (πόρος), which means both "passage" and "journey," which in turn became the Latin word *portus*, and the modern international *port*. For thousands of years, seaports have been vibrant centers of civilization involving trade and the exchange of currencies, commodities, and cultures. The world's most ancient port known as of today was recently discovered in the Egyptian coast of the Red Sea coast, 112 miles south of Suez, and dates back to 2500 BC at the time of Pharaoh Khufu (Davis 2013). The harbor's findings, carved anchors and man-made docks, verify a vivid port that served the country's exports of copper and other minerals.

Extensive archeological evidence verifies that humans built ships since at least the 11th millennium BC: "Papyrella" was a Mesolithic oar boat that dates back to 11,000 BC; it consisted of many fascicles of thin cane ("papyruses") tied together with ropes and was used for prehistoric fishing and trade, including the trade of obsidian stone (a semiprecious volcanic stone used as a cutting and piercing tool) in the Aegean sea (Hellenic Maritime Museum 2013). Numerous ancient ships and rock art depicting ships have been discovered in America, Australia, the Indus Valley, Scandinavia, the Netherlands, Nigeria, and so on. As for England, the ancient "tin islands," findings confirm that their prehistoric homes were made of whale bones, while they were actively involved in the export of tin and metals (Burns-Kokkinaki 2004).

A modern timeline of global port development encompasses four key eras and outlines a set of drivers: First, the era of national independence, whose grandeur signified the beginning of the industrial revolution and faded in the 1960s. Second, the era of containerization from the 1960s until the 1980s, a time when global trade had regained its pre-WWII level through manufacturing and trading of value-added goods. Third, the era of logistics, whose rapid expansion lasted from the 1980s to the 1990s. That era saw an intensive production of high value-added goods, while more complex production and trade networks and efficiency resulted in the most profitable ports operations. Fourth, globalization gained momentum in the 1980s and continued throughout the 1990s, by means of trade and financial growth around the globe. The author proposes a fifth era in the Post-New Economy era, signifying the aftermath of the 2008 global economic meltdown.

1.2.2 Advanced Thinking: National Ports, Diplomacy, and Economy

While examining history, the role of shipping in a country's diplomatic and economic strategy can hardly be exaggerated. To quote Sir Walter Raleigh (1552–1618), "whosoever controls sea trade, eventually controls the world itself." Four centuries later, his belief becomes more timely than ever: from the British Empire, to the American dream and the Chinese and Indian rapid development, the making of a global empire only seems feasible through controlling maritime trade, seaports, canals, and terminals.

The timeline of the customary Western history covers the premodern (pre-1500), early modern (1500–1850), modern (1850–1945), and contemporary eras (Pieterse 2012). The significance of ports was highlighted between the fifteenth and eighteenth centuries, when the economic doctrine of Mercantilism emerged in the Western Hemisphere. The theory advocated that a positive balance of trade and government control of foreign trade had to be achieved in order to safeguard the nation's sovereignty and military security. Building colonies as trading networks grew to become an appealing political strategy. Nations would retain their role of leadership and accumulate wealth as raw materials and finished goods were traded between colonies and the mother country.

The "First British Empire" era depicts the British maritime leadership that emerged in the seventeenth century, culminated in the eighteenth century, and diminished in the nineteenth century, during the Age of Revolutions. As facets of a potent British Navy, the North American colonies supplied mother England with precious metals, raw materials, and finished goods, such as ships and spare parts. The American Revolution occurred during the eighteenth century as 13 states united to break the British rule and establish their national sovereignty by growing their sea power in the Atlantic Ocean.

Essentially, this initiative altered the course of history, and in 1775, General George Washington, who was later to become the first President of the United States, privately launched a compact marine force against the mighty British sea power: *If we mean to be a commercial people, or even to be secure on our side of the Atlantic, we must endeavor as soon as possible to have a Navy.* "In the Service of the ministerial Army," General Washington directed the fleet to hunt down the globe's most indomitable naval power (Daughan 2011; Nelson 2008; Palmer 2013). This was the making of the American Empire and the American Dream, sealed by the Declaration of Independence in 1776, while laying the foundations of a mighty and resilient nation.

In the case of America, we observed how a nation's dynamic maritime activities can strengthen its economy and global role. At the same time, nations have also experienced the retarding effect of introverted economic policies that deprive a country from the pertinent advantages of sea trade.

China's history is a vivid example of the adverse results of commercial isolation, followed by a thriving economy in the twenty-first century.

Deng Xiaoping (1904–1997), China's leading economic reformer, helped the nation achieve the Four Modernizations by accomplishing specific foreign funds, management, production, and technological innovations, thus boosting its economic development. Maritime investment was endorsed through developing special trade and economic zones, where market liberalization was promoted. In his words:

> No country that wishes to become developed today can pursue closed door policies. We have tasted this bitter experience and our ancestors have tasted it. In the early Ming Dynasty in the reign of Yongle when Zheng He sailed the Western Ocean, our country was open. After Yongle died the dynasty went into decline. Counting from the middle of the Ming Dynasty to the Opium Wars, through 300 years of isolation China was made poor, and became backward and mired in darkness and ignorance. No open door is not an option (Ferguson 2012; Murphey 2007).

Modern China is a global economic and shipping magnet with an average annual growth of 9% between 2001 and 2012. Although the country has not been immune to the global shocks of 2008, it has enjoyed a tremendous growth since 2001, when it joined the World Trade Organization and liberated its stock exchange system. And yet, prior to the Industrial Revolution, China was alienated from global trade and adopted an introvert policy focusing in rice production, with severe effects on population growth, reduced earnings, health, and economic output. Among the lessons learned, what is certain is that a country that wishes to control its economy first needs to control its sea trade and global agreements.

1.2.3 Ports History, Planning, and Development

To investigate the significance of ports, we must comprehend the trade patterns throughout history. Civilizations have traditionally been using their seaports as tools for establishing diplomatic and trade relations with other nations, accumulating wealth, conquering geographic boundaries, and minimizing global distances. Through maritime history, ports and ships transform their world into a fusion of markets, commodities, and factors of production.

The ports selected in this subsection have one thing in common: they all have undergone great challenges, fluctuations, and yet they have thrived throughout with consistency and flexibility. Having to select among numerous, perhaps hundreds of ports that at some point in history achieved global leadership and recognition was not an easy task. Fortunately, there is a great number of leading ports that thrived and demonstrated exceptional achievements through genius, unity, endurance, and continuity. Many of these will be used as case studies throughout this book.

a. The Port of Houston, Texas

You'll never find oil here.

Calvin Payne
Standard Oil (Times Magazine 1901)

Houston's story is one of persistence and triumphantly overcoming adversities. In the 1820s, a new European settlement established an active shipping center (World Port Source 2013b). In 1836, the new Republic of Texas was established; the port's first dock was constructed in 1840, followed by the first railroad (1853). Subsequent to the American Civil War (1869), the Houston Ship Channel in the absence of funds for deepwater dredging focused on its regional railroad center. Since the 1890s, regional authorities and businessmen attempted to obtain funds to support the Channel's deepwater dredging activities. Their efforts were unsuccessful, until oil was discovered in 1901, in Spindletop, Texas (Museum of Houston 2013; Port of Houston Authorities 2013). Patillo Higgins, the "Prophet of Spindletop," partnered with Anthony Lucas, an inventive engineer, and advocated their salt dome theory to secure funds for drilling (World Port Source 2013b).

As the Port of Houston's cargoes shifted from timber and cotton to global oil trade, the ship channel needed deepwater dredging to accommodate larger vessels. US Representative Tom Ball, Mayor H. Baldwin Rice, and other business leaders proposed a revolutionary plan whereby the dredging cost would be equally shared between Houston and the federal government (Houston Ship Channel 50th Anniversary Collection 1926–1964). The US Congress unanimously accepted the "Houston Plan" proposal, and its concept was implemented in multiple US ports (Museum of Houston 2013). By 1902, over 500 Texas corporations relocated in Beaumont. The Port of Houston's channel was launched in 1914, and in 2014, it celebrates 100 years as a major global deepwater port (Port of Houston Authorities 2013).

Houston, Texas, is now established as the energy capital of the world, ranking first in US international commerce and has the 10th largest global port.

b. The Port of Shanghai

Shanghai in Chinese means "Climbing Above" (Shang) "The Sea" (Hai). The thousand-year-old history of Shanghai has been subject to volatile periods of decrease and abundance, thus resembling the history of China. During the Sung Dynasty (960–1279), the Administration of Mercantile Marine was established, with a toll house for merchandise. Between the tenth and thirteenth centuries, the port had gained importance as a "trading station," as indicated by its name change "ShangHai Chen" (Madrolle 1912), whereas Chen means harvester, farmer, and a gold collector (Oxford Advanced Learner's English Chinese Dictionary 2012).

During the Ming Dynasty (1368–1644), the Chinese financial system was manifesting signs of capitalism. Yet, Shanghai was not as developed as the industrial centers of Beijing, Nanjing, Yangzhou, and so on. The port's commercial development was restricted because of the ban of shipping activities and pertinent trade-restraining policies. The Ming seafaring expeditions ceased unexpectedly after 1433, the reasons being (a) the high cost of ocean-going journeys, (b) safeguarding against the Mongols' threats, and possibly (c) the perception of growth through sea trade and control of the seas was incompatible to the Dynasty's aspirations.

Shanghai evolved during the late Qing Dynasty (1644–1912) as one of China's principal trading ports (Murphey 2007). In 1684, the Shanghai port was authorized to trade with foreign vessels, as it obtained exclusive control over customs tariffs and charges for all international trade in Jiangsu Province. By 1735, the Port of Shanghai became the Yangtze region's most prominent seaport. The Canton System (1757–1842) served as China's method for regulating foreign trade in its own territory. In 1842, the Anglo-Chinese Treaty of Nanking declared Shanghai as the major port in China that handles foreign commerce, thus empowering the port to develop into a global trade center (Dermigny 1964). The treaty opened the Port of Shanghai, and other key ports for trade, that is, Ningbo, Canton, Xiamen, and Fuzhou (Madrolle 1912). British and American merchants united in 1863 to establish the International Settlement in the Port of Shanghai, while the French retained an independent French Concession. At the time, the port had six large docks and enjoyed a high accumulation of foreign ships. Its significance grew immensely as it engaged a crucial strategic location for trading with the West (Bullock 1884). By the early 20th century, Shanghai was the greatest seaport and city in the Far East. In 1949, with the People's Republic takeover, the financial regime had a debilitating impact on Shanghai's sea trade, infrastructure, and investment capital.

As soon as the central government granted Shanghai the power to implement financial reforms in 1991, the port has evolved at a soaring rate, becoming one of Asia's financial centers and the busiest container port in the world (World Port Source 2013c). Throughout modern Chinese history, the Port of Shanghai has been a considerable contributor of tax-related funds in China.

c. The Port of Piraeus, Greece
The ancient oracle and the Greek Shipping Legacy

The Port of Piraeus has been Greece's major seaport for over 2500 years, and this is a significant honor to bear. Being the Port of Athens, capital of Greece, it has enjoyed eras of remarkable prominence and glory, at times of both war and peace. While modern Greece struggles to survive from its Sovereign Debt Crisis pursuant to the 2008 financial crisis, the port's history is here to remind us that shipping will help the nation find its way out of the crisis.

Greece and prehistoric sea voyages of 13,000 BC

Greece is a small, yet extraordinary country of 50,942 square miles, whose size is one-fifth the size of Texas. The nation's intense, almost religious connection with the sea trade dates back from prehistoric times. This can be justified by its 6000 islands and an enormous coastline of 13,676 km that can only be compared to the coastlines of the greatest maritime nations on the planet, that is, the United States (19,924 km), China (14,500 km), and England (12,429 km) (CIA World Factbook 2013). To the Greeks, the seas have only been a path of growth and survival.

Archeological findings in Franchthi Cave (Argolis), between 1968 and 1976 under the leadership of Thomas W. Jacobsen of Indiana University (United States), have discovered a prehistoric reed boat made of papyrus (hence called Papyrella). In the cave, a vast selection of lava-generated obsidian stones was discovered, together with 500 tools. European laboratory analyses have determined it as deriving from the island of Melos, which is 80 miles away.

The primary obsidian stones from Melos island discovered in the Franchthi Cave are dated in the late Paleolithic era at 13,000 BC, as verified using modern techniques known as "obsidian hydration dating" and a most innovative method named "secondary ion mass spectrometry." The small reed-papyrus boat has been restored and exhibited in the Maritime Museum of Piraeus. A simulation of this prehistoric voyage reveals that the boat's speed was only 2 miles per hour. Taking into consideration the adverse currents encountered in the open seas, 52 hours were actually needed for this 80-mile voyage.

500 BC: Port of Piraeus and Its Long Walls: The First Ancient Maritime Security Measures

The Port of Piraeus was a major seaport that first gained glory during Greece's "Golden Century," that is, fifth century BC. In 480 BC, the Persian king Xerxes occupied Greece and the port was considered as a high-risk target because of its strategic location as a passage from Asia to Europe.

This is when the infamous "Long Walls" were constructed all around the city of Athens and the Port of Piraeus, securely linking the cities to the port at times of war.

These two port-and-city-protecting walls were 16 miles (26 kilometers) long and were designed to serve two purposes: (a) as military-defense barrier walls and (b) as protected logistics corridors where the military and food supplies would move safely and unobstructed. This remarkable construction was also enhanced with a double-gate system, to deter the enemy's entry. The purpose of this design was that even if attackers would concentrate their full force to the outer gate, they would in fact be trapped between the two gates, in the mercy of the Greek forces. The solid structure, the architecture, and overall concept of the long walls were a significant achievement in the history of Greece (Foucart 1887; Frazer 1898; Pausanias 2nd century AD, description of Greek, Book I: Attika).

This strategic construction of solid rock walls prevented the heartland of Greece from being surrounded by land or sea. The Port of Piraeus underwent reconstruction (i.e., modern-day retrofitting), and the Long Walls were fortified numerous times, to ensure national sovereignty.

Thucydides, an ancient general and historian (460–398 BC), named the Port of Piraeus "Greece's commercial heart, the major point of entry into the country."

Themistocles and the Oracle on the Persian Invasion

Themistocles was a highly skilled general and politician, who led the Greek military operations in view of the Persian invasion. It was customary for the political leaders in antiquity to consult the High Priestess in the Delphic Temple of Apollo, seeking for an oracle that would inspire their strategic and tactical military operations.

The Priestess' oracle for Themistocles was that only Wooden Walls would help the country avert an attack from the Persians, thus implying that the solid Long Walls made of stone were not sufficient.

Themistocles convinced his people that ships were the "Wooden Walls" of the oracle. Hence, he utilized the funds from the Lavrion mines and gathered donations from the aristocracy. He managed to build 400 ships with an equal amount of ship sheds situated next to the Roofed Arsenal and the Hippodamian market place (Foucart 1887; Ludlow 1883).

These ships were capable of averting the Persian invasion and saved the entire Europe from what would be a colonization of the West from the East.

Modern-day Piraeus reflects the glory and the prosperity brought by the Greek shipping tycoons of the twentieth century, like Onassis, Niarchos, Livanos, Carras, to name a few. Today, Piraeus remains one of the greatest ports in the oil-and-gas-rich Mediterranean basin. It is one of the largest container ports in Europe and a leading global destination for luxury cruise ships. Overall, the port serves over 24,000 ships each year, which is a remarkable statistical figure for a country on the verge of bankruptcy. COSCO shipping and the government of China have secured special privatization and concession agreements to utilize the port's unique strategic location, where Chinese cargoes are being handled and distributed in Europe through sea, land, and air.

The oracle, or rather Themistocles' interpretation of the oracle, is verified millennia later. In the turbulent times of modern Greece, the "wooden walls" oracle is more timely than ever:

The high seas are Greece's only way to financial prosperity, as, despite the crisis, the nation remains a leading global maritime power (second after Japan), carrying 16% of the global cargoes.

d. The Port of Istanbul (Constantinople), Turkey, and the Bosphorus Canal

Over millennia of rich history, this epic port still attracts tremendous commercial and diplomatic attention, because of its unparalleled geopolitical location: situated between the East and the West, this port still facilitates global trade between Europe, Asia Minor, Russia, and the landlocked, oil-rich Caspian Sea. The ancient port of Lygos was founded in thirteenth century BC as a Thracian colony. In 657 BC, it was colonized by Byzas of Megara (Athens). The port enjoyed both riches as a result of trade and political independence throughout the centuries, when Emperor Constantine I the Great (272–337 AD) moved the capital of the Roman Empire from Rome to the Byzantium port in 324 AD. Recent archaeological excavations have exposed the immense port ruins with dams, jetties, platforms, and anchors, and confirmed ancient and medieval trade and connections with the port of Alexandria and other Mediterranean key ports. For the next centuries, the official name of the city was Constantinople until 1930 when the Turkish Government changed it to Istanbul (Plummer 2006). Through the port's passage from antiquity to Christianity and finally the Ottoman Empire, it was Europe's largest and wealthiest city (Taylor 2006). In contemporary times, the Port of Istanbul is home to 35 Forbes World's billionaires (Forbes 2014; Oxford Business Group 2009). Since the mid-1990s, Istanbul's economy has been one of the fastest growing among OECD metro-regions (OECD 2008). As the only sea route bridging the oil-rich Black Sea and the Mediterranean, the Bosphorus is one of the busiest waterways globally; over 200 million metric tons of oil move across the strait annually, and the traffic on the Bosphorus exceeds three times that on the Suez Canal (Oxford Business Group 2009). Consequently, there have been plans to construct a new canal, referred to as "Canal Istanbul," parallel to the strait, on the European side of the city (Jones 2011). Istanbul hosts and coordinates several completed and proposed oil and gas pipeline projects, transferring energy between Eurasia's wealthiest nations, namely, Russia and the Caspian Sea, China, Israel, and Europe. For over 2500 years, UNESCO's monument of world heritage, or "Golden Horn" as it is called for its geopolitical significance, is still a bridge and a barrier for cultures, religions, and empires (World Port Source 2005).

e. The Port of Haifa, Israel

The Port of Haifa is the largest of Israel's three major international seaports, which include the Port of Ashdod and the Port of Eilat. The port lies on the shores of the Bay of Haifa on northern Israel's coast on the Mediterranean Sea. Since prehistoric times, Haifa Bay has been a refuge for mariners under the lee of Mount Carmel. Haifa harbor is mentioned for the first time in the year 104 BC when Ptolemy Lathyrus of Cyprus landed a huge force in Shikmona port in order to wage war on Alexander Yanai, the Jewish king (Port of Haifa 2013).

The Greek explorer Scylax in his "Seafarers' Manual" (Περίπλους, i.e., Periplus [350 BC]) referred to Haifa as the city on Mount Carmel (Baschmakoff 1948; Scylax 350 BC). When the ancient port became silted, the port was moved to a new site to the south. The Christian Bible mentions the Kishon River and Mount Carmel (Encyclopaedia Judaica and the Jewish Virtual Library 2013).

The Talmud (תַּלְמוּד), a central book of Rabbinic Judaism, mentions Haifa 100 times, as it was the home of prominent Jewish scholars. At this time, Greeks were engaged in trade on the coast near the Port of Haifa. The Port of Haifa flourished under Byzantine rule (330–1453 AD). The Persians conquered the Port of Haifa in seventh century AD. During the ninth century, the port established trade with ports in Egypt and the city contained some shipyards. By the eleventh century, the Port of Haifa was a prosperous mercantile center. When the Crusaders conquered Haifa in the year 1100, it became an important town and the main port for Tiberias, the capital of the Galilee. The port fell into disrepair during the Mamluk reign and acquired the reputation of a pirate lair in the eighteenth century (Port of Haifa 2013; World Port Source 2013a).

The modern Port of Haifa is home to one of Israel's two oil refineries. The refinery processes nine million tons of crude oil each year. The Port of Haifa is also home to the oldest and biggest business park in Israel, *Matam*, which houses manufacturing and research and development facilities for several high-tech companies including Intel, Microsoft, Google, IBM, and Yahoo, among others (World Port Source 2013a).

The first person to comprehend the tremendous possibilities of a port in Haifa was Theodor Herzl, the father of Political Zionism, who in 1898 wrote a prophetic description of the town in his book *Altneuland*. During the twentieth century, as the ships' size grew, draft restrictions necessitated dredging activities in order for the port to take advantage of its full geopolitical potential. Construction of the port began in 1922. The port allowed Haifa to blossom, and in 1936, the city had over 100,000 inhabitants. The port served as a gateway for thousands of immigrants to Israel after the Second World War (Samuels 1949). With Israel's strategic geographical position, Haifa served as a crucial gateway to the rest of the world and helped Israel develop into an economic power. Today, the port brings both passenger and cargo traffic to a bustling metropolis, much as Theodor Herzl predicted over a century ago. The UNESCO World Heritage Site Bahá'í World Centre is located in the Port of Haifa. In July 2013, the Israel Ports Company announced its plans to invest $2.2 billion equally shared to expand the infrastructure of the new ports in Haifa Bay and in South Ashdod.

f. The Port of Singapore

The globe's busiest port was established as a British trading harbor on the Malacca Strait in 1819. Singapore's outstanding position on the principal sea route among India and China concurs with its designation as "the gateway of the East" (Mongabay 2012).

The island's initial success resulted from its function as a convenient passageway, bunkering station and duty-free services for the three-way trade among China, India, and the rest of the Far East. This trade could well be an ancient sea route that flourished for at least 2000 years and expanded through the colonial and modern times. By the late nineteenth century, the British overlords of Singapore had exerted their influence or control throughout the Malay Peninsula, and Singapore took over as the outlet for Malaysia's tin and rubber, in addition to the gateway that supplied workers, spares, and services (Lane et al. 1922).

Singapore's independence in 1965 inflicted tremendous socioeconomic pressure to the port as income, trade, and 20% of local jobs were lost. These conditions led Singapore's leadership to aggressively promote export-oriented, labor-intensive industrialization via efforts designed to attract foreign investment.

In 1990, the economic environment of modern Singapore was consistently based on the colonial services, yet they were significantly enhanced and included global trade, export-oriented manufacturing, oil refining, shipbuilding, ship repairs, production of goods and services, and so on.

Singapore is still the ideal location for transshipment operations and replenishment transits through the Far Eastern Tropics, trading with the United States, India, China, Japan, and Western Australia. With a population of only five million people, it has remained a global leading port for a number of years. In 2012, the port handled over 30 million tons of cargo, ranking Singapore as the second busiest port in the world after Shanghai. According to Port Singapore Authorities, the sky is the limit: their capable workforce has set their eyes to even greater achievements.

g. The Port of Hamburg

The ancient port of "Hamma-Burg" (Hamma Fortress) was first cited during the ninth century as a small seaport with a moated castle and 200 residents. During the next centuries, the Vikings had burned the city down eight times, until in 1189 when King Frederick I Barbarossa selected the port for its strategic location to secure Germany's prosperity and power in Europe. The king granted Hamburg the right to maintain markets in an imperial charter and customs-free journeys to the North Sea.

In 1321, Hamburg joined the Hanseatic League, which was the most prominent merchant federation of the Middle Ages covering the North Sea and the Baltic Sea (Encyclopedia Britannica 2013). The Hansa ports gained advantages such as the control of shipbuilding and deterring piracy by employing convoy ships (Port of Hamburg 2013). Other benefits included trade agreements and naval operations in the region. Hamburg dominated trade in the Baltic ports with striking speed from the 13th century to America's discovery in the fifteenth century, throughout Hansa's decline in the seventeenth century.

The German reunification of 1871 recognized the port as Europe's leading hub for global trade and transatlantic passenger voyages. In 1862, the port modernized its multimodal operations, and in 1872, an efficient ship-to-rail transfer occurred at Kaiserkai's Imperial Dock. In 1888, Hamburg's Free Port attracted major traders of the time as a major hub port and warehouse with simplified customs clearance process (Port of Hamburg 2013). In the 1930s, the Hamburg port hosted the Hamburg–Harburg area oil refineries, shipyards, and warehouses. Hamburg shipyards suffered great losses during WWI and WWII, in particular during the allies' bombing attack "Operation Gomorrah" (1943) that totally destroyed the port. Pursuant to the fall of the Iron Curtain and the European Union's establishment, Hamburg reassumed its role as "gateway to the world" as more than 15,000 ships from over 100 countries call the port each year. The port is the world's largest roofed warehouse (Übersee-Zentrum), and the Waltershof container terminal is the largest in Europe.

1.3 PORT OWNERSHIP, STRUCTURE, AND ORGANIZATION

This section focuses on the characteristics of port ownership types, while it evaluates how ports have adjusted structural and organizational designs resulting from globalization, the radical market changes, and the competitive demands affecting the industry.

Over the years, ports have been affected by global trade and the shipping industry in terms of ownership, technology, and services provided. Port evolution was swayed by modifications in port ownership, management, and structural and operation patterns and brought about radical changes in labor recruitment, training, and production requirements.

As markets became progressively globalized, sea transport volumes soared. From the 1950s to the latest global economic crisis, the growth rate of global trade was virtually double compared to the economic activity as a whole. From 2000 to 2008, global trade expanded by at least 5.4% annually, while financial transactions, as assessed by the international GDP, grew by only 3% annually (World Ocean Review 2013).

1.3.1 Forms of Port Ownership, Structure, and Organization

The resulting changes in distribution of power, investment, and innovation have reshaped port ownership, structure, and organization. A new hierarchy surfaces among port authorities, governments, terminal operators, and ship-owning companies, in a complex supply chain network. This evolving relationship has the power to influence the way in which port managers and decision makers in the entire supply chain interact.

The *United Nations Conference on Trade and Development (UNCTAD) Handbook for Port Planners in Developing Countries* (1998) specifies the governmental powers of a nationwide port authority in the following manner, provided that the operating decision making will be undertaken regionally:

- *Monetary policy*: Authorization to establish common economic objectives for ports, such as investment policies and goals (as defined on a standard basis), with a common policy to local—as opposed to centralized—facilities funding, and informing the government authorities on loan requests
- *Tariff policy*: Capacity to regulate tariffs and charges as needed to safeguard the general public interest
- *Investment policy*: Ability to accept plans and projects for port investment opportunities in amounts exceeding a specific figure, under the condition that the suggested plans were largely in line with another national plan, sustained by the authority
- *Labor policy*: Authority to establish common hiring standards, a standard salary structure, standard qualifying criteria for professional advancement, and the ability to authorize standard labor union processes
- *Legal policy*: Capacity to represent the local port authorities as legal advisor
- *Licensing policy*: If applicable, authority to set up principles for accreditation of port workforce, brokers, or agents
- *Research and data analysis policy*: Authority to gather, evaluate, and distribute statistical data on port activity for common use, and to finance scientific, industrial research into port matters as needed

1.3.2 Port Governance

Port governance may be defined as the method by which power is exercised in the management of a nation's socioeconomic assets for growth (World Bank 1991). The key categories of port governance are as follows:

1. *Government/state ownership and administration.*
2. *Semigovernmental organization: Autonomous ports/public trusts.* A nonprofit administration manages a port entity and its specified services. This is a fairly typical arrangement during the 1980s, for example, London and Liverpool ports before privatization, as well as the French "Portes Autonomes." This port concept may be afflicted by budget deficit or overregulation.
3. *State/regional ownership*, for example, Rotterdam, Hamburg, and Yokohama. The significant benefit of this port model is the region's full engagement in meeting the port's requirements owing to the port's role in the regional economic

growth. Regional subsidies may also be provided, as well as other financial incentives, in order to promote regional prosperity. On the other hand, its draw-back is the region's disinclination to participate in any nationwide program.

4. *Privately owned ports.* Since 1947, at least one-third of British ports became public and pursuant to the privatization of the Port of Felixstowe in the 1980s, many other British trust ports pursued this path, driven by an estimated yearly revenue of at least 10 million dollars. In 1983, 19 ports under the Associated British Ports umbrella were privatized, thus increasing manual work productiv-ity by 40% (Associated British Ports [ABP] 2013, http://www.abports.co.uk). Privatization triggers efficient allocation of port assets, resulting in increasing its capital equity and revitalizing the regional economic climate.

Port systems are also distinguished in terms of the following characteristics:

- Geographical orientation, that is, global, national, or regional
- Regulatory and policy system (globalization, liberalization, and protectionism)
- Service arrangement, that is, private, public, or mixed
- Infrastructure ownership, that is, land, technical structures
- Superstructure and equipment ownership, that is, warehousing, cargo handling equipment, ship-to-shore handling equipment, outdoor sheds, and so on
- Management of stevedores (longshoremen); trade union members or nonunion arrangements

1.3.3 Port Ownership and Structural Types

Port typology is also segmented in terms of ownership, structure, and service arrange-ments, namely, segmentation between a fully public port, a tool type port, a landlord port, and a privatized port. For a port authority, the decision-making and selection pro-cess are crucial for the port's competitive edge, productivity, profitability, and regional development. In order for a port to maximize market share and compete in the global platform, it is necessary to select the tools that will determine its effectiveness and com-parative advantage. Figure 1.4 illustrates the port management typology and the different service options available.

1. *Public Service Ports*: Service ports possess a public character. Within this frame-work, the port authority is the employer and provides the entire spectrum of services needed for global port operations. There is a tendency for certain service ports to be converted into landlord ports, especially in Africa. Certain ports in developing countries continue to be governed in accordance with the service type. The port possesses, controls, and runs all assets, equipment and services available, whereas cargo operations are performed by port authority employees. Service ports are typically managed by national or state authorities and its lead-ership is composed of public officials, recruited by state or federal government.

Cargo handling is a primary service offered by service ports. In certain devel-oping economies, these services are carried out by independent public companies. Typically, these companies are also managed by the same state or government authorities, just like the ports. A modern corporate challenge entails managing port authorities and related service providers of contradicting interests.

2. *Tool Ports*: Within a tool port model, the port authority is the owner, developer, and handler of the port's infrastructure and superstructure, whereas it leases the port's superstructure, including cargo handling equipment, to cargo handling companies.

   Assignments are allocated through divided operational duties, as stipulated in their contractual obligations. While the port authority owns and operates the cargo handling equipment, the private cargo handling firm does not have the authority to entirely control the cargo handling operations; hence, it typically liaises with the shipowner or cargo owner. Over the past few years, there has been a fine line between tool ports and service ports: in an effort to increase efficiency and eliminate discord among port authorities and cargo handling firms, certain port authorities permit the operators to employ their own gear, hence gradually becoming less of a tool port and more of a service port. Other common elements between tool ports and service ports include the financing arrangements and their public character.

   Modern arrangements have increased the role of private terminal operators, thus resulting into power struggles and a complex or inconsistent decision-making protocol. In general, a port's function and its impeding success depend on power delegation, as well as the allocation of services and liabilities between the port authorities, the cargo handling companies, and private terminal operators.

   On the other hand, the tool port model is an appealing option for ports that wish to benefit from a public–private coalition, in particular when the port authorities wish to minimize the risk entailed with an initial capital investment and may not rely entirely on private ownership. Tool ports are also an attractive option during the initial formation of a port, owing to the relative simplicity and time efficiency of legal and regulatory framework. During the initial stage of a port's reform, no state resources need to be allocated to the privately owned business; hence, tool port models are a simpler and faster model to follow.

3. *Landlord Ports*: Landlord port models combine a public–private alignment: the port authority serves as a legislative and administrative structure, whereas the private sector provides its own superstructure (i.e., cargo handling equipment, cranes, derricks, etc.) and other port operations. In addition, private industries, for example, chemical plants, oil refineries, and cargo and liner ship terminals, lease infrastructure and space from the landlord port. Typically, this port type is the leading model for the larger hub ports and busy medium-sized ports, such as Houston, New York, Antwerp, to name a few. Typically, a contract stipulates terms and conditions for leasing time, use of superstructure (i.e., offices, cargo storage, and maintenance areas), and assets' ownership and land repossession. The annual lease typically involves a lump sum payment per square meter, providing for an annual estimated inflation rate. The contract will also define stevedores (longshoremen) and other labor provisions, both union and nonunion, as provided by either a port employment arrangement or private terminal operators.

4. *Fully Privatized Ports or Private Service Ports*: Under the full port privatization model, the port's ownership and public policy-making is fully transferred from the state to the private entity. The benefits from this option include government support through private investments, generating commercial growth and increase of employment, as well as technological innovation and modernization of equipment. This model is encountered mostly in British and New Zealand ports.

Characteristics	Port types			
	Public service port	Tool port	Landlord port	Privatized port
Port management	Public	Public	Public	Private
Navigational management	Public	Public	Public	Public
Navigational infrastructure	Public	Public	Public	Private
Port infrastructure	Public	Public	Public	Private
Superstructure (equipment)	Public	Public	Private	Private
Superstructure (buildings)	Public	Public	Private	Private
Cargo handling activities	Public	Private	Private	Private
Pilotage	Public and private	Public and private	Public and private	Private
Towage	Public and private	Public and private	Public and private	Private
Mooring services	Public and private	Public and private	Public and private	Private
Dredging	Public and private	Public and private	Public and private	Public and private
Other functions	Public and private	Public and private	Public and private	Public and private
Ownership/management characteristics	- Unity of command - Overstructured management	- Terminals/private cargo handling company do not entirely control cargo operations. - Power struggles due to private entity's limited funding contribution, administrative and equipment usage issues.	- Long-term contracts - Terminal's loyalty to port - Terminals/private cargo handling company owns and operates cargo handling equipment.	- Limited government authority and interference. - Government loses financial control and ability to benefit from future profit or development. - Regional development may not be a priority to the private sector. - Any future gains mainly benefit the private sector. Long-term benefits from port's previous clients and supply chains. - Serious security concerns. - Risk of speculation. Private company free to resell, redevelop, or lease to third party, with huge profits, no government control or interference. - Strategic location encourages private company to expand business activities. - Risk of monopolistic behavior, as tariffs are decided by the private sector. Port tariff regulator may be needed to avoid overcharge, conflict with regional/national interests and supply chain. - Motivation to invest and take high financial risks. - Jobs creation may not be a priority to the private sector.

Private sector role	Limited	Moderate	Positive partnership	High
Flexibility	Limited	Moderate	High; most adaptable to industry requirements.	High; most adaptable to industry requirements.
Stability	Yes	Moderate	High, long-term contracts	N/A
Problem solving potential	Limited	Moderate	High	N/A
Innovation, modernization	Limited	Limited; private entity acts as labor pool, limited innovation incentive.	High	High
Access to public funds	Limited	Limited; private entity does not own equipment; hence, there is limited investment incentive.	High	High
Dependence on government budget and support	High	High	Moderate	Limited
Incentives for growth	Limited	Moderate	High	High
Internal conflict/competition	Limited	Yes, because of split cargo handling operations.	Limited	High conflict potential pertaining to social responsibility, and all the above factors.

FIGURE 1.4 Port management typology and predominant service arrangements. (Courtesy of M.G. Burns, based on World Bank, 2007, "Alternative Port Management Structures And Ownership Models," Port Reform Tool Kit, Module 3 [accessed May 12, 2013]; World Bank, 1991, Managing Development—The Governance Dimension, Washington, DC; UNCTAD, 2012, UNCTAD Global Economic Outlook. Available at http://www.unctad.org [accessed May 10, 2013]; UNCTAD, 1998, Guidelines for Port Authorities and Governments on the Privatization of Port Facilities. United Nations Conference on Trade and Development. Distr. General UNCTAD/SDTE/TIB/1, September 23, 1998.)

This is often regarded as the most risky type of port arrangement owing to high exposure of the national interests, especially pertaining to issues of national sovereignty, security, social responsibility, trade balance, and financial liability. Once the sale agreement is finalized, sellers have the right to engage in any commercial activities, including land property speculation, and resale the land to any government or privately owned business, hence raising sensitive diplomatic issues.

Some characteristic examples of the potential complications involved with privatization are as follows:

Panama Terminal acquisitions by Hutchison Port Holding group, China

Since 1997, the port terminals of Balboa and Cristobal were operated by the Panama Ports Company, a member of the Hutchison Port Holding (HPH) group. Certain maritime groups regarded this strategic purchase as China's attempt to control the canal. Hence, the HPH CEO had to reassure the US government that this acquisition served purely trade purposes and no further involvement with the canal's activities was intended.

US Ports management contracts with the Dubai Ports World, UAE

The Dubai Ports World controversy took place in post-September 11 America and became the dominant national security discussion in the country. The debate entailed the port management contracts of six major US seaports to a United Arab Emirates-based company. DP World acquired the Peninsular and Oriental Steam Navigation Company and subsequently took over the full management of terminals in New York, Baltimore, Miami, New Orleans, and Philadelphia, together with 16 freight handling facilities and stevedoring operations at nine US ports. Pursuant to diplomatic action taken by the US Government, Dubai Ports World ultimately sold the assets to the Global Investment Group, the asset management division of the American International Group.

1.3.4 Port Privatization

Privatization pertains to the private sector ownership, that is, the transfer of property ownership from the public to the private sector or the utilization of private investment capital to finance ventures in port facilities, machinery, infrastructure, and superstructure (UNCTAD 1985 and UNCTAD 1998). Over the past few years, there is an increasing trend for seaports to pursue privatization. The contract types pertaining to privatization depend on the type of privatization, the time duration, and the degree of control and investment obligations on behalf of the private entity. Hence, the most commonly used privatization agreements, with the prevailing forms of privatization, and their respective contractual agreements are as follows:

1. *Full privatization* pertains to the full port ownership passing onto a private entity that eventually becomes the owner of all the land, water, infrastructure, superstructure, and generally all properties and assets within a port's domain.

   An *asset sale agreement* is applicable for the outright sale of a terminal facility.
2. *Partial privatization* pertains to an arrangement where only a fragment of the property, assets, or operations is purchased by the private sector, for example,

the superstructure is sold to the private sector, or the permit issued by a public port authority to a private business to develop and control a terminal, berth, or a designated port service (Trujillo and Nombela 1999). This type of privatization involves different forms of agreements, reflecting the time of lease or concession, the element of permanency, and the obligations of the private sector to invest, develop, and manage the segment of port facilities stipulated in the contract. These contractual categories include the following:

- *Concession agreement*: Long-term facility lease, typically for 20–40 years.
- *Service or leasehold contracts*: A private operator performs specific operational tasks, whereas the port authority retains ownership of the facility and equipment.
- *Build–Operate–Transfer (BOT), Build–Own–Operate (BOO), and Build–Own–Operate and Transfer (BOOT) arrangements*. These agreements stipulate that as a prerequisite for operating a port segment, the private sector must participate in financing, building, and managing of port facilities. Upon expiration of the agreement, the ownership is shifted back to the public sector.

On the basis of these models mentioned, it is worth noting that there is a great variation in different privatization agreements, options, and the extent of ownership of different assets, services, and operations within the entire port area.

1.4 PORT WORKFORCE: PRODUCTIVITY, GROWTH, AND EMPOWERMENT STRATEGIES

To reestablish our sense of maritime economic growth, labor, and their effect on productivity, we need to disassociate them with the global impact and the national product and observe their correlation within an industrial framework; ports to be precise. Productivity is an essential component of cost efficiency; it serves as an efficiency quantifier of a worker, software and piece of equipment, a process, an entire port, or even the supply chain as a whole.

Because of the inherent complexity and overlapping functions of different segments of the supply chain within a port, and the dissimilarity of ports' strategies, resources, functions, and operations around the world, several anomalies arise in measuring port productivity and growth.

The main functions and utility of seaports are to facilitate the movement of ships and commodities; eliminate bottlenecks; support cargo loading and unloading, bunkering, shipbuilding, and ship repair operations; and serve as hubs for inland trade and intermodal and multimodal activities (see Chapter 4). In addition, they accommodate trade zones and support hundreds or thousands of large, medium, and smaller businesses that are located in the ports' vicinity or belong to their national and global supply chain (see Chapter 2). Ports collaborate with the oil and gas industry by hosting refineries, while feeding offshore platforms and deepwater drilling operations (see Chapter 11). In their warehouses and terminal facilities, they store cargoes and container boxes. They are the tolerant and gracious hosts of numerous professionals from all these industries, including Coast Guard officers, bankers, surveyors, inspectors, auditors, insurance brokers, and a legion of seafarers. On a 24/7 basis, their emergency response team, in collaboration with the Coast Guard (Port State Control), will address any security, safety, or environmental challenge within the inner and outer port limits.

The positive output of the above functions created by entrepreneurs, workers, capital, and assets is called productivity growth, whereas any bottlenecks, delays, loss of life or property, or damage of the environment are counterproductive factors that hinder growth.

1.4.1 Measuring Productivity, Throughput, and Growth

In port management, productivity measures may be investigated within the entire supply chain or the entire port entity, or within departmental segments, that is, port sectors, ship types, and terminals. It is important to distinguish the three different indicators of growth: productivity is a combination of throughput and cycle time.

On the basis of the scientifically approved concept that productivity includes entrepreneurship, capital, and labor, it can be further divided into man-made and technological. This distinction will help us measure its two key elements: technology (output vs. input) and labor (output vs. input). Productivity is calculated by dividing the average output per period by the overall resources utilized or costs incurred during that time: output or the end service and/or product are typically measured in business earnings and inventories, whereas inputs comprise three (out of four) factors of production, that is, entrepreneurship, labor, and capital.

While measuring productivity, it is worth noting that in the modern era of service providers, productivity may also encompass intangible, long-term benefits that in accounting principles may be considered as "accrued revenue," such as establishing positive clients' feedback, building company reputation, and attracting new business, all of which ensure business continuity.

For port authorities, economic growth is stimulated by investment capital, technology optimization, and increase in the volume and quality of labor. The shipping industry's need for growth is a powerful incentive for technological advancements in ports and ships alike, which in turn elevate the quality of human labor and amplify the possibilities for productivity. The industry's emerging concepts of promoting sustainability and regulatory compliance will eventually introduce new growth factors, that is, sound processes to promote occupational health, safety, security, environmental protection, social responsibilities, and quality. The industry's motto—"Zero accidents, zero incidents, zero non-conformities"—suggests that regulatory compliance can also be measured and assessed and will eventually be included as inputs of production and components of productivity.

UNCTAD recommends two main types of port performance indicators: (a) macro performance, evaluating aggregate port outcomes on growth and financial activity, and (b) micro performance, assessing input/output ratio measurements of port operations (UNCTAD 1999).

Consequently, there are three major port productivity indicators:

1. *Cycle time or physical factors*, where service is measured in cycle time, such as a ship's turnaround time, which includes waiting time owing to port traffic plus cargo loading time, and so on. Certain models include multimodal synchronization time, that is, time counts when the port interacts with other transportation modes or components of the supply chain, for example, cargo dwell time from entrance to port until it is loaded onboard the ship.

2. *Factor productivity indicators*, which comprise labor and capital input during ships' stay at the port, for example, input for loading and unloading. (i) *Labor productivity* displays the employee's overall performance and measures the individual's value added in the production and sale of the output. (ii) *Capital productivity* pertains to the value added per dollar. It is measured from calculating the output after maintenance of port equipment, as well as the labor's skills that contribute to the value added throughout the process. The author wishes to add a third factor productivity indicator: (iii) *Entrepreneurship*, which, together with labor and capital, generates a by-product that is crucial to productivity and growth: (iv) *Value added* signifies the income generated from the port's performance, including facilities and services provided. In this labor–capital arrangement, the port's growth is produced by the merged efforts of its employees (labor) and those who supply the capital (government, shareholders, terminal operators, etc.). Value added has a higher percentage in the earnings of integrated port activities, for example, assembly lines, shipbuilding, ship repairs, and maintenance, and a reduced percentage of earnings to less integrated port activities, such as commercial and operational functions. Value added is distributed among capital and labor (key factors of production), and this distribution may frequently be subjective or inconsistent. Value added also pertains to the supplementary functions and quality of performance, services, or labor, which exceed the client's usual expectations and offer more with minimum or no extra charges. Most important, value-added services provide a port with the competitive edge that can significantly enhance growth and profitability.

3. Financial indicators pertaining to ships' traffic and cargo volume at any given time (World Bank 1999). For instance, operating surplus or total revenue and costs associated with the ships' charge, for example, charge per 20-foot equivalent unit (TEU) for container ships, or cargo volume, or ships' gross tonnage (GT)/net tonnage (NT).

Port economic impact indicators can be assessed to evaluate a port's socioeconomic impact at a regional, state, or global level.

Total factor productivity is frequently considered as the actual growth driver inside the market or an industry segment; although labor and capital are essential contributing factors, up to 60% of economic growth is due to total factor productivity.

While the methodologies and aforementioned indicators provide accurate evaluations of a port's overall performance, inconsistent or inaccurate findings could derive from poor measurements of accountability, process sustainability, management/supervision, and performance control.

As an example, induced employment has been increasing considerably, to the detriment of direct port employment. Considering that certain inputs have shifted from the regional economy or are acquired at an extremely low cost, the added value becomes increasingly reliant on direct and induced labor. This has a twofold impact in measuring port productivity: first, it affects the validity of measurements, as input labor calculations may not be accurate or consistent. Second, it distorts the overall port growth and productivity indices. As reflected in Figure 1.5, the factors of production, that is, entrepreneurship, land, labor, and capital are utilized as the input that will determine the port's growth, productivity, and output.

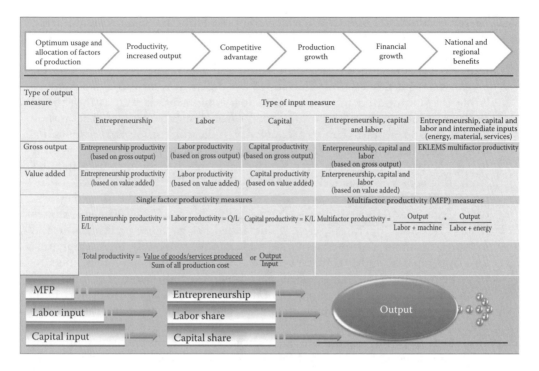

FIGURE 1.5   Port growth and main productivity models. (Courtesy of M.G. Burns.)

1.4.2 The Econometrics of Labor and Production

In the previous section, we established the significance of labor as a factor productivity indicator.

A simple aggregate production function is used to estimate how inputs produce goods and services:

$$Y = AF(K, L) \tag{1.1}$$

which takes into consideration four variables, whereby Y is the level of aggregate output, K represents capital, L is labor, and A measures total factor productivity. A is hereby considered to be exogenous to capital and labor inputs. Hence, any increases in the input of K or L will result to constant, decreasing, or increasing returns to scale. F signifies the functional association whereby Y, that is, the level of economic output, is modified by input variations of capital (K) and labor (L).

The translog production functions can also be expressed as:

$$y = \alpha + \alpha k + \beta l + \gamma g \tag{1.2}$$

which can be expanded in a geometric infinite distributed model as follows:

$$y = \alpha + \alpha_1 k + \beta_1 l + \gamma_1 g + \alpha_2 k^2 + \beta_2 l^2 + \gamma_2 g^2 + \psi_{kl}^{kl+} \psi_{kg}^{kg+} \psi_{lg}^{lg} \tag{1.3}$$

The time element can also be introduced among the production variables to capture the dynamic nature of production and exogenous parameters, such as port traffic, extended cargo loading time owing to cargo volume, and so on:

$$Y = F\left(K\left(t\right), A\left(t\right) L\left(t\right)\right) \tag{1.4}$$

Other control variables may be included as input in the above equations, such as investment, the business cycle, and so on, which are also considered to have a positive correlation with output together with capital and technological innovation.

The above calculations focus on the level of aggregate output, and yet a port manager will also need to compare input to output, that is, estimate the productivity of each terminal, and on the basis of their findings, calculate the overall port productivity. A port is a complex corporate entity, composed of a multitude tnd exogenous factors that will hinder productivity, that is, a strike, or act of God.

The following regression models are calculated:

Terminal Productivity:

$$\ln(Q/L)_t^T = \alpha_0 + \alpha_1 \ln(S)_t^T + \alpha_2 \text{Strike}_t^T + \alpha_3 \text{Weather}_t^T$$
$$+ \alpha_{4\%} \Delta \text{Cargo}_t^T + \alpha_5 \text{Terminal size}_t^T + \alpha_6 \text{time} + ut \tag{1.5}$$

Port Productivity:

$$\ln(Q/L)_t^P = \beta_0 + \beta_1 \ln(S)_t^P + \beta_2 \text{Strike}_t^P + \beta_3 \text{Weather}_t^P$$
$$+ \beta_{4\%} \Delta \text{Cargo}_t^P + \beta_5 \text{time} + ut \tag{1.6}$$

where
   $\ln(Q/L)$ = natural log of labor productivity as measured by the output per worker within the port.
   $\ln(S)$ = natural log of a salary rate index.
   Strike = a dummy variable, 1 for strike years, 0 otherwise.
   $_\%\Delta$Cargo = % alternation in the cargo value, that is, imports and exports, domestic and global, to evaluate the outcomes of the market cycle and cargo value on labor productivity, both locally and nationally.
   T = terminal size = average annual output per terminal, that is, a proxy for diminishing returns.
   P = port size = average annual output per seaport, that is, a proxy for diminishing returns.
   time = a linear time trend, designed to record any benefits of technological innovation on port labor productivity.
   ut = unit root tests are useful to evaluate if trending data should be first differenced or regressed on deterministic time functions to render the data stationary. Unit root tests can be used to estimate which pairs of assets seem to exhibit mean-reverting behavior.

The suggested model of port labor productivity is calculated by the average outcome of labor and entails factors such as work effort, the market cycle, or cargo volume and port traffic, while downsizing marginal returns and technological change.

1.4.3 Port Growth, Productivity, and Empowerment

While observing today's globalized businesses in the "Era of Abundance," the collapse of powerful maritime companies and degradation of ports, despite their affluent technological and capital resources, can only be described as a paradox. On the other hand, organizations with inspired leaders and empowered employees seem to stand out in performance, production, and quality, by using the company's assets to the maximum. And this can only be achieved by utilizing the main driving force an organization has: human resources.

As discussed in Section 1.4.1, port productivity is measured in terms of (a) cycle time; (b) factors of production, which include entrepreneurship and labor productivity; and (c) financial indicators pertaining to ships' traffic and cargo volume. In observing these three aspects of productivity, it becomes apparent that the human factor is the determining factor that can substantially influence all the other parameters, that is, machinery and capital, and control the cycle time element, that is, the turnaround time, and so on. The human factor is also the catalyst in terms of value-added services or products, which, again, is distributed among capital and labor.

Figure 1.6 demonstrates the timeline of productivity and the impact of global workforce:

- The first era covers the 1920s and the pre-World War I, till the post-World War II era of the 1950s: "Taylorism" was a factory management system introduced in the late nineteenth century, with the purpose of maximizing productivity through establishing a protocol for the industrial processes and increasing efficiency by assigning to each worker very specific, repetitive functions. These principles enjoyed a high demand in 1928, as they seemed to match with the prevailing "Fordism" theories, which favored an industrialized, mass-production working environment.
- In the 1950s and 1960s, the Fordist production and logistics systems brought about the mass production assembly lines. A focus on cost reduction and the introduction of robotics and computerization commenced in the 1970s, as a result of a radical increase in oil price from $18 per barrel to $40+ per barrel, which increased industrial and domestic expenditure.
- From the 1970s till the early 1990s, the market's focal points were quality standards and optimum productivity. The oil market that typically serves as an industrial productivity and cost measurement fluctuated from $85 in 1979 to a gradual collapse of $20 through 1986. During this time, computerization, a wide use of satellite communication systems, cell phones, and the first tailor-made maritime software were introduced to replace the telex machine, challenging ship-to-shore communication and manual calculations for the ships' loading, trim, and stowage measurements.
- The 2000s impressive technological advancements introduced large-scale automation and networks that not only enabled the wider use of advanced software and port handling equipment, but also facilitated a booming market and the radical increase of global waterborne freight.

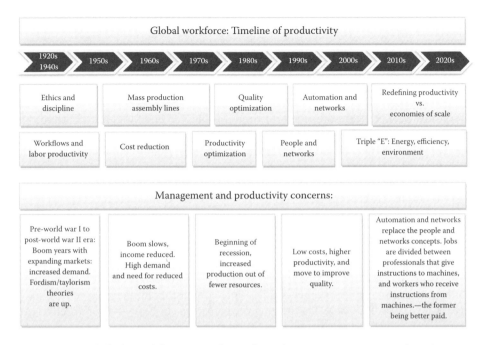

FIGURE 1.6 Global workforce: timeline of productivity. (Courtesy of M.G. Burns.)

- Finally, the 2008 economic crisis served as a wakeup call on the global markets, as productivity had to be redefined through Triple E: Energy efficiency, Economies of scale, and Environmental integrity. This trend is expected to last through the 2020s.

While econometric formulas can be used to estimate the actual human factor's input in terms of labor and entrepreneurship, social scientists would focus on empowering employees, in order to achieve optimum productivity.

Employee empowerment means different things in different organizations, based on culture and work design, all of which are based on the concepts of job enlargement and job enrichment. Employee empowerment also means giving up some of the power traditionally held by management for time and efficiency purposes. It does not mean that management relinquishes all authority, but it may allow port operations to run with sustainable accountability, even in the absence of the top management. It requires a significant investment of time and effort to develop accountability and add to individuals' capabilities, while developing clear agreements about roles, responsibilities, risk taking, and boundaries.

As modern ports are increasingly becoming diverse, multiethnic, and multicultural, it has become crucial to identify the core values that surpass national boundaries and cultures.

In today's business, excessive competition and economic pressures require employees to take initiatives, be inventive, and offer a great deal of their time, skills, and imagination to their business. To meet these new demands, micromanagement must be replaced by self-management and employee motivation. As the rigid corporate pyramid structure tends to fade, empowerment is considered as a key motivator and a vital tool for all organization members.

REFERENCES

Associated British Ports (ABP). 2013. Available at http://www.abports.co.uk. Accessed May 11, 2013.

Baschmakoff, A. 1948. *La Synthèse des Périples Pontiques: Méthode de précision en paléoethnologie. (based on ancient texts by Flavius Arrianus; Scylax.* Flavius Arrianus; Scylax. Series: Etudes d'Ethnographie, de Sociologie et d'Ethnologie, 3. Éditions Geuthner, France.

Bullock, C.J. 1884. Yangtse Kiang: Shanghai. *China Sea Directory*, Volume 3, 2nd edition. London: Admiralty Hydrographic Office, Great Britain.

Burns-Kokkinaki, M. 2004. The significance of navigation during prehistory, Presented at the European Anthropological Association's Annual Conference, Greece.

Burns-Kokkinaki, M. 2012. The currency factor: Underlying trends in the decline of International Trade and Economy, Asian Logistics Round Table, Canada Conference. Managing Connectivity in the Supply Chain—Human Resources, Sustainability and Security in the Presence of Global Financial Risk.

Butler, R. 2012. The history of Singapore. Mongabai, Indonesia. Available at http://www. mongabay.com/reference/country_studies/singapore/history.html. Accessed April 6, 2013.

Daughan, G.C. 2011. If by sea: The forging of the American Navy—From the Revolution to the War of 1812. *First Trade*. Basic Books (AZ), 1st edition.

Davis, C. 2013. Wadi El Jarf sit reveals oldest harbor, papyri ever found in Egypt. Huffington Post, Egypt State Information Service, Sorbonne University.

Dermigny, Louis. 1964. La Chine et l'occident. Le commerce a Canton au XVIIIe siecle, 1719–1833, Volume I of III. Paris Imprimerie Nationale, France.

Encyclopaedia Brittannica. 2013. Available at http://www.britannica.com/EBchecked/topic/254543/Hanseatic-League. Retrieved August 14, 2013.

Encyclopaedia Judaica and the Jewish Virtual Library. 2013. Available at http://www. jewishvirtuallibrary.org/jsource/judaica/ejud_0002_0016_0_16164.html. Retrieved June 5, 2013.

Ferguson, N. 2012. *Civilization, the West and the rest*, 2nd edition. Penguin Books Ltd., London.

Forbes. 2014. The world's billionaires list, 2014. Available at http://www.forbes.com/billionaires/list/.

Foucart, P. 1887. Les fortifications du Piree, Bulletin Correspondences Hellenique II, pp. 129–144 (French).

Frazer, J.G. 1898. *Pausanias's Description of Greece*. Translated with a commentary by J.G. Frazer. London: Macmillan and Co. Limited, New York: The Macmillan Company.

Hellenic Maritime Museum. 2013. Prehistoric and ancient age, Greece. Available at http://www.Hmmuseum.Gr. Accessed April 6, 2013.

Hoekman, B. and Messerlin, P. 1999. Liberalizing Trade in Services: Reciprocal Negotiations and Regulatory Reform. World Bank, July 1999. Available at http://siteresources.world bank.org/INTRANETTRADE/Resources/SER-brook-fin.pdf. Accessed in May 8, 2013.

Houston Ship Channel 50th Anniversary Collection, 1926–1964. Available at http://www. lib.utexas.edu/taro/uhsc/00004/hsc-00004.html. Accessed April 6, 2013.

Jones, S. 2011. Istanbul's new Bosphorus canal to surpass Suez or Panama. *The Guardian*. Accessed April 29, 2013.

Lane, S.A., Trimmer, G.W., Gibbons, V., James, F.S., Wellwood, W.P., Tomlin, F.L. and Williams, P.L. 1922. A short history of the port of Singapore, With particular reference to the undertakings to the Singapore harbor board. Fraser & Neave, Singapore.

Ludlow, T.W. 1883. The harbors of ancient athens. *The American Journal of Philology* 4:192–203.

Madrolle, C. 1912. *Madrolle's Guide Books: Northern China, The Valley of the Blue River, Korea*. Hachette & Company, France.

Mongabay. 2012. The history of Singapore. Available at http://www.mongabay.com/refer ence/country_studies/singapore/HISTORY.html. Accessed April 6, 2013.

Murphey, R. 2007. *East Asia: A New History*, 4th edition. Pearson Longman, England.

Nelson, J. 2008. *George Washington's Secret Navy: How the American Revolution Went to Sea*, 1st edition. International Marine/Ragged Mountain Press, Maine, USA.

OECD Territorial Reviews: Istanbul, Turkey. 2008. *Policy Briefs*. The Organisation for Economic Co-operation and Development.

Oxford Advanced Learner's English Chinese Dictionary. 2012. Oxford University Press, 7th Edition.

Oxford Business Group. 2009. *The Report: Turkey 2009*. Oxford, Eng.: Oxford Business Group.

Palmer, M. 2013. *The Navy: The Continental Period, 1775–1890, Naval History and Heritage Command*. Available at http://www.history.navy.mil. Accessed June 2, 2013.

Pieterse, J.N. 2012. Periodizing Globalization: Histories of Globalization, New Global Studies, Volume 6, Issue 2, University of California, Santa Barbara.

Port of Haifa. 2013. Available at http://www.haifaport.co.il. Retrieved June 5, 2013.

Samuels, G. (1949-08-21). From Munich to Haifa: Journey into the Light; For present-day Israeli immigrants, the trip is a dawn of hope after many years of dark tragedy. *The New York Times*. Retrieved June 5, 2013.

Scylax. 350 BC. «Περίπλους» Periplus, a Seafarers' Manual (Ancient Text).

Statistical Communiqué of the People's Republic of China on the 2012 National Economic and Social Development. 2013. China NBS.

Taylor, I. 2006. Lost treasures of Constantinople test Turkey's twenty-first-century ambition. *The Guardian*. Available at http://www.guardian.co.uk/world/2006/jan/25/turkey.iantraynor. Accessed April 6, 2013.

The American Association of Port Authorities. 2013. Available at http://www.aapa-ports.org. Accessed April 6, 2013.

The International Association of Ports and Harbors (AAPA). 2013. Available at http://www.iaphworldports.org. Accessed May 12, 2013.

The Museum of Houston. 2013. Available at http://www.museumofhouston.org/index.php/home/exhibits_podcast. Accessed April 23, 2013.

The Port of Hamburg. 2013. Available at http://www.hafen-hamburg.de. Accessed May 12, 2013.

The Port of Houston Authorities. 2013. Available at http://www.portofhouston.com/about-us/history/. Accessed May 12, 2013.

The Times Magazine. 1901. Spindletop gusher amazes oilmen. January 10, 1901, Newspaper Edition.

The World Factbook 2013–14. Washington, DC: Central Intelligence Agency, 2013. Available at http://www.cia.gov/library/publications/the-world-factbook/index.html. Accessed August 30, 2013.

Trujillo, L. and Nombela, G. 1999. Privatization and regulation of the seaport industry. University of Las Palmas, Canary Islands and World Bank, Spain. Available at http://info.worldbank.org/etools/docs/library/64583/2181seaport.pdf. Accessed June 12, 2013.

UNCTAD. 1985. Port Development. *A Handbook for Planners in Developing Countries*, 2nd Edition. United Nations Conference on Trade and Development, Geneva 1985. Publishers: UNCTAD, New York.

UNCTAD. 1998. Guidelines for Port Authorities and Governments on the Privatization of Port Facilities. United Nations Conference on Trade and Development. Distr. General UNCTAD/SDTE/TIB/1, September 23, 1998.

UNCTAD. 2012. UNCTAD Global Economic Outlook. Available at http://www.unctad. org. Accessed May 10, 2013.

US Army Lt. Col. Comer Plummer III. 2006. Ancient History: Walls of Constantinople. *Military History Magazine*. Available at http://www.historynet.com. Accessed May 18, 2013.

US Bureau of Economic Analysis. 2013. Available at http://www.bea.org. Accessed June 1, 2013.

World Bank. 1991. Managing Development—The Governance Dimension, Washington D.C.

World Bank. 2007. Alternative Port Management Structures and Ownership Models. Port Reform Tool Kit, Module 3. Accessed May 12, 2013.

World Ocean Review. 2013. Available at http://worldoceanreview.com/en/wor-1/transport/global-shipping/. Accessed May 14, 2013.

World Port Source. 2005. The Port of Istanbul. Available at http://www.worldportsource. com/ports/review/TUR_Port_of_Istanbul_3090.php. Accessed May 14, 2013.

World Port Source. 2013a. The Port of Haifa. Available at http://www.worldportsource. com/ports/review/ISR_Port_of_Haifa_246.php. Retrieved June 5, 2013.

World Port Source. 2013b. The Port of Houston. Available at http://www.worldportsource. com/ports/review/USA_TX_Port_of_Houston_60.php. Accessed May 14, 2013.

World Port Source. 2013c. The Port of Shanghai. Available at http://www.worldport-source.com/ports/review/CHN_Port_of_Shanghai_411.php. Accessed May 14, 2013.

USEFUL READING

Dictionary of Military and Associated Terms. 2005. US Department of Defense.

Gallup, J.L., Sachs, J.D. and Mellinger, A.D. 1998. Geography and Economic Development. National Bureau of Economic Research, NBER Working paper series.

Pausanias, 2nd Century AD. Description of Greece, Book I: Attika. 1918. Translated by W.H. Jones Edited by Capps. E., Page T.E., Rouse, W.H.D., The Loeb Classical library. William Heinemann, London.

Shirk, S.L. 1993. *The Political Logic of Economic Reform in China*, 1st edition. University of California, Bekerley and Los Angeles.

Smith, A. 1776. *An Inquiry into the Nature and Causes of the Wealth of Nations*, Book 1, Chapter 3, That the Division of Labor is limited by the Extent of the Market.

UNCTAD, Review of Maritime Transport. 2011. UN Symbol: UNCTAD/RMT/2011.

UNCTAD, Review of Maritime Transport. 2012. UN Symbol: UNCTAD/RMT/2012.

Wachsmuth, op.cit. I. pp. 306–336, pp. 1.176.

Weintrit, A. and Neumann T. 2013. *Marine Navigation and Safety of Sea Transportation: STCW, Maritime Education and Training (MET), Human Resources and Crew Manning, Maritime Policy, Logistics and Economic Matters*. CRC Press, Taylor & Francis Group, USA.

CHAPTER 2

Connecting Hub Port Gateways to the Inland Infrastructure

> It is upon the sea-coast and along navigable rivers,
> that industry of every kind naturally begins to subdivide and improve
> ... not till a long time after that, those improvements
> extend themselves to the inland parts of the country.
>
> Adam Smith (1723–1790)

Ports play a pivotal role within the supply chain system, as sea transport represents approximately 90% of the global trade. Progressively, ports' sizing and characteristics have been transformed, favoring the ones situated in the areas of the new extended trade routes, that is, areas with low-cost factors of production, large markets, large populations, significant economic capacity, and sizeable innovative activity. As of 2013, there are approximately 40 mega-hub zones globally, most of which are formed by hub cities growing outward and into one another. Nevertheless, permanency is not an attribute of modern hub ports; their commercial peak depends on the value of the region they represent, trade routes, markets, and technologies used. History can demonstrate multiple examples of global hubs that flourished for a number of years and gradually declined.

The elementary factors that define the magnitude of a port's influence within a supply chain network are the following:

a. Supply chain integration and the necessity of market visibility (see Section 2.1)
b. Geography and the need for market accessibility (see Section 2.2)
c. Traffic and a port's need to eliminate congestion and bottlenecks (see Section 2.2)
d. Port competition and substitution, and factors that weaken a port's competitive edge and enable the overlapping of market areas or supply chain services (see Section 2.3)
e. Factors of production, that is, labor, capital, assets, land use, technology, and innovation, as a means of gaining a competitive edge (see Section 2.3)

Ports are major players in the global trade, serving as the transport nodes that facilitate the flow of cargoes across global supply chains. Hub ports were created pursuant to a soaring growth of the global trade, which has quadruplicated in the past 40 years in an effort to connect the seaborne commodities with the inland infrastructure. Some of the key reasons for this soaring of global commodities include globalization, technological innovations, and the intensification of trade agreements (see Chapter 7). Nevertheless, the obstacle that had to be overcome involved fast turnaround times for cargo handling and storage. Containerization, which emerged in the 1960s, but has boomed since the 1990s, facilitated the large-scale carriage of goods by sea, land, and air, through multimodal/intermodal transportation (see Section 4.7).

Table 2.1 shows that in 1970, global trade in sea, land, and air amounted to 2605 millions of tons, whereas in 2011, it amounted to 8748 millions of tons loaded (UNCTAD 2012).

Furthermore, Figure 2.1 and Table 2.2 clearly demonstrate the steady growth in global trade.

A port's significance is directly related to its location and the niche it may represent. Gradually, the power was shifted from the ports as strategic alliances and partnerships were formed among terminal operators and liner companies. At a deeper level, the influence ports exert depends on political strategies at a regional and national level: it may be to a nation's or state's best interests to promote specific ports and regions, the reasons being security, growth of manufacturing activities, unemployment, and so on. Political support may provide the leverage while forming powerful strategic alliances within the supply chain. Furthermore, it may be to a region's interests to segment regional ports' span of operations, in order to develop unique competitive advantages and minimize regional competition.

The US networking system applies "the five corners' strategy," whereas the nation's major seaports, trade routes, and logistic systems are clustered in the nation's five corners. Interestingly enough, the country's oil and gas reserves, including the recently found shale gas reserves, are conveniently located nearby existing logistics corridors (see Figure 2.2). By 2016, the United States will be a major oil and gas exporter; hence, the location of the new shale gas reserves is rather opportune.

The Network Theory clearly demonstrates why hub ports are more opulent than other players of the network systems. Supply chains aggregate in the nodal points where markets, economies, and infrastructures meet: hubs are the focal points of opportunity, growth, and innovation.

TABLE 2.1   Development in International Seaborne Trade, 1970–2011 (Millions of Tons)

Year	Oil and Gas	Main Bulks	Other Dry Cargo	Total (All Cargoes)
1970	1440	448	717	2605
1980	1871	608	1225	3704
1990	1755	988	1265	4008
2000	2163	1295	2526	5984
2005	2422	1709	2978	7109
2006	2698	1814	3188	7700
2007	2747	1953	3334	8034
2008	2742	2065	3422	8229
2009	2642	2085	3131	7858
2010	2772	2335	3302	8409
2011	2796	2477	3475	8748

Source: UNCTAD, Review of Maritime Transport (2012), based on data supplied by reporting countries and as published on the relevant government and port industry web sites. Main bulk products include iron ore, grain, coal, bauxite/alumina and phosphate.

Note: The data for 2006 onward are based on various issues of the Dry Bulk Trade Outlook, produced by Clarkson Research Services.

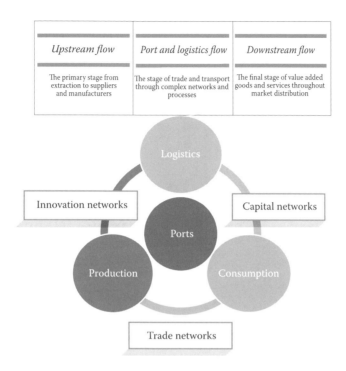

Upstream flow	Port and logistics flow	Downstream flow
The primary stage from extraction to suppliers and manufacturers	The stage of trade and transport through complex networks and processes	The final stage of value added goods and services throughout market distribution

FIGURE 2.1   The integration of hub ports in the global supply chain. (Courtesy of M.G. Burns.)

TABLE 2.2   World Seaborne Trade in 2006–2011, by Type of Cargo (Millions of Tons)

	World Cargo Loaded				World Cargo Unloaded			
Year	Total	Crude	Petroleum Products and Gas	Dry Cargo	Total	Crude	Petroleum Products and Gas	Dry Cargo
2006	7700.3	1783.4	914.8	5002.1	7878.3	1931.2	893.7	5053.4
2007	8034.1	1813.4	933.5	5287.1	8140.2	1995.7	903.8	5240.8
2008	8229.5	1785.2	957.0	5487.2	8286.3	1942.3	934.9	5409.2
2009	7858.0	1710.5	931.1	5216.4	7832.0	1874.1	921.3	5036.6
2010	8408.9	1787.7	983.8	5637.5	8443.8	1933.2	979.2	5531.4
2011	8747.7	1762.4	1033.5	5951.9	8769.3	1907.0	1038.6	5823.7

Source: UNCTAD, Review of Maritime Transport (2012), based on data supplied by reporting countries and as published on the relevant government, port industry web sites, and other specialist web sites and sources.

As transportation nodes are handling increasingly larger cargo volumes, port authorities have been asked to measure and maximize their performance in terms of ships' turnaround time, efficiency, cargo operations, congestion, and market concentration through their regional clients. An efficient port structure that establishes gateways to the inland and the hinterland necessitates the collaboration of the industry's key players, including terminal and liner operators, manufacturers, and cargo traders. A hub port's successful performance and connectivity require large amounts of capital investment,

(a)

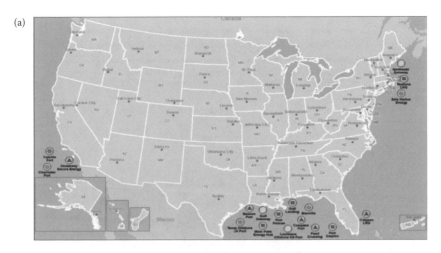

(b)

FIGURE 2.2 US ports and logistics network. (a) US deepwater ports. (From US Maritime Administration (MARAD), US Department of Transportation. Available at http://www.marad.dot.gov/ports_landing_page/deepwater_port_licensing/dwp_map/dwp_map.htm [accessed on June 15, 2013].) (b) US major ports. (From 101 Shipping. 2013. Available at http://101shipping.com/shippingseacost.html [accessed on June 15, 2013].) (c) US tonnage on highways, railroads, and inland waterways. (From US Department of Transportation, Federal Highway Administration. 2013. Available at http://www.ops.fhwa.dot.gov/freight/freight_analysis/nat_freight_stats/tonhwyrrww2007.htm [accessed on June 15, 2013].) (d) US national oil and gas production. (From Penn State University. 2013. Available at http://www.personal.psu.edu/uxg3/blogs/coal/assets_c/2009/08/US%20Oil%20and%20Gas%20Reserves%20Map-62158.html [accessed on June 15, 2013].) (e) US national pipeline mapping system. (From US Department of Transportation, Gas Transmission and Hazardous Liquid Pipelines. 2012. Available at http://www.phmsa.dot.gov/staticfiles/PHMSA/ImageCollections/Images/Pipeline%20Map.jpg.)

(c)

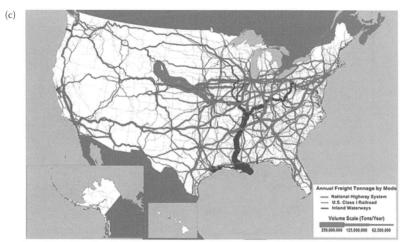

Sources: Highways: U.S. Department of Transportation, Federal Highway Administration, Freight Analysis Framework, Version 3.4, 2012. Rail: Based on Surface Transportation Board, Annual Carload Waybill Sample and rail freight flow assignments done by Oak Ridge National Laboratory. Inland Waterways: U.S. Army Corps of Engineers (USACE), Annual Vessel Operating Activity and Lock Performance Monitoring System data, as processed for USACE by the Tennessee Valley Authority; and USACE, Institute for Water Resources, Waterborne Foreign Trade Data. Water flow assignments done by Oak Ridge National Laboratory.

(d)   Oil and natural gas production in the United States
(Derived from Mast. et al. 1995)

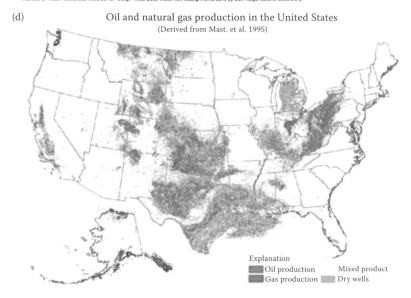

FIGURE 2.2 (Continued) US ports and logistics network. (a) US deepwater ports. (From US Maritime Administration (MARAD), US Department of Transportation. Available at http://www.marad.dot.gov/ports_landing_page/deepwater_port_licensing/dwp_map/dwp_map.htm [accessed on June 15, 2013].) (b) US major ports. (From 101 Shipping. 2013. Available at http://101shipping.com/shippingseacost.html [accessed on June 15, 2013].) (c) US tonnage on highways, railroads, and inland waterways. (From US Department of Transportation, Federal Highway Administration. 2013. Available at http://www.ops.fhwa.dot.gov/freight/freight_analysis/nat_freight_stats/tonhwyrrww2007.htm [accessed on June 15, 2013].) (d) US national oil and gas production. (From Penn State University. 2013. Available at http://www.personal.psu.edu/uxg3/blogs/coal/assets_c/2009/08/US%20Oil%20and%20Gas%20Reserves%20Map-62158.html [accessed on June 15, 2013].) (e) US national pipeline mapping system. (From US Department of Transportation, Gas Transmission and Hazardous Liquid Pipelines. 2012. Available at http://www.phmsa.dot.gov/staticfiles/PHMSA/ImageCollections/Images/Pipeline%20Map.jpg.)

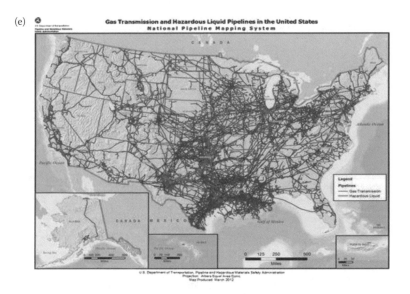

FIGURE 2.2 (Continued) US ports and logistics network. (a) US deepwater ports. (From US Maritime Administration (MARAD), US Department of Transportation. Available at http://www.marad.dot.gov/ports_landing_page/deepwater_port_licensing/dwp_map/dwp_map.htm [accessed on June 15, 2013].) (b) US major ports. (From 101 Shipping. 2013. Available at http://101shipping.com/shippingseacost.html [accessed on June 15, 2013].) (c) US tonnage on highways, railroads, and inland waterways. (From US Department of Transportation, Federal Highway Administration. 2013. Available at http://www.ops.fhwa.dot.gov/freight/freight_analysis/nat_freight_stats/tonhwyrrww2007.htm [accessed on June 15, 2013].) (d) US national oil and gas production. (From Penn State University 2013. Available at http://www.personal.psu.edu/uxg3/blogs/coal/assets_c/2009/08/US%20Oil%20and%20Gas%20Reserves%20Map-62158.html [accessed on June 15, 2013].) (e) US national pipeline mapping system. (From US Department of Transportation, Gas Transmission and Hazardous Liquid Pipelines. 2012. Available at http://www.phmsa.dot.gov/staticfiles/PHMSA/ImageCollections/Images/Pipeline%20Map.jpg.)

state-of-the-art technological innovation, an aggressive marketing strategy, but foremost powerful alliances, that is, governmental support and major clients.

Supply chains increasingly pursue a decentralized strategy in order to reduce production costs associated with central ports. This tendency is mirrored in the major players' efforts to merge, vertically integrate, or conclude long-term agreements.

The following sections will address issues to be resolved within the supply chains, including integration, location, and ports' repositioning within a supply chain.

2.1 LOGISTICS INTEGRATION OF PORT ACTIVITIES

The aim of this section is to investigate the ports' transformation inside integrated supply chains and consider challenges and opportunities in a rapidly changing—yet much promising—industry.

Logistics integration may be defined as "the integration of key business processes from end user through original suppliers that provides products, services, and information that add value for customers and other stakeholders" (Lambert et al. 1998). The benefits of integrated transport include cost efficiency and time efficiency between transshipment nodes, shared risk, distributive arrangements of investment and profit-sharing, reduced risk of cargo damage, pilferage or loss, the multimodal carrier retains network and coordination flexibility, efficient management of goods transported, efficient information and funds flow along their overall value chain, simplified formalities, and documentation processes.

Key objectives of integration include (a) benefits from economies of scale, based on joint resources; (b) visibility throughout the entire supply chain; (c) meeting—and exceeding—the final customer's product needs; and (d) strengthening the supply chain's performance and enhancing their marketing edge, as they compete with other supply chains.

It entails the broadest visions of holistic supply chain management that links the activities of supply chain partners as a source of customer value and competitive advantage.

From a ports' perspective, integration encompasses three levels:

a. Port operations, that is, from cargo handling to storage
b. Port facilities planning, monitoring, controlling, and maintenance
c. Holistic integration at a corporate level, where the port reaches out to the entire supply chain for the exchange of information, resources, and corporate goals

In essence, integration is (a) internal, that is, commences from within a port, or any company within the supply chain; and (b) external, that is, expands throughout the supply chain:

a. *Internal integration* in the shipping industry entails the following key characteristics:
   • Corporate leadership and teamwork
   • Interdepartmental goals, tasks, and performance
   • Getting the credit, sharing the risk
   • Exchange of customer information and market knowledge
   • Sharing innovation, talent, and resources to achieve economies of scale
   • Efficient IT, interdepartmental sharing of knowledge
   • Cross-functional planning
b. *External integration* in the shipping industry pertains to the following elements:
   Supply chains may collaborate like pools of companies, sharing compatible goals, ethics, and plans at a micro and macro level.
   • IT networks and software that encompass the entire supply chain and enable visibility, cargo, and carrier tracking
   • Alignment of leadership and strategies within the network
   • Exchange of customer information and market knowledge with selected client
   • Performance evaluation to be applied through the entire network
   • Performance adjustments to be applied horizontally (supply chain) and vertically (port authorities)
   • Getting the credit, sharing the risk
   • Sharing innovation, talent, and resources to achieve economies of scale
   • Cross-functional planning

It is worth noting that the maritime industry employs state-of-the art computer software, satellite systems, and technology that can be employed in every aspect and department, for example, security and safety monitoring, cargo operations, maintenance, emergency response, and so on. Furthermore, because of the industry's global nature, maritime professionals undergo continuous training and are more attuned to working in an interactive international environment. Therefore, the benefits of integration can be greater compared to other industries.

2.1.1 The Five Stages of Integration for the Maritime Industry

Globalization and technological innovations have enabled integration to transform supply chains to a great extent. Almost 25 years ago, integration was distinguished into four different stages within a supply chain. The outcome of integration will greatly depend upon the company's marketing strategy, structure, and operational process (Stevens 1989).

In this book, the stages of integration need to be redefined and restructured in order to encompass the modern global reality for the maritime industry and ports in particular:

Level I: The supply chain consists of different corporate entities and divisions; port strategy and operations are formed on the basis of the company's resources. Lack of information exchange creates lack of visibility throughout the logistics process, causing delays and poor outcome. Information breakdowns cause inability to follow a lean and agile process; lack of cohesion (Port Strategy 2013). The company misses opportunities of growth, owing to poor information exchange and poor market visibility. In order to prevent the port's collapsing, its only option will be to pursue integration, initially internal, and eventually external.

Level II: The organization applies internal integration, aiming at profit maximization and obtaining low-cost factors of production. The company lacks long-term visibility, and communication is not used effectively in order to attune to the supply chain's goals and objectives.

Level III: Corporate integration is focused on the port's territories. The port interacts with clients and its supply chain partners only in regard to operations, time, services, and activities that affect the port. Interdepartmental collaboration and information exchange have been achieved, and the port has achieved a holistic strategy. The port's vision and objectives pertain to annual market forecasts and growth. Because of the port's introvert attitude, it fails to reach out to global market opportunities and fails to deeply comprehend its clients' strategic goals and therefore fails to meet them.

Level IV: The port attains external integration throughout the entire supply chain. This new corporate attitude enables the port to expand its vision throughout the markets, manufacturers, terminal and liner operators, and other clients. The port now has a new, interactive approach and information sharing that enable it to share the risk and resources and benefit from economies of scale. It now functions as an efficient component of the supply chain that adds value to its partners' activities.

Level V: The external integration expands from a supply-chain level to a regional, national, or global level. The port authorities aim to increase the port's bargaining power by designing long-term strategies. With a keen understanding that the port's power within the current supply chain may be influenced by other, competing ports, the port extends its vision and marketing activities to other supply chains and new clients.

2.2 STRATEGIC LOCATION AND MARKET ACCESSIBILITY FOR EXISTING AND EMERGING SEAPORTS

The previous section addressed, among others, a supply chain's risk of power imbalance, that is, an imbalance between demand and supply that may grant certain key players excessive market power over others. As the global sea trade routes are constantly changing depending on the demand for certain commodities, it is imperative to discuss the significance of a port's strategy of controlling or enhancing its commercial leveraging.

Port managers can increase their business scope by strategically pursuing leverage through (a) utilizing their geopolitical power, (b) integration, (c) efficiency and the ability to handle large volumes of cargo and traffic with limited delays and interruptions, (d) eliminating competition and substitution through establishing competitive advantages, and (e) optimum utilization of resources and the port's factors of production, as these will determine the strategy's success.

Among the above factors, geography is the most rudimentary element in the economic, commercial, and foreign policy of nations, owing to its lasting power.

A nation's global power greatly depends on its seaports and its maritime vigor. In this sense, the author strongly believes that a port's strategy is not simply a commercial act but a political instrument. President Thomas Jefferson* (1743–1826) verified the political and socioeconomic significance of controlling commerce and navigation, during his speech at the US Congress. In his words, *the marketing of our productions will be at the mercy of any nation which has possessed itself exclusively of the means of carrying them; and our policy may be influenced by those who command our commerce.*

Strategy can be defined as "the art and science of utilizing during both peace and war, of all of the national forces, through specific objectives, concepts and resources, in order to ensure the objectives and concepts of national policy" (Burns 2014; *Dictionary of Military and Associated Terms* 1985). Strategy pertains to securing and maintaining a competitive edge over competitors, even when the resources available are limited.

The modern global market is characterized by cutthroat competition; hence, it would be appropriate to utilize military principles when considering a port's strategy: The two main thresholds of a port's tactics should be (a) operational, that is, based on existing contracts, capacities, and performance; and (b) force developmental, that is, market pressure points are employed for designing strategies at a macro level and are triggered by potential or actual threats, desired goals, and market prerequisites. While ports' assets, infrastructure, and resources are essential to sustain a strategy, the strategy itself has to be resourceful and inventive enough to be able to be attained even with the least of resources.

2.2.1 Ports' Success Factors

A capacity that discerns seaports as tools for regulating global trade is both consistent with the shipping industry's views and reflected in a nation's trade and economic policies. While national interests aspire to cash in on the ports' strategic geopolitical locations and competitive advantages, in practice the maritime industry is highly volatile, characterized by dynamic fluctuations and complex supply and demand networks. Ports, just like ships, are competing in a highly antagonistic global environment, which on top of the laws of

---

* The third President of the United States, an American Founding Father and the principal author of the Declaration of Independence (1776).

supply and demand are strongly influenced by politics, trade agreements, and customers' preference, in addition to currency wars, volatile commodity trade prices, scarce commodities, safety and security threats, and so on. These are some of the factors that formulate global trade patterns and sea routes and eventually determine a port's potential for productivity, employability, and economic growth.

Based on the above, factors that may influence a port's strategic position include the following:

1. Global capital markets (United States, Japan, Western Europe, China. etc.)
2. High production, demand, or supply hubs and regions (Shanghai, Busan, Houston, Hong Kong, S. Louisiana, etc.)
3. Value-added trade centers (i.e., of services, fuel refineries, agriculture, and manufacturing)
4. Transit areas (Panama Canal, Suez Canal, etc.)
5. Supply/replenishment areas (e.g., Singapore, Malta, Cyprus, etc.)
6. Free ports and free trade zones: US Virgin Islands (United States), Eilat (Israel), Singapore (Singapore), Malta (Malta), Hamburg (Germany), Colon "Zona Libre" (Panama), Suez Canal Container Terminal (Egypt), over 30 free trade zones of Dubai (UAE)
7. Shipbuilding, ship repair zones: China, Japan, South Korea, United States, Germany, United Kingdom, and so on

Interestingly enough, a port's ability to withstand the above challenges is restricted by its geographic and national boundaries. While ports are typically bound by geological and meteorological variables, regional and national regulations, taxation regime, and government budget allocation, ships enjoy the privilege of resilience, for example, flexible trade routes, commodities, converting type and design, and even regulatory compliance through flags of convenience (UN ECE/Trans/210, 2010).

Based on the above, the port selection criteria encompass a multitude of factors, including the following:

- Port physical and technical infrastructure (nautical accessibility profile, terminal infrastructure and equipment, hinterland accessibility profile)
- Geographical location (vis-à-vis the immediate and extended hinterland and vis-à-vis the main shipping lanes)
- Port efficiency
- Interconnectivity of the port (sailing frequency)
- Reliability, capacity, frequency, and costs of inland transport services by truck, rail, and barge (if any)
- Quality and costs of auxiliary services such as pilotage, towage, customs, and so on
- Efficiency and costs of port management and administration (e.g., port dues)
- Availability, quality, and costs of logistic value-added activities (e.g., warehousing)
- Availability, quality, and costs of port community systems
- Port security/safety and environmental profile of the port
- Port reputation

The question to be answered herein is to what extent geography plays a role in a port's significance. Table 2.3 demonstrates how the leading economies and global trading centers have evolved over the past decade.

TABLE 2.3 Global Economies and Seaports, 1999–2011 Global Trading Centers

	1999				2011		
Country	Overall Rank	Size Rank	Interconnectedness Rank	Country	Overall Rank	Size Rank	Interconnectedness Rank
Germany	1	2	2	China	1	1	1
United States	2	1	6	United States	2	1	3
France	3	3	2	Germany	3	3	2
Japan	4	3	5	Netherlands	4	6	3
United Kingdom	5	5	2	Japan	5	4	8
Netherlands	6	8	1	France	6	5	6
Italy	7	7	7	Italy	7	7	7
Canada	8	6	12	United Kingdom	8	8	5
China	9	9	8	Belgium	9	9	11
Belgium	10	11	9	South Korea	10	10	10

Source: International Monetary Fund (IMF), Finance and Development (2011).

Note: (1) Weighted average of the size and interconnectedness rankings using a 0.7/0.3 weight breakdown, respectively, and (2) excludes links representing less than 0.1% of each country's GDP.

In the year 1999, it seems that the major trade routes involve countries of the Western Hemisphere, namely, North America and European Union member-states. In 10 years, China became the world's leading trader. It is worth noting that in 1999, China restructured its foreign trade policy, after which, the nation's most profitable companies joined the New York Stock Exchange, while in 2001 China joined the World Trade Organization.

Figure 2.3 reflects the world's leading ports from 2003 to 2011 shift from the Western Hemisphere ports toward the eastern countries. In less than a decade, Asia's leading ports seem to have doubled in total cargo volume, whereas the Western Hemisphere's trade growth has been relatively slow.

According to the International Monetary Fund (IMF), trade in the 1970s was primarily limited to a few developed nations, particularly the United States, Germany, and Japan, which collectively made up over a third of international trade. By the 1990s, world trade became broad enough to encompass a number of emerging economies of the Far East. By 2010, China became world's second most significant trading partner after the United States, overpowering Germany and Japan (2011).

The uprising of rapidly developing countries like China, India, and Brazil has resulted in an increasing growth of cargo volume. As seen in Table 2.4 and Figure 2.4, the new

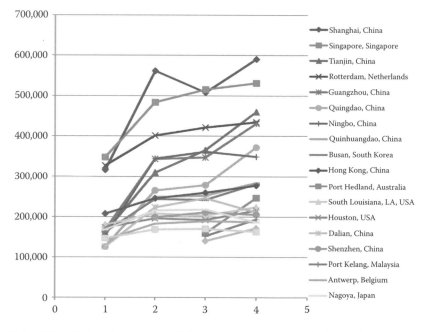

FIGURE 2.3 World's leading ports: 2003–2011. (Courtesy of the author, based on data from World Shipping Council. 2013. Available at http://www.worldshipping.org [accessed on June 15, 2013]; IMF. 2013. *World Economic and Financial Surveys.* World Economic Outlook, April 2013, Hopes, Realities, Risks. International Monetary Fund Publication Services, Washington; Agência Nacional de Transportes Aquaviários—ANTAQ, Brazil. 2013. Available at http://www.antaq.gov.br [accessed on June 21, 2013]; Institute of Shipping Economics and Logistics. 2012. Containerization International Yearbook 2012. ISBN-13: 978-1843119906; U.S. Army Commerce Statistics Center, Secretariat of Communications and Transport, Mexico. 2013; Waterborne Transport Institute, China. 2013. Available at http://www.wti.ac.cn [accessed on June 12, 2013]; AAPA Surveys. Available at http://www.aapa-ports.org [accessed on June 21, 2013]; and various port Internet sites [2003–2013].)

TABLE 2.4   World Ports Ranking, Total Cargo Volume, 2003–2011

	Ports	2003	2007	2008	2011
1	Shanghai, China	316,210	561,450	508,000	590,439
2	Singapore, Singapore	347,694	483,616	515,415	531,176
3	Tianjin, China	161,460	309,460	365,163	459,941
4	Rotterdam, Netherlands	326,958	401,181	421,136	434,551
5	Guangzhou, China	167,720	343,250	347,000	431,000
6	Qingdao, China	125,620	265,020	278,271	372,000
7	Ningbo, China	153,980	344,000	361,163	348,911
8	Qinhuangdao, China		248,930	252,000	284,600
9	Busan, South Korea	162,460	243,564	241,683	281,513
10	Hong Kong, China	207,612	245,433	259,402	277,444
11	Port Hedland, Australia			159,391	246,672
12	South Louisiana, Louisiana, USA	180,493	207,785	203,157	223,633
13	Houston, USA	173,320	196,014	192,473	215,731
14	Dalian, China	126,020	222,860	246,000	211,065
15	Shenzhen, China		199,000	211,000	205,475
16	Port Kelang, Malaysia			152,348	193,726
17	Antwerp, Belgium	142,875	182,897	189,390	187,151
18	Nagoya, Japan	168,378	215,602	218,130	186,305
19	Dampier, Australia			140,375	171,844
20	Ulsan, South Korea	146,940	168,652	170,279	163,181

Source: M.G. Burns, based on data from World Shipping Council. 2013. Available at http://www.worldshipping.org (accessed on June 15, 2013); Agência Nacional de Transportes Aquaviários—ANTAQ, Brazil. 2013. Available at http://www.antaq.gov.br (accessed on June 21, 2013); Institute of Shipping Economics and Logistics. 2012. Containerization International Yearbook 2012. ISBN-13: 978-1843119906; US Army Commerce Statistics Center, Secretariat of Communications and Transport, Mexico. 2013; Waterborne Transport Institute, China. 2013. Available at http://www.wti.ac.cn (accessed on June 12, 2013); AAPA Surveys. Available at http://www.aapa-ports.org (accessed on June 21, 2013); and various port Internet sites (2003–2013).

supply chains and ports geography mainly involves Asian ports close to China, that is, Singapore, South Korea, and Malaysia.

In this new reality, the distance between ports has been expanded, and yet because of low-cost benefits, technological advances and economies of scale, the global supply chains have been positively affected. Focusing exclusively on the geographic parameters and distance between ports, the impact on the supply chains is hereby examined.

A probabilistic gravity model of spatial interaction was formulated by Huff (1963) in order to analyze market areas for retail outlets (see case study by Kim et al. 2011).

Since this section deals with the geographic significance of ports within a supply chain, the author has hereby modified Huff's model to evaluate the geographic significance of a port's location, owing to its vicinity with the suppliers' and consumers' markets.

The extended gravity model reflects the likelihood of a supply chain selecting a specific port and is stated in Equation 2.1:

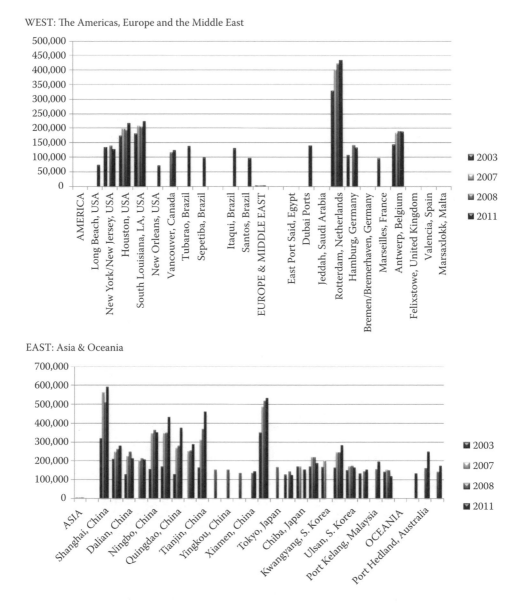

FIGURE 2.4　World's leading ports by hemisphere, 2003–2011. (a) West: The Americas, Europe, and the Middle East. (b) East: Asia and Oceania. (Courtesy of the author, based on data from World Shipping Council. 2013. Available at http://www.worldshipping.org [accessed on June 15, 2013]; IMF. 2013. *World Economic and Financial Surveys.* World Economic Outlook, April 2013, Hopes, Realities, Risks. International Monetary Fund Publication Services, Washington; Agência Nacional de Transportes Aquaviários—ANTAQ, Brazil. 2013. Available at http://www.antaq.gov.br [accessed on June 21, 2013]; Institute of Shipping Economics and Logistics. 2012. Containerization International Yearbook 2012. ISBN-13: 978-1843119906; US Army Commerce Statistics Center, Secretariat of Communications and Transport, Mexico. 2013; Waterborne Transport Institute, China. 2013. Available at http://www.wti.ac.cn [accessed on June 12, 2013]; AAPA Surveys. Available at http://www.aapa-ports.org [accessed on June 21, 2013]; and various port Internet sites [2003–2013].)

$$P_{ij} = \frac{S_j/K_{ij}^{\lambda}}{\sum S_j/K_{ij}^{\lambda}} \qquad (2.1)$$

where

P_{ij} = likelihood that supply chain i selects port at location j

S_j = size of the node at location j, where the node refers to any component of the supply chain, for example, supplier, manufacturer, buyers' market, and so on

K_{ij} = knots, that is, distance per time required to access location j. In cases where $\lambda > 1$, distance per time has a more significant impact, whereas if $\lambda < 1$, the supply chain's node size has a greater impact

On the other hand, a port's or nation's geographic remoteness may act as a deterrent to trade and transport. A formula for evaluating remoteness is often employed to evaluate the multilateral trade conditions of national imports and exports while considering factors of trade remoteness. This is frequently estimated as follows:

$$\text{Rem}_i = \sum_j \frac{\text{dist}_{ij}}{\text{GDP}_j \ \text{GDP}_w} \qquad (2.2)$$

The formula measures a country's average weighted distance from its trading partners, where weights are the partner countries' shares of world GDP, herewith signified as GDP_w. (UNCTAD and WTO 2013).

Having established that a port's geography provides it with a competitive edge, a port's success, to a great extent, lays in its strategic position, that is, (a) a port's geopolitical and socioeconomic importance at a macro-level, which also has the greatest duration over time, and (b) a port's location in the vicinity of significant trade routes, which benefits the port at a short term.

Various formulas have been developed to measure a nation's average weighted mileage from its trading partners, yet these econometric models are inaccurate as distance cannot be considered as the only barrier of trade (Anderson and van Wincoop 2003). Unless the port offers some unique competitive advantages, its geographical position alone will not be beneficial in the long term. To measure the nodes of a specific trade within a supply chain, a three-tiered evaluation is required: (a) a Regional Assessment will be required in order to choose a geographic market within a continent or a number of nations; (b) Market and Trade Assessment will be used to appraise the trading area and the market clusters; and finally (c) Port Evaluation will determine the port's ability to meet—if not exceed—the supply chain's purposes.

Consequently, despite the significance of a port's geographical location, there are alternative factors that could enhance the position of a competing port. The key factors that are considered significant by shipowners, terminal operators, and supply chains include the following:

- Broader strategic geographical location, that is, port is a part of a significant trade route, or market, and belongs to an influential supply chain
- Port efficiency and reputation

- Port's size, infrastructure, and superstructure, that is, ability to handle cargo
- Value for time and money (quality of service and output, i.e., cargo handling, warehousing)
- Interconnectivity: intermodal reliability and supply chain connections
- Satisfactory levels of regulatory compliance, that is, conformity of environment, safety, security, and quality (extreme levels of compliance, i.e., too low or too high, are considered by shipowners and terminal operators as undesirable)

As observed in the Major Global Ports' facts and figures as stated above, within a decade, the global trade routes of major commodities have significantly changed, pursuant to the rapid growth in production of countries in the Far East. Shanghai, for example, enjoys the global leadership of trade, since in the port's vicinity, it hosts 36 industrial zones. The particularity of China is that its rising salaries are no longer as competitive as they used to be in the 1990s boom. The country's competitive advantage and sustainability derive from a combination of advantages that offer the country's production reliability, a stable political regime, central government controlling as to the ports' cargo specialization and the trade zones' products specialization, efficient telecommunications, infrastructure, and supply chain networks (Harney 2008). Other advantages that attract foreign investors include lower costs for land, government incentives, and less stringent environmental and safety regulations that enable the country to produce at a faster pace (e.g., coal is used as a leading source of industrial energy, whereas its use in the Western Hemisphere has been banned for environmental reasons).

A port's geographical significance should always be evaluated under the prism of derived demand: because of the fact that the demand for ports is a derivative of its markets, trade zones, and so on, ports' commercial significance is directly related to the commodities they trade.

Having said that, it is worth clarifying that the geographical and geopolitical comparative advantages are not strictly focused on a single port but pertain to a wider region that enjoys similar comparative advantages. Nevertheless, even the regions of utmost geopolitical significance are affected by competition. The Panama Canal is such an example:

Case Study: The *Panama Canal* is a strategic transit area. As an efficient route of strategic geopolitical significance to global trade, it facilitates navigation between the Atlantic and the Pacific Oceans, that is, trade within the US East Coast–West Coast (saving between 4500 and 11,500 nautical miles), as well as the trade from US West Coast to Europe and West Africa. Since its inception, the canal offered substantial cost financial savings to intercoastal trade. By 1940, the United States' national revenue was approximately 4% higher than it would have been without transiting the canal.

The $5.25 billion expansion of the Panama Canal is due to be completed by 2015 (Sabo 2013). It is anticipated to increase its geopolitical significance with enormous advantages for the authorities. This expansion is expected to double the size of ships previously transiting the canal; the expanded capacity will actually be as high as 13,200 containers, which will travel on ships as large as 1200 feet long and 160 feet wide. The Panama Canal has been competing with the US Intermodal system, for example, the NAFTA Corridors, the Pan-American Highway within the continent, and very remotely with the Suez Canal in the South and Southeast Asia–United States sea routes. Figure 2.5 demonstrates the history of the Panama Canal, its centennial celebrations, and significant images pertaining to its 2015 expansion.

In 2013, the Hong Kong-based Chinese company HK Nicaragua Canal Development Investment Co. Ltd. (HKND Group) was granted a 100-year concession by the Nicaraguan

FIGURE 2.5 Dry bulker transiting south toward Pedro Miguel Locks. (Courtesy of the Panama Canal Authority.)

Government, to build a canal joining the Atlantic and Pacific oceans. The concession does not guarantee that the new canal will be built in the foreseeable future. Disputes in the San Juan River between Nicaragua and Costa Rica have been intensified with the possibility of a future canal. Furthermore, its construction will take at least a decade, that

CASE STUDY: THE PANAMA CANAL, CELEBRATING 100 YEARS OF GROWTH

- The lifeblood of Panama and the global trade.
- Global Leader in maritime industry services.
- Foundation of the global trade and transport practices.
- An eminent structure of innovation, reliability, and visibility.

A NEW ERA AHEAD

The Panama Canal (Figure 2.5) is roughly 50 miles (80 kilometers) long along the Atlantic and Pacific oceans. This seaway is a passage across the Continental Divide, that is, an isthmus that connects North and South America. The canal utilizes an efficient system of locks, that is, chambers with entry and exit gates. The locks serve as water lifts that raise vessels to 81.3 feet (26 meters) from ocean level to the Gatun Lake level.

The Panama Canal's history extends back virtually to the first explorers of the American continent. Upon the end of the 19th century, scientific innovations and the need for global sea transport induced the canal's construction.

The Hay–Pauncefote Treaty signed in November 1901 stipulated England's consent for the United States to construct and operate the canal. In 1902, US President Roosevelt paid $40 million for the legal rights to the Panama concession. The Republic of Panama was established in 1903 pursuant to its separation from Colombia. The Panama Canal's construction was concluded in 1913 and officially launched in 1914 (see Figure 2.6a and b).

FIGURE 2.6   Historical photos of the canal's construction and in use. (Courtesy of the Panama Canal.)

The water utilized to elevate and lower ships in every single lock originates from the Gatun Lake by gravitational pressure; it arrives to the locks via a mechanism of major culverts that expand beneath the lock chambers through the sidewalls and the center wall. The Panama Canal Expansion Program is a planned state-of-the-art technological marvel with 3 million cubic meters of concrete used in the construction of the new lock processes for the Expansion Program. The significance of the Panama Canal's expansion in global trade and transport can hardly be exaggerated, as more enticing, optimum services will not be found elsewhere.

With a total manpower of roughly 10,000 people and 24/7 operations throughout the year, the Canal is the lifeblood of Panama and the Americas. Figures 2.7 through 2.12 illustrate additional features, and forthcoming features, of the canal.

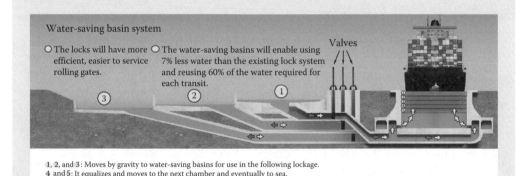

FIGURE 2.7   The canal's water-saving basin system explained. (Courtesy of the Panama Canal.)

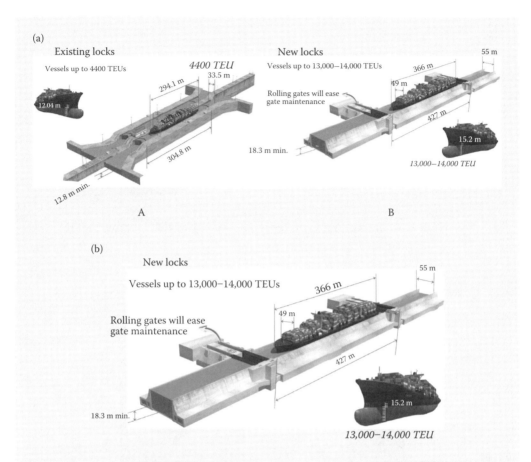

FIGURE 2.8 The existing locking system and the new locking system. (a) Existing locks: vessels up to 4400 TEUs; (b) new locks: vessels up to 13,000–14,000 TEUs. (Courtesy of the Panama Canal.)

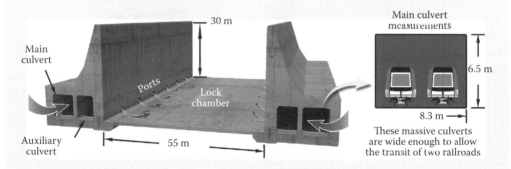

FIGURE 2.9 A profile of the new lock chamber and lateral culverts. (Courtesy of the Panama Canal.)

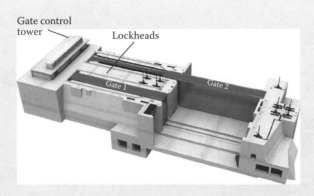

FIGURE 2.10   The Rolling Gate recess. (Courtesy of the Panama Canal.)

(a)

(b)

FIGURE 2.11   A rendering of the third set of locks on the Atlantic side looking north and a rendering of third set of locks on the Pacific side looking north. (a) Rendering: Third set of locks on the Atlantic side looking north. (b) Rendering: Third set of locks on the Pacific side looking north. (Courtesy of the Panama Canal.)

FIGURE 2.12 An overview of the canal expansion program. (Courtesy of the Panama Canal.)

is, at a time when the Arctic opening to sea trade will be another attractive alternative for certain global routes.

As an antipode, the Suez Canal has no easy alternatives. In 1961, the Israeli Government designed an overland Suez Canal deviation by an Israeli company, transporting Japanese goods to Eilat, Israel's Free Trade Zone, and move them overland to Haifa for further transshipment to its final discharging node, in Italy (JTA 1961). The inland option is frequently an attractive alternative, yet seaports, owing to the ships' economies of scale, achieved by size and low cost of fuel, so far exceed any train or truck alternative.

On the basis of this presumption, the likely region for a future canal to be built connecting Eastern Europe and the Middle East would have to pass through Israel, Syria, Iraq, Iran, or Turkey. The fragile political climate in the region for a number of years now would not encourage such a risky endeavor. This brings us to the other alternative: when the Suez Canal closed in 1973 during the Yom Kippur War (5 days war), ships' only other alternative was to navigate around Africa, resulting in significant global supply chain disruptions, time lost, and fuel oil consumption. This offered Durban and other South

African ports a great advantage, as they were used as bunkering ports, as well as repairs and replenishment stations.

The above case studies have taught us that the maritime industry evolves and discovers new alternatives for geography, and all the factors of production.

One of the more significant findings to emerge from this section is that in a rapidly growing and changing global trade, ports are still bound by geographical restrictions, yet it is other factors such as cost, time, and quality that may contribute to the change of trade routes. Technology seems to have outgrown geography, as IT and advanced satellite telecommunications have eliminated distances. A port's geopolitical significance may be permanent, that is, a nation's location, or temporary: depending on supply and demand fluctuations. A port's significance is also related to the regional factors of production and trade agreements that determine the country's trade growth and trade balance (import/export ratio).

The next section will examine the principal factors that affect a supply chain and the role of ports in different network formations and will seek for solutions.

2.3 SUPPLY CHAIN OPPORTUNITIES, COMPETITION, AND CONFLICT PREVENTION

This section highlights the changing role of ports and port managers as supply chains and global trade change. The three fundamental issues to be addressed are as follows: first, in which manner did the logistics networks and nodal connection affect the global maritime industry? Second, who are the new industry's key players within this redefined industry? And third, how can ports eliminate conflict and competition and benefit from the industry's opportunities?

The significance of port connectivity became magnified over the past few years, when globalization and outsourcing required the transport of large sea trade volumes throughout expanded global sea routes. According to the latest UNCTAD annual statistics (2012), the world cargo volume has almost doubled in 13 years, that is, from 27 billion in cargo ton-miles in 1999 to over 45 billion in cargo ton-miles in 2012. Ninety percent of this cargo involves sea trade, which means that the world's 2000 seaports, 10,000 terminals, and 50,000 ships were employed to accommodate this trade growth. With the globalization of cargoes and markets, modern ports needed to restructure their operations and networking systems, in order to enhance connectivity throughout their supply chain associates. Although the focal point of supply chains remains the reliable transportation and distribution from the ports to the markets, the new supply chain networks became more complex and lengthy in processes.

Designated transshipment hub ports have been developed owing to an increasing demand for gateway ports (OECD 2008). As of 2013, at least a quarter of the world's largest ports serve as transshipment hubs, whereas most of the global hub ports' container throughput is being distributed as transshipment. The following is a case study of Gulf Winds International Inc., a typical third-party logistics company serving global cargoes in Texas, USA.

Furthermore, larger ship sizes such as Maersk's Triple E type aim to achieve economies of scale on behalf of the shipowners, which also suggests that less global ports will be used to serve the most profitable trade routes. This will eventually lead to fewer global ports of call, the concentration of cargoes in fewer, selected regions, and increased competition among domestic ports.

CASE STUDY: GULF WINDS INTERNATIONAL

A 3PL, THIS IS ABOUT MORE THAN THE MOVE

In our modern era of mega-ports, mega-ships, and a global trade volume of 8 billion tons, which is expected to quadruple in the next few decades, a seaport's magnitude can only be sustained with a robust and reliable supply chain network that will offer customized drayage, storage, transloading, and long-haul transportation in a dependable, timely, efficient, and cost-effective manner.

This is the job of third-party logistics (3PL) companies that have gained a competitive edge by offering tailor-made solutions in the complex global logistics networks and market accessibility, eliminating port traffic and bottlenecks. Efficient 3PL companies are the bloodline of global, national, and regional trade and transport: they strengthen a port's efficiency, eliminate competitors and transport substitution from other regions, and as a result reinforce a nation's marketing edge at a global level.

Gulf Winds Intl. is a 3PL company with over 250 trucks and 2 million square feet of warehouse space. Having opened its seventh warehouse in 2012, GWI offers industry-leading services focused on the following:

- Container drayage and management
- Warehousing: storage and distribution, transloading
- Logistics: transportation management, truck brokerage, port logistics
- Other services:
  - Oilfield logistics
  - Cost management
  - Customs examination station
  - Out-of-gauge cargo
  - Local and long-haul transportation

Now, when the 3PL provider is headquartered in a strategic maritime location such as Houston, Texas, it is easy to understand why GWII's Chairman, Steve Stewart, was named as "Maritime Person of the Year," and why August 20 was proclaimed as "Steve Stewart Day" in Houston.

GWI's mission is to provide world class logistics services through continual investment in their "people, clients, community and the world that we live in." Not only does Gulf Winds strive to live out this mission in everything they do, they make it their personal motto to consistently be "More than the Move."

The company's strategic locations adjacent to port and rail facilities in Houston and Dallas combined with state-of-the-art information systems provide real cost savings opportunities for their global clients.

Tedd and Steve Stewart awarded with the Harrisburg Rotary Club for charitable work.

Source: http://www.gwii.com/

The supply chain's performance prerequisites remain the same, for example, what still matters is time, value for money, reliability, productivity, and quality (OECD 2009). And yet, what changed were the distribution paradigms.

The industry is mostly familiar with the two leading networks, that is, point-to-point system and the hub-and-spoke system, yet global supply chains and their distribution paradigms have evolved gradually. As seen in Figure 2.13, the supply chain networks have evolved over time, in a manner that affects the players' bargaining power.

Stage 1: The first supply chain networks adopted the point-to-point distribution models. This linear process entailed linear types of negotiations; hence, the power distribution within the entire supply chain was more even. As can be seen in Figure 2.13, more points were involved and negotiations in this linear process were also performed in a point-to-point method. At this stage, port authorities were placed in the advantageous position of coordinating the factors of production along the goods movement process.

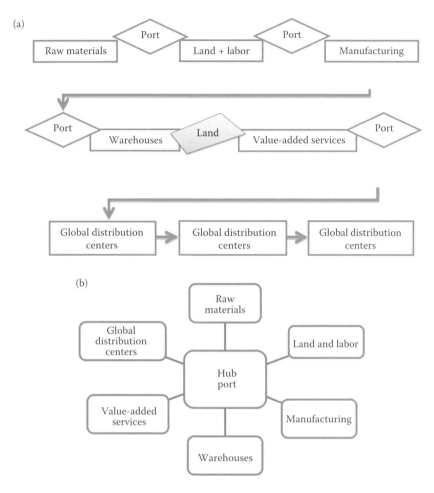

FIGURE 2.13 The evolution of supply chain distribution paradigms. (a) Type 1: point-to-point network: even distribution of power. (b) Type 2: hub-and-spoke network: the hub has increased bargaining power over the spokes. (c) Type 3: interactive hub-and-spoke networks: terminal operators, shippers, and manufacturers merge, and coordinate multiple networks. (d) Type 4: holistic networks: a large-scale lean and agile system.

Stage 2: In an effort to benefit from economies of scale, supply chains became more centralized. The multiple global distribution centers merged and were relocated in the vicinity of mega-ports. Hence, they evolved into hub-and-spoke distribution paradigms, where the central hub has the bargaining power over the spokes. As can be seen in Figure 2.13, hub ports are the focal point of the entire network, whereas smaller ports served as spokes, that is, as feeders for inland connectivity and cargo distribution. Each spoke is assigned to a different shipper and terminal operator, which again shifts the power toward mega-ports.

Stage 3: As terminal operators, shippers and manufacturers merge or form strategic alliances; they coordinate with multiple networks. Efficiency systems need to be designed, for the supply chain to achieve economies of scale. The large-scale mergers lead the industry toward the next arrangement.

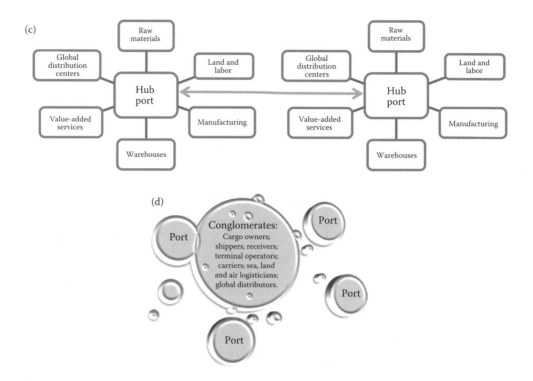

FIGURE 2.13   (Continued) The evolution of supply chain distribution paradigms. (a) Type 1: point-to-point network: even distribution of power. (b) Type 2: hub-and-spoke network: the hub has increased bargaining power over the spokes. (c) Type 3: interactive hub-and-spoke networks: terminal operators, shippers, and manufacturers merge, and coordinate multiple networks. (d) Type 4: holistic networks: a large-scale lean and agile system.

Stage 4: This is the industry's most recent arrangement: a holistic, large-scale lean and agile system, where optimum efficiency is achieved, in terms of resource allocation and economies of scale. The market power is increasingly shifting from the ports to the multinational conglomerates that own a large part of the supply chain, mostly shipping lines (see Chapter 4) and terminal operators (see Table 2.5). In order to increase their market share, the key players among the supply chains focus on the big picture, that is, controlling the trade flow from a holistic perspective, without permanently promoting particular routes.

On the basis of this principle, the key supply chain players can now achieve low costs and efficiency by endless network combinations, where ports and hinterlands compete, and thus become "the weakest link of the chain" (UNECE 2010). The power is hereby shifted from the hub ports to the global conglomerates that own the vast majority of the spokes within the supply chains. This concentration of power within the new structures makes no port irreplaceable, a fact that intensifies competition among global and national ports.

These major players of the maritime industry share some common strategies, such as (a) geographical and operational diversity, (b) intensive financial and investment activities, (c) strategic alliances within the maritime, energy, logistics, and offshore industries, and so on. Quite frequently, a single conglomerate controls the vast majority of the supply chain and all the factors of production, that is, from the raw materials, to manufacturing, the sea

TABLE 2.5 World's Top 10 Terminal Operators: Corporate Profile and Equity-Based Throughput, 2011

Terminal Operator in Market Power Sequence	Corporate Profile	Strategic Location of Terminal Operations, Logistics, and Other Port Activities						Million TEUs	% Share of Global Throughput
		Americas and Panama Canal	Asia	Africa	Europe	Oceania	Middle East and Suez Canal		
1. PSA International	PSA is the world's largest conglomerate in investments and finance, focused in the maritime and hotel business. PSA Marine services are actively involved in port operations and management, support vessels for the offshore oil and gas industry, heavy-lift, oil spill response, pilotage, port and terminal towage, and salvage services. Its network of operations includes 29 ports and 17 countries, mainly in Asia and Europe, as well as in Panama, South America, and the Middle East.	F/IT, O, E, L, C, A	Δ, F/IT, O, E, L, C, A		F/IT, O, E, L, C, A		F/IT, O, E, L, C, A	47.6	8.1
Ownership countries: China and Singapore									
2. Hutchison Port Holdings, Subsidiary of Hutchison Whampoa Limited (HWL)	HPH is a leading conglomerate, focused in Port and Terminal Investments, development and operations. Its network of port operations comprises 320 berths in 52 ports, spanning 26 countries mainly in Asia, Panama, and South America, as well as in the Middle East, Africa, Europe, North America, and Australia.	F/IT, A, L, C, A	Δ, F/IT, A, L, C, A	F/IT, A, L, C, A	F/IT, A, L, C, A	F/IT, A, L, C, A	F/IT, A, L, C, A	43.4	7.4
Ownership country: Hong Kong, China									

(continued)

TABLE 2.5 (Continued) World's Top 10 Terminal Operators: Corporate Profile and Equity-Based Throughput, 2011

Terminal Operator in Market Power Sequence	Corporate Profile	Strategic Location of Terminal Operations, Logistics, and Other Port Activities						Million TEUs	% Share of Global Throughput
		Americas and Panama Canal	Asia	Africa	Europe	Oceania	Middle East and Suez Canal		
3. DP World	DP World Cargo Services; Container	T	T	T	T	T	Δ	33.1	5.6
Mostly owned by Dubai World	Rail Road Services Private Limited (CRRS/DP World Intermodal); P&O	C	C	C	C	C	T		
	Maritime Services (former British-	L	L	L	L	L	C		
	owned Peninsular and Oriental Steam	F/I	F/I	F/I	F/I	F/I	L		
Ownership country:	Navigation Company). Its network of	A	A	A	A	A	F/I		
United Arab Emirates	port operations entails 60 terminals						A		
	throughout six continents; most of its								
	business is focused in Asia (India and								
	China) and Australia, as well as the								
	Suez Canal and Europe. Its new								
	strategic acquisitions are located in								
	Africa, Western Europe, and Brazil. Sea								
	and land container handling generating								
	approximately 80% of its earnings,								
	whereas 20% of its activities involve								
	shipping and logistics services.								

					32.0	5.4
4. APM Terminals and APM Terminals operate in 68 countries	F	F	F	F	F	
Maersk Shipping Lines with interests in 69 port and terminal	T	T	T	T	T	
Maersk Drilling facilities and over 170 Inland Services	O	O	O	O	O	
Maersk Supply Services operations. The *Maersk Line fleet*	C	C	C	C	C	
comprises more than 500 vessels,	L	L	L	L	L	
Ownership country: whereas the group of companies has	F/I	F/I	F/I	F/I	F/I	
Denmark, HQ in the expanded to oil and gas drilling, short	A	A	A	A	A	
Netherlands sea shipping, global supply/						

procurement services, warehousing and
distribution networks. Their activities
have expanded mainly in the Americas,
Asia, and Europe, followed by strategic
locations in Africa, the Middle East,
and the Suez Region. APM expands its
inland services in India, a rapidly
growing economy.

(continued)

TABLE 2.5 (Continued) World's Top 10 Terminal Operators: Corporate Profile and Equity-Based Throughput, 2011

Terminal Operator in Market Power Sequence	Corporate Profile	Strategic Location of Terminal Operations, Logistics, and Other Port Activities						Million TEUs	% Share of Global Throughput
		Americas and Panama Canal	Asia	Africa	Europe	Oceania	Middle East and Suez Canal		
5. COSCO Group	COSCO consists of six listed companies and owns over 300	F	ΔP	F	F	F	F	15.4	2.6
		T	F	T	T	T	T		
Ownership country: Republic of China	subsidiaries globally, offering services in terminal operations, ship building, freight forwarding, ship repairs, container manufacturing, financing, real estate, trade, and IT. Its owned fleet has now expanded to over 800 ships. COSCO owns the world's largest fleet of commodities ships, followed by the Japanese company Nippon Yusen K.K., Japan. As of 2013, COSCO's terminal operations have expanded in Greece, whereas its shipyards and offshore drilling activities have also expanded. Interestingly enough, other global strategic regions in Asia and Africa have been heavily invested by "China Merchants Holding International," China's largest public port operator, also owned by the Government of China (CMHI 2013).	O	T	O	O	O	O		
		E	B	E	E	E	E		
		IT	R	IT	IT	IT	IT		
		L	O	L	L	L	L		
		F/I	E	F/I	F/I	F/I	F/I		
		C	IT	C	C	C	C		
		A	L	A	A	A	A		
			F/I						
			C						
			A						

6. Terminal Investment Limited (TIL)	Terminal Investment Limited (TIL) is a port, harbor, marine, and terminal manager and operator that invests in container terminals in five continents, often in joint ventures with other major terminal operators. Its geographically diverse activities focus on the Western Hemisphere, namely, Western Europe, the Americas, the Suez Canal, the Mediterranean Sea, and Africa. In addition to its terminal operations activities, TIL is also presently active in the development of several strategic alliances such as global Greenfield sites. TIL is closely affiliated with Mediterranean Shipping Company S.A. (MSC), one of the world's leading container lines, as it handles a large segment of MSC's volumes.	T A	Δ T A	T A	T A A	2.1 / 2.1
Ownership country: The Netherlands						
7. China Shipping Terminal Development	China Shipping (Group) Company is a shipping enterprise owned by the People's Republic of China. It is primarily involved in terminal management, logistics, finance, and investment, ownership and management of oil tankers, passenger and container ships, engineering, trading and IT.	F T L IT I A F/I	Δ F T P L IT I A F/I		F T L IT I A F/I	7.8 / 1.3
Ownership country: China						

(continued)

TABLE 2.5 (Continued) World's Top 10 Terminal Operators: Corporate Profile and Equity-Based Throughput, 2011

Terminal Operator in Market Power Sequence	Corporate Profile	Strategic Location of Terminal Operations, Logistics, and Other Port Activities						Million TEUs	% Share of Global Throughput
		Americas and Panama Canal	Asia	Africa	Europe	Oceania	Middle East and Suez Canal		
	China Shipping Terminal Development Co. Ltd is one of the group's subsidiaries principally engaged in investment and finance, development and management of terminals in the oil, chemicals, and containers sectors. Among others, the company owns 141 container ships, has holdings in 13 port facilities in China, and shares in terminals within the United States (LA and Seattle) and Egypt (APM Terminals 2013a,b).								
8. Evergreen	Evergreen Marine Corporation is a global shipping company involved in ports and terminal operations, containerized freight, supply chain management, and logistics. Its principal trading routes include all continents. It operates terminals in the United States (LA and Washington), Thailand, Italy, and so on. It also operates four leading transshipment hubs, two of which are in Taiwan's major ports, one in Panama and one in Italy. It owns over 150 container ships, and a number of logistics-related companies (Evergreen Marine 2013).	F T P L A	Δ F T P L A		F T P L A			6.9	1.2
Ownership country: Taiwan									

| 9. Eurogate | EUROGATE is Europe's leading shipping line-independent container-terminal group, actively involved in terminal operating, intermodal transport, containers, and logistics. They operate a major container terminal in Germany, and they are involved in three joint ventures with significant global leaders: (a) the NTB North Sea Terminal is co-owned by Maersk Line, (b) the MSC GATE is a joint venture with MSC-Mediterranean Shipping Company, and (c) in partnership with Contship Italia, they operate sea terminals throughout Europe, from the North Sea to the Mediterranean and the Atlantic. They are specialized in "box"-related operations, cargo-modal logistics, container-depot services, container servicing and container repair, and customized IT logistics and engineering services. | A T L IT A | 6.6 | 1.1 |
| Ownership country: **Germany** | | | | |

(continued)

TABLE 2.5 (Continued) World's Top 10 Terminal Operators: Corporate Profile and Equity-Based Throughput, 2011

Terminal Operator in Market Power Sequence	Corporate Profile	Strategic Location of Terminal Operations, Logistics, and Other Port Activities						Million TEUs	% Share of Global Throughput
		Americas and Panama Canal	Asia	Africa	Europe	Oceania	Middle East and Suez Canal		
10. HHLA	Hamburger Hafen und Logistik AG (HHLA) was established in 1885. It is a partly privatized company actively involved in four segments: (a) terminal operator, (b) logistics, (c) intermodal transportation, and (d) real estate. The firm operates three container terminals in Germany and Ukraine and extends its cargo handling and transport services by sea, rail, and road (Hamburger Hafen und Logistik AG, Germany 2013).				△ T A L E C			6.4	1.1
Ownership country: Germany									

Source: M.G. Burns, based on data from Drewry Maritime Research (2012). Available at http://www.drewry.co.uk (accessed on June 21, 2013); PSA International. 2013. Available at http://www.internationalpsa.com (accessed on June 21, 2013); HPH. 2013. Hutchison Port Holdings. Available at http://www.hph.com/webpg.aspx?id=87 (accessed on June 15, 2013); Dubai Ports World. 2013. DP World. Available at http://www.dpworld.com (accessed on June 15, 2013); APM and Maersk. 2013; China Ocean Shipping, COSCO Group. 2013. Available at http://www.cosco.com (accessed on July 3, 2013); Terminal Investment Limited (TIL). 2013. Available at http://www.tilgroup.com (accessed on July 3, 2013); CN Shipping. 2013; Evergreen Shipping, USA. 2013. Available at http://www.evergreen-shipping.us/ (accessed on July 3, 2013); Eurogate, Denmark. 2013. Available at http://www.eurogate.de/live/eg_site_en/show.php3?id=1; Hamburger Hafen und Logistik AG, Germany. 2013. HHLA. Available at http://hhla.de/en/ (accessed on July 3, 2013); CMHI. 2013. Available at http://www.cmhi.com.hk. China Merchants Holding International China's largest public port operator; and Bloomberg. 2013. Available at http://www.bloomberg.com/news/2013-07-02/cosco-chairman-wei-retires-after-building-biggest-bulk-fleet.html; China Shipping Terminal Development. Available at http://www.cnshipping.com. Accessed July 4, 2013; Feport, Belgium. Available at http://www.feport.be. Accessed July 3, 2013. Feport is connected with Terminal Investment Limited (TIL), http://www.tilgroup.com, http://www.feport.be/index.php?cmd=operators&view=detail&ID=63.

Note: △, Region of origin; A, Strategic alliances; B, Ship building; C, Cargo handling; E, Real estate; F, Fleet ownership/management; F/I, Financing/investment IT; IT; L, Logistics and warehousing; O, Oil and gas, offshore drilling activities; P, Ports/hubs; R, Ship and/or container repairs; T, terminals.

and land carriers, the logistics company, the terminal, the warehouse, the cargo handling facilities and stevedores (longshoremen), and the distribution centers. This can be clearly demonstrated in Table 2.5, which highlights the corporate profiles of the top 10 global terminal operators that manage over 35.90% of the global terminals in the industry.

The table further shows each company's geographical diversification and corporate activities, such as ports and terminals management, investment, ship building and repairs, ownership of cargo or fleet, IT activities, logistics and warehousing, oil and gas, offshore drilling activities, container boxes ownership, distribution and repairs, and so on.

As an antipode to these conglomerates and the strategic alliances within supply chains, the vast majority of global ports seem to lose bargaining power at a regional, national, and global level.

This tendency generates further market imbalance, with the ports losing their competitive edge, while competing with neighboring ports. At the same time, controlling these routes has been a major subject of distributive negotiations and conflict among the major industry players, owing to the strategic significance of hinterland hubs as contacts between global cargoes and regional markets.

In a nutshell, the parties at a disadvantage are the supply chain players that directly benefit from and rely on a specific port (i.e., industries, customers, local authorities, markets), whereas winners in this power shift are the global players that are mostly interested in a cost-effective cargo transport, and they achieve this through alternative routes and transportation mode substitutes, for example, emerging ports, alternative inland routes, and so on. In this respect, the traditional role of ports is expected to change in the next few years, toward a specialization between global gateway ports and domestic transshipment hubs. Industry analysts consider that provided the sustainability of demand growth, the option of hub ports may not be plausible in the long run, since direct service costs are more competitive compared to transshipment hubs.

Port competition typically has two contradicting effects in the industry: healthy antagonism, which leads to improved services, increased efficiency, quality, and productivity, and reduced time delays at port. On the other hand, internal competition (i.e., regional and national) may lead to an input increase, for example, on behalf of the ports, without the analogous profit. To survive, modern ports may be forced to lower port tariffs, without eliminating the threat of loss of business from substitute ports.

The standard response for a port to reestablish its market power is leverage, integration, and the formation of strategic alliances.

Regardless of a port's size, the following factors will determine a port's competitive edge:

- National and regional trade agreements
- Trade barriers: protectionist policies, cabotage restrictions, and so on
- Regional economic potential; vicinity to markets
- Interconnectivity potential; vicinity to logistic networks
- Port facilities, development, superstructures, services, infrastructure, berths
- Port's ability to handle cargo volume and type
- Port tariff and overall cost
- Port's traffic, berth occupancy
- Labor reliability: strikes, trade unions
- Regulatory compliance. Port, legal framework, regional/national
  - Safety: weather conditions, for example, temperatures, hurricanes
  - Security: terrorism, piracy, and so on

- Environmental protection
- Political climate/stability
- Quality

Opportunities always multiply as they are seized, and ports can further increase their comparative advantages in numerous ways, as discussed in the next chapter.

REFERENCES

101 shipping. 2013. Available at http://101shipping.com/shippingseacost.html. Accessed June 15, 2013.

AAPA Surveys. Available at http://www.aapa-ports.org. Accessed June 21, 2013.

Agência Nacional de Transportes Aquaviários–ANTAQ, Brazil. 2013. Available at http://www.antaq.gov.br. Accessed June 21, 2013.

Anderson, J.E. and van Wincoop, E. 2003. Gravity with gravitas: a solution to the border puzzle. *American Economic Review* 93: 170–192.

APM Terminals. 2013a. Company profile. Available at http://www.apmterminals.com/. Accessed July 2, 2013.

APM Terminals. 2013b. Available at http://www.apmterminals.com/uploadedFiles/corporate/Media_Center/Press_Releases/130313%20China%20Shipping%20Terminal%20stake%20in%20APM%20Terminals%20Zeebrugge.pdf. Accessed July 2, 2013.

Bloomberg. 2013. Available at http://www.bloomberg.com/news/2013-07-02/cosco-chairman-wei-retires-after-building-biggest-bulk-fleet.html.

Burns, M. 2014, Intermodal Transportation and Marine Transportation: Strengths, Weaknesses, Opportunities, and Threats Analysis. Transportation Research Board of the National Academies. 93th Annual Meeting. January 12–16, Washington D.C. Available at: http://pressamp.trb.org/aminteractiveprogram/EventDetails.aspx?ID=28944. Accessed February 12, 2014.

China Ocean Shipping, COSCO Group. 2013. Available at http://www.cosco.com. Accessed July 3, 2013.

China Shipping Terminal Development. 2013. Available at http://www.cnshipping.com. Accessed July 4, 2013.

CMHI. 2013. China Merchants Holding International China's largest public port operator. Available at http://www.cmhi.com.hk.

Dictionary of Military and Associated Terms. 1985. Washington: US Department of Defense. p. 244.

Drewry Maritime Research. 2012. Available at http://www.drewry.co.uk. Accessed June 21, 2013.

Dubai Ports World. 2013. DP World. Available at http://www.dpworld.com. Accessed June 15, 2013.

Eurogate, Denmark. 2013. Available at http://www.eurogate.de/live/eg_site_en/show.php3?id=1.

Evergreen Marine. 2013. Available at http://www.evergreen-marine.com/CorporateProfile.jsp. Accessed July 3, 2013.

Evergreen Shipping, USA. 2013. Available at http://www.evergreen-shipping.us/. Accessed July 3, 2013.

Feport, Belgium. 2013. Available at http://www.feport.be. Accessed July 3, 2013.

Hamburger Hafen und Logistik AG, Germany. 2013. HHLA Available at http://hhla.de/en/. Accessed July 3, 2013.

Harney, A. 2008. *The China Price: The True Cost of Chinese Competitive Advantage.* Penguin Books Ltd, Penguin Books (USA) Inc.

HPH. 2013. Hutchison Port Holdings. Available at http://www.hph.com/webpg.aspx?id=87. Accessed June 15, 2013.

Huff, D.L. 1963. *A Probabilistic Analysis of Consumer Spatial Behavior*, W.S. Decker (ed.). Emerging Concepts in Marketing, Chicago: American Marketing Association, pp. 443–461.

Institute of Shipping Economics and Logistics. 2012. Containerization International Yearbook 2012. ISBN-13: 978-1843119906.

International Monetary Fund (IMF). 2011. Changing patterns of global trade. IMF's Strategy, Policy, and Review Department. Available at http://www.imf.org. Accessed July 1, 2013.

International Monetary Fund (IMF). 2011. Finance and Development. Accessed June 15, 2013.

International Monetary Fund (IMF). 2013. World Economic and Financial Surveys. World Economic outlook, April 2013 Hopes, Realities, Risks International Monetary Fund Publication Services, Washington D.C. ©2013.

Kim, P.-J., Kim, W., Chung, W.-K. and Youn, M.-K. 2011. Using new Huff model for predicting potential retail market in South Korea. *African Journal of Business Management* 5(5): 1543–1550. DOI: 10.5897/AJBM10.396. ISSN 1993-8233 ©2011 Academic Journals.

Lambert, D.M., Cooper, M.C. and Pagh, J.D. 1998. Supply chain management: Implementation issues and research opportunities. *The International Journal of Logistics Management* 9(2): 1–19.

OECD. 2008. Joint Transport Research Centre, Round Table, April 10–11, 2008, Paris. Port Competition and Hinterland Connections, Summary and Conclusions.

OECD. 2009. Port Competition and Hinterland Connections. OECD, *International Transport* Forum. Published by: OECD Publishing, Publication date: July 6, 2009.

Penn State University. 2013. Available at http://www.personal.psu.edu/uxg3/blogs/coal/assets_c/2009/08/US%20Oil%20and%20Gas%20Reserves%20Map-62158.html. Accessed June 15, 2013.

Port Strategy. 2013. Chinese terminal operators expand. Available at http://www.port strategy.com/news101/world/asia/chinese-terminal-operators-expand#sthash.vF78nTMm.dpuf.

PSA International. 2013. Available at http://www.internationalpsa.com. Accessed June 21, 2013.

Sabo, E. Bloomberg. 2013. Panama canal expansion fuels 7% growth in first quarter. Available at http://www.bloomberg.com/news/2013-06-14/panama-s-economic-growth-slowed-to-7-in-first-quarter.html. Accessed July 5, 2013.

Secretariat of Communications and Transport, Mexico. 2013. Available at http://www.sct.gob.mx. Accessed June 20, 2013.

Stevens, G.C. 1989. Integration of the supply chain. *International Journal of Physical Distribution and Logistics Management* 19(18): 3–8.

Terminal Investment Limited (TIL). 2013. Available at http://www.tilgroup.com. Accessed July 3, 2013.

The Global Jewish News Source (JTA). 1961. Available at http://www.jta.org/1961/08/11/
    archive/israel-to-test-overland-transportation-in-competition-to-suez-canal.
    Accessed July 1, 2013.

UNCTAD and World Trade Organization. 2013. A practical guide to trade policy analy-
    sis. Available at http://vi.unctad.org/tpa/web/docs/ch3.pdf. 6 chapters, Available at
    http://vi.unctad.org/tpa/web/docs/ch6.pdf. Accessed June 15, 2013.

UNCTAD, 2012. UNCTAD Ad Hoc Expert Meeting on Assessing Port Performance Room
    XIV Palais des Nations Geneva, Switzerland. 2012. Port Performance Indicators
    A case of Dar es Salaam port by Mr. Hebel Mwasenga Principal Planning Officer
    Tanzania Port Authority. Available at http://unctad.org/meetings/en/Presentation/
    dtl_ttl_2012d11_Mwasenga.pdf. Last accessed: February 8, 2014.

United Nations Economic Commission for Europe. 2010. *Hinterland Connections of
    Seaports*, ECE/Trans/210. New York and Geneva: United Nations.

US Army Commerce Statistics Center. 2013. Available at http://www.fedstats.gov. Accessed
    June 21, 2013.

US Army Engineer Institute for Water Resources (IWR). 2013. Available at http://www.
    iwr.usace.army.mil. Accessed June 21, 2013.

US Department of Transportation, Gas Transmission and Hazardous Liquid Pipelines.
    2012. Available at http://www.phmsa.dot.gov/staticfiles/PHMSA/ImageCollections/
    Images/Pipeline%20Map.jpg.

US Department of Transportation, Federal Highway Administration. 2013. Available at
    http://ops.fhwa.dot.gov/freight/freight_analysis/nat_freight_stats/tonhwyrrww2007.
    htm. Accessed June 15, 2013.

US Maritime Administration (MARAD), US Department of Transportation. Available at
    http://www.marad.dot.gov/ports_landing_page/deepwater_port_licensing/dwp_map/
    dwp_map.htm. Accessed June 15, 2013.

Waterborne Transport Institute, China. 2013. Available at http://www.wti.ac.cn. Accessed
    June 12, 2013.

World Shipping Council. 2013. Available at http://www.worldshipping.org. Accessed June
    15, 2013.

CHAPTER 3

Port Management and Economic Growth

Nothing contributes so much
to the prosperity and happiness of a country
as High Profits.

David Ricardo (1772–1823)

3.1 ESTABLISHING A PORT'S COMPETITIVE EDGE IN A NICHE WORLD

Ports are a nation's links to prosperity. Since a chain is only as strong as its weakest link, a port's constitutional weakness within a supply chain will eventually affect a country's economic sovereignty. After all, ports of all sizes and types are important tools to build a country's competitive advantages, as long as they aim toward a niche market. Despite this compelling theoretic approach, in reality, the performance of a country's ports is not always synchronized; neither is it focused on comparative advantages that are based on opportunity cost. Perhaps this is the reason why today's trade has experienced an increasing disparity among harbors that have adjusted to the new technological and supply-chain needs and harbors that have not reconciled with this new reality.

As discussed in Chapter 2, competition among ports is beneficial for the supply chain and the market end users, as they obtain tariff discounts and optimum productivity, quality, and quantity of services. Another equally significant element of competition is that it enhances innovation and corporate breakthroughs: to survive, ports are forced to differentiate, keep low costs, save time and resources, and improve techniques and processes. While ports thrive for success, they surpass their own limitations and invent ways of doing things better and faster. In fact, this concept verifies Charles Darwin's theory on *The Origin of Species* (Darwin 1859), which supports that evolution is a result of natural selection whereby external or performance traits improve, enabling an individual to get accustomed in a changing environment. Evolution is therefore the positive side of competition, that is, exactly what modern ports need in order to regain (a) their niche, (b) their comparative advantage, and (c) their opportunity cost.

3.1.1 Comparative Advantage, Competitive Advantage, Absolute Advantage and Niche: Better, the Best, or Simply Different?

In the twenty-first century's rapidly growing and ever-evolving global trade and transport, it seems difficult to clearly define the difference between a port's absolute advantage, as opposed to a comparative advantage or a niche. In order to better understand the evolution of socioeconomic thoughts with a logical sequence, these theories are hereby presented (a) in a chronological series as they were developed and (b) in connection with the factor of production that better pinpoints the magnitude of growth. It is worth noting that for the past two centuries or so, these theories are still dominant: modern scholars have only managed to extend them but never to replace them:

a. An *Absolute Advantage* is based on *higher workforce productivity*. It enables a nation or a port to offer a specific commodity or service at a cheaper price on the basis of increased labor efficiency. In port management, it suggests that the port can either (a) utilize the same amount of labor compared to other ports, and deliver a greater output, or (b) employ a lower amount of workers and be equally productive to other ports that offer similar services though they employ more workers.

   Adam Smith (1776) developed the *absolute advantage* theory in his publication *An Inquiry into the Nature and Causes of the Wealth of Nations*. During the ages of the powerful British Empire and the concept of mercantilism, Adam Smith contended that, in practice, mercantilism could not bring profit and growth to all countries, as during an export/import transaction, one country's profit was based on another country's low cost and limited gains. Hence, Smith proposed that only free trade and specialization would enable countries to develop their absolute advantage, that is, fields of specialization, where all trade and transport parties involved would profit.

b. A *Comparative Advantage* focuses on *lower opportunity cost*. It establishes the capabilities of growth and profitability from industrial and commercial specialization. Opportunity cost is the value of a choice—in terms of service or production—that must be relinquished so as to engage in a priority production that will potentially bring more profit. Corporate strategies and the decision-making process of port managers are based on measuring the opportunity costs of two or more options and comparing the benefits of each option at a micro and macro level.

   The principles of *comparative advantage* were conceptualized by the British economist David Ricardo in 1821, that is, 45 years after Smith's absolute advantage theory (1772–1823). Applying his principles in the maritime industry, a port's value and profitability for the services rendered are analogous to the degree of obstacles encountered during the input process. Alternatively, earnings could also increment owing to advancements in production or achieving an increase in output, while retaining the same level of input (Ricardo 1821). A port with a *comparative advantage* indicates its potential to deliver services at a reduced opportunity cost compared to other ports. When this occurs, not only can a port's tariffs be more competitive as a means of attracting more clients, but a better allocation and controlling of resources can help the port grow.

c. A *Market Niche*, finally, is a specialized market segment; a unique service or commodity that is offered to a market that has an unmet demand.

Although a niche strategy is a widely recognized way of maximizing profits, there is no long-term guarantee of market share retention or sustainable success. New entrants can always claim their market share, or the demand for the product or service may diminish (Noy 2010). Hence, port strategies should be reviewed and reconsidered on the basis of market fluctuations, annual balance of trade statistics, government trade agreements, and so on.

Ports of any size, type, or geographical location can offer a niche when they have studied their supply chain's market segments and have identified the different supply and demand trends. Niches are not readily available but are strategically designed by identifying this specific market demand and supplying the commodities or services required.

The "niche" principle was presented as a component of an economic theory as industry segmentation and niche generic strategy (Claycamp and Massy 1968; Porter 1980; Smith 1956).

d. A *Competitive Advantage* resembles the benefits of a comparative advantage and a market niche. It positions a company as an industry's leader, in terms of either lower cost or product differentiation advantages. Michael Porter has recognized two principal kinds of competitive advantage: (a) cost advantage, where the company is in a position to offer the same products or services as its competitors, yet at a decreased cost, and (b) differentiation advantage, when the company creates superior advantages or value within their products or services that surpass those of their competitors.

A nation's economic and business objectives can be reshaped when it learns to develop, identify, and evaluate the comparative and absolute advantages or niche markets connected to its ports and supply chains. Once a port's advantages are recognized, it may prompt other regional ports to diversify and therefore grow in size and influence, as opposed to consuming their resources in regional competition.

A port may seek for an absolute advantage through higher productivity, may gain a comparative advantage through pursuing lower opportunity costs, or enter a niche market through service differentiation. This means that a port can still be better, be the best, and be different, on the basis of its geography, socioeconomic and legal factors, demography, markets, and all of its factors of production. The ports' matrix reflecting potential niche markets and advantage areas could be based on the following criteria:

a. Geographic profile: infrastructure and interconnectivity profile, for example, island-nations, coastal areas, and landlocked areas
b. Regulatory framework: safety, security, environmental laws, and so on, strict or lenient
c. Legal and governance framework: tax reform/incentives, government subsidies, liability, social responsibilities
d. Demographic profile: gender; age, ethnicity, education, employment status, income range, and socioeconomic structure of population
e. Financial and economic framework: banking and financing systems; trade, manufacturing, transport incentives; balance of trade; currency strength; market-based economies versus command-based economies; national debt; consumer spending; inflation
f. Factors of production—cost-based profile: land, labor, technology and entrepreneurship. Adam Smith's factors of production

g. Factors of production—innovation profile: technology, entrepreneurship, labor (e.g., innovation vs. brain drain)

h. Commodities profile: raw materials, energy, value-added goods, commodities and factors of production

When economic theories are extended to fit the principles of port management, it becomes evident that a port can achieve advantage, specialization, or niche through four principal paths: (a) by delivering a service that is hard to provide, (b) through innovation, (c) through lower cost, and (d) through efficient allocation of resources:

a. *Delivering services that are difficult to supply*

As an example, Port of Anchorage is Alaska's lifeline, handling over 90% of the state's cargoes (Prokop et al. 2011).

The significance of Alaska's ports is increasingly highlighted as a result of Alaska's vast oil, gas, and shale gas resources. Ongoing scientific researches and plans by national and local authorities, USCG, the US Army Corps of Engineers, and so on, aim at the extraction and transportation with the Port of Anchorage as a focal logistics point (Prokop et al. 2011).

The recent economic and energy developments require that Alaska plays a leading role as a national and global energy exporter in the decades to come.

The constraints the energy industry and the port face include weather and geographic restrictions, inadequate infrastructure, difficulties in securing the high levels of investment required through public or the private sector, and developing innovative engineering and architecture designs that would fit the Arctic, sub-Arctic, and Alaskan inland waterway particularities (Northern Economics, Inc. 2011).

Hence, Alaska's comparative advantage (or niche) will be achieved under the condition that a highly strategic and innovative energy and transport plan will be developed to encompass port functions, inland waterways, and global networks.

Port planners must also look into the future, as global warming will open new, unpredictable Arctic shipping routes by 2050 (Zabarenko 2013). Scientists have discovered at least three alternative sea passages that will shorten vessels' navigation by at least 5000 nautical miles, or 14.88 days, calculated based on an average vessel speed of 14 knots (nautical miles per hour). The new sea routes will connect Alaska, Northern Europe, and Russia to the Far East markets of China, Japan, South Korea, and so on, which suggests that other ports along the Arctic route will benefit as well.

b. *Innovation*

An innovation pertains to novel ideas that offer increased value and usage to a port's, company's, or nation's factors of production, that is, labor, technology, processes, services, and products. For a method to be considered as innovative, it should offer a novel perspective of service or product that exceeds a simple "improvement." The Global Innovation Index was developed through a collaboration of the United Nation's World intellectual property organization, Cornell University, and INSEAD.

Its design is based on two sub-indexes, the Innovation Input and the Innovation Output, which are both established on the five input pillars of innovation. These five pillars reflect components of the global economy that facilitate pursuits of innovation: (1) organizations, corporation; (2) think-tanks, workforce, research; (3) infrastructure systems, (4) industry growth and refinement; and (5) business enterprise growth. The output of innovation is reflected through two pillars that establish proof of innovation, namely, supporting evidence of (1) technology, intelligence, and information; and (2) creative designs and progressive outputs (Global Innovation Index 2013).

This is one of the most widely accepted indexes for measuring and evaluating innovation that modern ports and the maritime industry can use.

c. *Cost minimization*

A port's costs consist of direct and indirect costs. *Direct costs* are allocated for financing the input of a specific task, product or service, salaries for technical work, port labor, research and development, and transportation expenses directly allocated for existing business. *Indirect costs* include the overhead (administrative and facilities) costs, administrative salaries, maintenance, procurement and spare part purchases, and traveling for marketing and future business.

Cost estimation techniques include linear regression, high–low method, account analysis, scatter graphs, and so on.

By estimating the production function of a port, one can estimate its cost function. Based on the Cobb–Douglas production–function model:

$$P_{it} = A^{\gamma} \left(\frac{\theta U_{it}}{1 - \theta} \frac{\text{MRe}}{\text{MRi}} \right) N_{it}^{1 - \theta} \tag{3.1}$$

where P_{it} represents the port's real production over a specific time, $(1 - \theta)$ and θ signify the parameter of share coefficients in terms of the real port output, A is the technology input, γ reflects the capacity to modify the port's technological performance, and N represents the number of employees. To ensure the port's cost efficiency, the MRe represents the marginal revenue of employment at the port, whereas MRi represents the amount of investment.

Capital cost

A port's capital costs are typically estimated by the investment price, that is, cost of assets such as buildings, storage facilities, equipment, and so on. Ungo and Sabonge (2012) proposed a capital cost estimation that is computed by calculating variables pertaining to a new asset's cost. Depending on the source of investment, contractual arrangements, and financing options, port managers may estimate the asset's residual value, interest rates, and amount of years financed. The capital cost per annum signifies the yearly loan repayment for the total sum of the purchased asset's value with a residual price upon repayment of the loan.

The capital cost is estimated as follows:

$$CC = \text{NAC} \left[\frac{i - (1 + i)^n}{(1 + i)^n - 1} \right] - R \left[\frac{i}{(1 + i)^n - 1} \right] \tag{3.2}$$

where NAC is the new asset cost, i represents the annual interest rate as appropriate, n is the year's number of loan repayment, and R is the residual asset's value.

Energy consumption

The energy consumption factor is a critical expenditure for ports and terminals, as the cargo handling equipment, infrastructure, and superstructure are heavily dependent on energy costs and fuel efficiency. Costs also depend on other parameters such as maintenance costs, life expectancy for the equipment, and so on. Hence, the estimation for the energy consumption per year is calculated by the equation:

$$EGC = \frac{CC}{LT_n} + MC + EC \qquad (3.3)$$

where EGC is the energy consumption, CC is the capital cost, LT_n is the asset's lifetime expectancy estimated in years, and MC and EC reflect the maintenance cost and energy cost, respectively, for the period.

Cost minimization in a capital intensive industry such as the maritime industry should by no means be achieved at the expense of quality, reliability, and productivity. Cost minimization can be achieved when the budget allocation and the expenditure process are characterized by the following principles: (i) utility, that is, each expenditure should bring benefit to the port's production and a tangible result to the business output; (ii) sensibility and shrewdness in expenditures and the management of economic affairs; and finally (iii) consistency should be applied in order to comply with the port's expenditure plan.

A port's commitment to cost minimization may employ techniques of computing input and costs of production in order to develop the minimum costly output, that is, services or products. Among the factors of production, labor and technology are the two factors that are directly related to time and money and therefore monitoring and controlling the performance of labor and machinery can bring optimum results.

Technological performance
A port planning strategy is heavily dependent on technology and machinery; hence, technological performance, availability, and dependability warrant a port executive's serious consideration. As this chapter deals with the port's competitive advantage, the measurement of a port's dependable services should also be evaluated. A modified version of the Hidden Markov Model has been hereby developed in order to evaluate the system's functionality and dependability not restricted to a single unit machinery but for the functionality of all port terminals. The estimation subdivides the port system's reliability into mechanical integrity and cargo handling process. Hence, the equation verifies to which extent a port's technological performance offers the requested level of reliability in terms of cargo handling and distribution services. Since a modern port's performance is typically measured by evaluating the overall performance of different segments, the following formula can be used to estimate dependability per berth or terminal, in order to isolate the different divisions and functions, or for the port as a whole:

$$R(t) = \left(\prod_{e=1}^{n} Re(t) \right) Rch(t) \qquad (3.4)$$

where $R(t)$ reflects the reliability of all the equipment within a port; $Re(t)$ represents the reliability and availability of the port's equipment to meet the demanded production; n reflects the sum of the equipment used at the port, that is, per berth or terminal; and Rch(t) represents the reliability of the equipment's subsystem, such as distribution functions, cargo handling auxiliary equipment, warehousing and logistics subsystems, and so on.

Labor performance

Although shipping is a global industry, costs typically vary at the regional, national, and local levels, based on factors such as national balance of trade, currency strength, inflation, cost of labor and unemployment, national debt and taxation, consumer spending, and so on.

China's wages serve as an example of cost minimization and how a nation's competitive edge may fluctuate over the years. According to the World Bank, a decade ago, that is, in 2003, cost of living and the factors of production, including salaries, were five times lower in China compared to the United States (World Bank 2013). From the middle of 2005 to late 2008, cumulative appreciation of China's currency, that is, the renminbi, against the US dollar exceeded 20%; however, the rate of exchange was still pegged to the US dollar from the 2008 world economic crisis to June 2010, when a progressive currency appreciation took place. As of 2013, China is second to the United States in the value of services it produces, although China's per capita income is below the global average (CIA 2013). When these economic figures are combined with a population of 1.349 billion people (World Bank 2013; CIA 2013 estimate), it gives China an unprecedented competitive edge.

"The soft three dollars" is a figure of speech commonly used by Asian traders over the past decade to describe that for every single dollar of productivity input, there were three dollars pertaining to trade commissions and profit through Asia's global exports.

Interestingly enough, after the 2008 global crisis, salaries in most of the world, for example, in Europe and the United States, have significantly dropped after four consecutive years of diminishing productivity. At the same time, salaries in Asia are increasing. In 2012, about half of China's workforce experienced a salary increase of at least 10%, thus making China the nation with the highest salary rate (Forbes 2012). The laws of supply and demand have proved that history repeats itself. The Western economics have outsourced to the East, and now they both expand their outsourcing activities primarily toward Africa and the less developed Asian economies, followed by South America and Eurozone's collapsed economies, such as Greece, Italy, Spain, and Ireland. The geography of outsourcing is extensively highlighted herewith, as a reminder that ports and maritime activities will grow where manufacturing, energy sources, and raw materials are located.

d. *Through efficient allocation of resources*

Finally, a port can achieve advantage, specialization, or niche through optimum allocation of resources, which implies that diversification is promoted in order to eliminate internal competition.

Excellent examples of business diversification that leads to growth and development can be seen from the Port of Virginia Authorities, USA:

i. The Norfolk International Terminals (NIT) is the largest facility of VPA, with remarkable infrastructure outlets, that is, railroad service and intrastate and interstate highways. NIT has evolved from a surplus Army base in the

1960s to a valuable supply-chain partner to Nissan (which has now moved to NNMT) and Evergreen, a major trans-Pacific shipping line in the 1980s. Decades later, it is a leading facility with expanding container, break-bulk, and car-carrier activities (Virginia International Terminals 2013).

ii. Portsmouth Marine Terminals represent the second largest facility of VPA. It is a highly automated port with multiple railroad outlets. Its container terminals have undergone technology-intensive and capital-intensive development over the past few years. APM Terminals Virginia has operated Portsmouth's container facility for over 30 years under a lease agreement, with its last contract of 20 years' lease signed in 2010. APM's $540 million investment in the facility is one of the most significant private ventures of all time in Virginia and signifies APM's strategy to unite all container terminals in the Hampton Roads. Regardless of future developments that may change the terminals' operational structure, VPA has secured a 20-year contract that generates high revenues and will help attain an even larger market share (APM 2013; Port of Virginia Authorities 2013).

iii. The Virginia Inland port is the third largest facility in size. Its niche is to serve as a commodity outlet in seven neighboring states in the East Coast and the heartland of America, that is, Washington, DC, West Virginia, Delaware, Maryland, New York, Ohio, and Pennsylvania.

iv. The Newport News Marine Terminal is the fourth largest (but an equally significant) facility, which specializes in break-bulk operations and warehousing, as well as container storage. It hosts the Nissan Import Auto Operations facility (Nissan of Newport News 2013).

Indeed, today's mega-ports achieve growth and profit maximization through one or more of these methods.

3.2 ECONOMIC GROWTH AS THE SPACE BETWEEN STIMULUS AND RESPONSE

> There is no security on this earth;
> there is only opportunity.
>
> **Douglas McArthur**
> *General of the US Army (1880–1964)*

Sustainability is a corporate juggle of staying power and prosperity and has reasonably become one of the most euphonious words in the port managers' ears, as what it really implies is "sustainable growth." It is one of the stronger indicators of economic, quality, and innovative influence, and because of this reason, it is found in most of the ports' web sites, port plans, and annual reports. Economic growth is perceived as a nation's or industry's capacity increase over time and the ability to accelerate the output. The measurement of economic growth is either in nominal terms, where inflation is included, or in real terms, which provides for inflation adjustments. Economic stimulus on the other hand pertains to a nation's or the port's capacity to trigger financial growth.

Global maritime frontrunners are aware that sustainability eventually results in profitability. The American Association of Port Authorities (AAPA) has developed and currently implements a sustainability platform for all its members. Sustainability within the maritime and port management practices also embodies the Leadership in Energy and Environmental Design (LEED) as well as Green Strategies for sustainable ports (AAPA 2013; EPA 2013a,b; Northeast Diesel 2008; USGBC 2013).

Sustainable ports embrace a holistic culture for their port operations. Environmental integrity commences with clean air and cleaner fuels, and technological innovations are offered to ensure the port's sustainability and market prominence.

Each individual port has a distinctive package of topographical, socioeconomic, national, environmental, operational, technical, and financial particulars that structure and determine its growth pursuits. In fact, it is those distinctive characteristics that will help a port differentiate and find its niche markets, such as

i. Its management type, for example, landlord, service port, and so on
ii. The regional regulatory framework for safety, security, the environment, social responsibilities, and so on
iii. The industrial zones, warehouses, and customers that will define its market segments
iv. Its cargo specialization, for example, oil and gas, containers, automobile industry, passenger ships, and so on
v. Its infrastructure, connectivity and accessibility to rail, intrastate and interstate highways for hinterland connection, airports, inland ports, and so on

3.2.1 Physical versus Strategic Growth

While governments achieve growth by increasing government subsidies and implementing currency, financial, or fiscal policy reforms, ports and entrepreneurs monitor and evaluate growth in two distinguished categories: (1) organic or physical growth, and (2) strategic or long-term planning growth.

3.2.1.1 Organic or Physical Growth

Organic or physical growth reflects a port's tangible and short-term growth. This is a tactical growth type that in military terms resembles a battle (as opposed to a port's strategy, which is the war itself). It is more easily recognizable by the industry and the majority of the media and the communities, because of its tangible, short-term, and nonconfidential nature: it is practically easier to measure a port's expansion in terms of assets, cash flow, investments, and current business compared to a strategic alliance of the next category, whose long-term benefits and initial confidential process cannot be easily assessed.

A port's physical growth is based on the accumulation, expansion, or development of assets and services such as (a) land and assets acquisition, (b) capital investment, (c) strengthening links between port and hinterland, (d) enhancing ports' energy efficiency, (e) dredging and expanding deepwater ports, and (f) bridging the maritime industry with oil and gas offshore activities. Namely:

a. *Land and assets' acquisition*, for example, the purchase of land, offices, terminals, berths; cargo handling equipment such as cranes and derricks; infrastructure and superstructure; port development and efficient land planning and land management within the port, while also monitoring and controlling the land development in the port's vicinity.

Land capacity is a critical factor for the port's planning and growth strategies. A port's future is determined by the land availability and its pertinent restrictions that are usually of regulatory, regional competition, financial, or community nature. Some of the major limitations to the growth of a port and its supply chain pertain to the limited availability of port land, which may become a demotivator to investors and potential clients and eventually a growth inhibitor. It is these restrictions that can prevent ports of considerable land capacity, from expanding its warehouse and multimodal or infrastructure plans. Conversely, there is no assurance that a port's land capacity expansion, investment, and assets acquisition will attract new contracts and will enhance its market share (UNCTAD 2004b).

The lesson learned from numerous global ports is that the measurable and tangible demand for a port's enhanced services should always precede the supply, that is, a port's plans for physical expansion. In cases where ports adopted "pull production" strategies based on an intangible forecast demand, that is, invested in an enlarged port with new terminals and berths, increased warehousing, and new services in the hope of attracting new customers, their plans were not successful. On the other hand, when ports implemented a "push production" strategy, based on actual consumers' demand, they managed to grow and increase their market share.

Some of the land use strategies commonly used in global ports include the following:

- *Reinforcing policy making frameworks* in order to obtain guidance and assistance for long-term port development. When port planning entails land acquisition and enlargement of the port–land interface, it is of critical significance to recognize the regulatory framework prior to the process of planning and implementation. Harmonization and understanding of any dissimilarities among regulatory and port policies should be proactively addressed, in order to eliminate future obstacles in the growth of port or land.
- *Optimum land utilization; ameliorating the land–port interface* among the port terminals and multimodal transportation, for example, rail, trucks. The port authorities should collaborate with various stakeholders in order to forecast the market and make plans on the hinterland accessibility, land development, market segments, and services to be offered. These plans should be monitored well in advance, in view of potential partnerships with the private sector. Efficient land utilization is a crucial port growth factor, to ensure the efficient handling, storage, and distribution of commodities from the port to local, intrastate, interstate, and global destinations.
- *Time efficiency*; identifying effective transportation routes that reduce travel mileage and relieve road traffic.
- *Initiating dynamic partnerships* with neighboring corporations, in order to support the port's land expansion plans.
- *Establishing a port buffer strategy* to tackle any interface considerations involving the port, commercial land functions and alternative land uses. Redesigning and developing land segments to serve as buffers to refineries, trade zones, or infrastructure connections with the hinterland (AAPA 2013; Geelong Port 2013a,b; Port of Rotterdam 2006).
- *Contractual stipulations, wording, and compliance.* This stage pertains to the assurance that the development project will be completed punctually by

the due date. Special contractual provisions should safeguard the port's interest, for example, the monitoring of the construction process or the signing of back-to-back contractual agreements between the port and third parties that may be partially or fully assigned the project's completion.

The port administration should meticulously plan the landside expansion and logistical integration in terms of both long-term architecture and short-term functional stages, in order to eliminate any contradicting visions among national or state authorities, policy makers, urban planners, private companies, and so on. Adopting and encompassing these different strategic perspectives with different key players are necessary in order for ports to be in compliance and promote growth, while eliminating any future obstacles.

b. *Capital investment*, cash flow, shares, and so on. From 1946 through 2005, that is, within 60 years, capital investment for US public port development for improvements to port facilities and infrastructure exceeded $30.1 billion, which amounts to an annual funding of $501 million. Over the past decade, the median annual investment leaped to $1.5 billion annually, while they were allocated for the development of new port facilities and the modernization of existing ones. By 2016, US port authorities and their private-sector partners, that is, marine terminals, will invest $46 billion to enhance terminals and connected infrastructure. Underinvestment in linking port and land infrastructure can be detrimental for the regional trade growth equally affecting a country's imports and exports.

A port's efficient market positioning is necessary to enhance accessibility to the regional and global markets (AAPA 2013).

Over the past few years, the private sector is increasingly investing in public ports' terminals. For example, American International Group acquired terminal leases in six US ports from DP World, along with the operations of Marine Terminals Corporation. A Deutsche Bank subsidiary has acquired the Port of New York and New Jersey's largest container terminal, that is, Maher Terminals. Furthermore, Goldman Sachs obtained a 49% share in Carrix Incorporated, parent of SSA Marine. Ports and terminals' investment is an appealing funding alternative for all concerned, because of the high returns on investment. The major modern obstacle for infrastructure funding pertains to the complexity and time-consuming process, such as the creditors increasing time for due diligence and evaluation process (MARAD 2013).

The investment benefits remain promising and lasting over time, as a result of (i) the steady economic recovery after the 2008 global crisis; (ii) the increasing global and US trade growth, which necessitates bigger ports and more terminals to accommodate the increased volumes of cargo; (iii) the recent oil and gas reserves discovered in the United States, Australia, and Western Africa (the Gulf of Guinea), which will create increased global trade flows; for example, based on a US Geological Survey, 30% of the globe's untapped gas reserves and 13% of its undiscovered oil are found in the Arctic region (USGS 2008); (iv) new trade routes combined with critical passages such as the expansion of the Panama Canal in 2015 and the opening of the Arctic Circle by 2030 because of global warming; and (v) sea ports and terminals offering reliable and sustainable returns on investment.

c. *Strengthening links between port and hinterland*; internal and external port integration; establishing long-term partnerships with influential supply chains that strengthen the connection between port, city, state, national, and global networks.

According to UNCTAD (2004a), a port's strategy should focus on assertively marketing the opportunities for expansion of supply chain networks and commodity flows to the hinterland. This will offer to ports the competitive advantage of the architecture as to the design and influence over the new networks with the hinterland. Starting from the port-interface clients such as terminal managers or transport companies that may expand their role in the supply chain, the network may eventually expand to include new players in the hinterland commodities flow, such as industries, manufacturers, commodity brokers, and so on. The collaboration between the public sector, which is typically interested in regional growth and prosperity, and the private sector, which more assertively pursues the return on investment, will bring a win/win situation, where both the industry and the community gain.

d. *Enhancing ports' energy efficiency* through efficient consumption, environmental initiatives, and energy mix, that is, alternative energy options. Global ports increasingly depend on green power or alternative energy in order to tackle new environmental goals. The US West Coast is one of the areas with many seaports establishing and meeting particular targets for the acquisition and consumption of renewable energy.

Certain seaports have intensely invested in green energy as a prerequisite of new port expansion projects or in joint venture with their regional communities. The installment of alternative energy including solar, photovoltaic, wind power, or generation machinery on port facilities is in the process of execution, typically in joint venture between port authorities, terminal operators, and energy providers. These joint ventures profit the local cities along with the port authorities (Port of Portland 2010).

Modern ports are dedicated to minimizing traditional energy use and reliance upon fossil fuel-generated power to lower emissions, boost efficiency, and minimize costs. Ports' energy planning processes aim to progress the ports' energy goals, such as stimulating energy conservation and green port engineering. Modern ports are committed to boost the ratio of alternative energy in their energy mix.

- *Biomass energy*: Biomass waste-to-energy facilities are constructed. Typically, the port obtains green energy certificates (RECs) and carbon dioxide offsets, and they will be able to receive municipal solid waste and building and demolition waste and transform it to energy using an environmentally friendly gasification system. Ports' participation in such environmental projects not only protects the environment but also enables the marketing of an innovative and promising technology that might be reproduced in other market sectors to minimize greenhouse gas emissions and provide green energy.

- *Fuel cells*: Fuel cells typically produce energy on-site, therefore lowering the tower's need for electrical power from the power company. The alternative energy produced by fuel cells supports energy-dependent building systems and equipment, consequently enhancing the port's energy security.

- *Wind energy*: Modern ports establish wind energy farms. Typically, port authorities cohost the venture with terminal operators whereas private developers build and operate the wind farm. As an example, the Port of New York/New Jersey has acquired a farm that contains up to nine wind turbines in a position to supplying up to 50,000 megawatt-hours of electricity, which is comparable to powering roughly 6000 homes (Port of New York and New Jersey 2013).

The level of a port's energy consumption and the energy type determine a port's commitment to protect the environment. A port's energy efficiency policy pertains to reducing the utilization of electric power, which compensates by minimizing pollution levels and financial cost, thus achieving port sustainability (Port of Helsinki 2013). Modern ports' planning and energy efficiency policies in the ports' five-year plans typically set annual objectives of decreasing ports' energy consumption by a specified percentage.

Investment in alternative energy pays off in the long run. Proactive ports that had invested in renewable energy equipment prior to the 2008 global crisis greatly benefited during the economic downturn and were able to handle power shortages. For instance, Indian ports like Gangavaram Port and Visakha Container Terminal Private Limited (VCTPL) are converting to green energy systems with the aim of reducing costs. Gangavaram Port has invested in solar panels and LED navigation lamps and has thereby saved 15% on its power bills. Furthermore, VCTPL has succeeded to cut back approximately 40% energy by enhancing energy efficiency in illumination and making an investment in energy-efficient, green technology (Indian Ports Association 2013).

e. *Dredging and expanding deepwater ports* to accommodate increasing mega-ship sizes. A port's financial prosperity is closely determined by the nation's ability to remain competitive in a global market. As larger vessels with deeper drafts are considered to provide significant economies of scale, regional economies will only benefit from these opportunities if they construct considerably deeper ports. Port dredging is a technology-intensive, capital-intensive process that requires proper funding, management, and compliance to an environmental regulatory framework, owing to the severe environmental issues generated therefrom. Another significant factor to consider is that the future trade patterns and sea routes are almost unpredictable (Marine Board Commission, NRC 1985). A port authority's planning strategy should therefore exercise canniness and due diligence in evaluating their current and future contract structures and securities prior to investing in dredging activities.

Dredging material is considered as highly polluted regardless of whether they are disposed in upland or in enclosed locations. For financial and environmental reasons, numerous seaports employ technological innovation to retain sediments in an effort to reduce or postpone dredging activities.

As an example, the Maritime and Port Authority of Singapore formed a strategic alliance with a regional engineering company to produce an innovative technology. Hence, polluted dredged components and commercial waste material are treated and recycled into ecofriendly, harmless building materials. This minimizes or eradicates removal and possible contamination problems as a result of dredging and maritime pollutants such as oil sediments and copper debris (Maritime and Port Authority of Singapore 2013).

f. *Bridging the maritime industry with oil and gas offshore activities*, developing offshore terminals. Ports increasingly serve as a land platform for offshore oil support services, thus bridging the maritime, oil, and gas activities. Their principal business is discharging international crude oil from tanker ships, cargo storage, terminal facilities, and distribution through joining pipeline systems to refineries. For example, America's first deepwater port, the Louisiana Offshore Oil Port in Baton Rouge, ensures the supply of offshore US crude oil through the entire Gulf Coast and Midwest (LOOP Port 2013). (Chapter 11 of this book is dedicated to ports' offshore activities.)

3.2.1.2 Strategic or Long-Term Planning Growth

Strategic, long-term planning serves as a ship's rudder: focusing on the port's long-term vision helps the organization materialize its short-term plans. In military terms, strategic growth depicts the war itself. It is established in confidentiality and entails long-term plans, and its benefits, such as its revenue, may be spread over numerous activities and time span, making it hard for the outsiders to measure and evaluate them.

Over time, ports are modified because of changes in cargo volumes and types, ship types and sizes, and customers' preferences, leading to changes of utilization in the port's land, infrastructure, and accessibility. Hence, port managers should strategically forecast the long-term trends and act in a manner that better serves short-term port development, while decreasing the probability of improper port's future development.

This is a strategic (as opposed to tactical) growth type that aims at enhancing a port's market power by profitable contractual agreements; acquiring a strategy; obtaining a market niche or specialization; gaining a key role in a regional supply chain; introducing a groundbreaking innovative technology; forming a strategic alliance with terminal operators and liners, oil majors, or pipeline consortiums; enhancing the port's hinterland and transshipment activities; obtaining long-term cost advantages through economies of scale; and so on.

A port's long-term strategy entails five principal alternatives for port reform:

1. *Reengineering, or business process redesign*, that is, undergoing an organizational restructuring through focusing on the ground-up architecture of the business process.
2. *Deregulation, liberalization*, which is frequently associated with a gradual shift from a public port toward a landlord model, with the potential of transferring certain services or terminals to the private sector.
3. *Privatization*, which entails the shift from a landlord type to a fully privatized port model. This port management option enables the sale of superstructural assets. Another privatization aspect is the development of joint ventures with logistics companies and private-sector operators that will facilitate the hinterland accessibility and networks.
4. *Corporatization*, which pertains to the restructuring of state or municipal establishments, that is, port authorities, into corporations, so as to acquire financial and management tools from the public sector. In ports' corporatization, the government controls most of the corporation's shares. Quite often, corporatization is the forerunner to partial or full privatization, in which case the private sector resumes increasing control over the port.
5. *Commercialization*: assignment of designated authorities and obligations from the central government to ports. A balance needs to be established between the port's control by the central government, and its antipode, which is the port's autonomy and operation as an independent entity. Since the 1960s, the concept of a gradual port autonomy became increasingly popular. However, the globalization between the late twentieth and the early twenty-first centuries demonstrated that the corporations can only thrive though unity, mergers, and the formation of larger conglomerates. It is the author's opinion that partial autonomy is a highly beneficial option that will make the port more flexible in their commercial negotiations, and yet government control will offer them the strength in unity that is offered in negotiating and establishing its multimodal logistics network.

A port's long-term strategy is frequently misinterpreted by the industry and the average market observer owing to its intangible, long-term, and confidential nature. In fact, in the history of shipping and global trade, very few things have been as misunderstood as a strategist's tactics, the major reasons of public misjudgment being the lack of inside information about the port's partnerships and different profitability options, lack of understanding on how budget allocation can generate increased funds, misconceptions about the tariffs' calculation, and which ship types or services generate more profit for the port.

A port's market share is calculated by the proportion of the global market's overall revenue that is acquired over a given time period. Market share is assessed by using the port's sales over the time period and dividing it by the total industry's revenue within the same time period. The outcome will offer a broad understanding of the port's market share compared to the market and its competition.

$$\text{Port's market share} = \frac{\text{Port's Cvm} \times t}{\text{Competitors' Cvm} \times t} \qquad (3.5)$$

where Cvm is the total cargo volume and t is the specified time period.

Alternatively, the port's market share may be estimated by the actual cargoes' value:

$$\text{Port's market share} = \frac{\text{Port's Cvl} \times t}{\text{Competitors' Cvl} \times t} \qquad (3.6)$$

where Cvl is the total cargo value and t is the specified time period.

3.2.2 Trade, Protectionism, and Free Trade

A nation's balance of trade, that is, the import/export ratio, is strongly affected by domestic and global socioeconomic alliances such as trade agreements, which consequently determine the growth of seaports within the country-members of the agreements. At the same time, policies of protectionism or free trade significantly determine the framework in which the agreements and port revenues can grow.

Globalization has enabled international economies of all growth levels to adopt free trade as a tool for promoting market diversity and benefiting from each country's cost-efficient factors of production. On the other hand, advocates of protectionism, frequently triggered by retaliation, would rather control their domestic balance of trade through a number of tools such as tariffs, import quotas, subsidies, currency manipulation, and so on.

In observing different national policies, one can discover endless combinations and degrees of applying these rules, each one of which can bring positive or negative effects to a country's socioeconomic system.

Trade agreements between two or more countries are governed by unilateral barriers, that is, where one country exports its goods to another with certain trade restrictions such as (a) quotas as to the imported volume of goods, (b) government-imposed trade prohibitions, (c) tariffs, or (d) nontariff barriers to trade (NTBs), and so on. Namely:

a. *Quotas* in global trade pertain to the government-enforced restrictions of the quantity imported, or under certain instances the monetary value of the commodities or services performed. There are typically two types of quotas: (i) an *import quota* is an NTB, which entirely eliminates the import of a specific commodity, whereas (ii) *a tariff quota* allows a commodity to be imported duty-free or at a reduced duty fee up to a specific quantity. Once the quantity surpasses the quota agreement, an increased duty rate applies. Quotas serve as a catalyst in reducing trade activities, especially when the national demand for a product is not responsive to price increases. Just like a financial weapon, quotas are typically more unsettling to global trade compared to tariffs, since the quota outcomes cannot be counterbalanced by export subsidies or by foreign currency depreciation (Britannica 2013).

b. *Government-imposed trade prohibition or export sanctions* refer to the limitation of free trade where one economy cannot purchase the commodities of another economy except if specific prerequisites are met, or requirements are satisfied. Trade prohibition is one of the least effective methods of trade control, as its consequences are much more severe compared to its benefits. Some of the repercussions of trade prohibition include (i) retaliation; (ii) trade fraud, as the exporting country may use intermediate countries as transiting stations, where the goods' documentation is reissued and a new country of origin is stated; and (iii) because of the commodity's artificial scarcity, the demand increases, leading to an increase in prices.

c. *Tariffs* restrict global growth as they disable countries to benefit from each other's competitive advantage. Tariffs create price distortions and minimize a nation's revenue maximization. Tariffs imposed also disable countries to expand their commercial or manufacturing activities. Once the tariff barriers are eliminated, any expenditure related to a nation's productivity shift from one commodity to another is short term, whereas the long-term benefits from the shift toward a free market are greater by far (OECD 2005).

d. *NTBs* enforce import control without applying the common tariff techniques, for example, rules of commodities' origin, import licenses, import bans and other regimes, subsidies, overvalued currency, and foreign exchange market control. In certain NTBs, once the measures are ratified, they may lead to an outcome similar to tariffs. For example: (i) antidumping measures: pursuant to their enforcement, trade tariffs apply on foreign exporters; and (ii) countervailing duties pertain to subsidies that enable the commodities to be sold at a lower price than the average market price. Global market sales enable the producer to minimize the prices in nations and markets that do not obtain government subsidies. Several NTBs are indirectly connected to foreign monetary policies, but their effect on foreign fiscal activity and global trade is tremendous. Certain NTBs are granted under certain conditions, when they are essential to safeguard safety, health, hygiene, scarce natural resources, and so on.

In accordance to OECD research, the monetary gains from the elimination of remaining trade barriers will be considerable, both in terms of "static" gains and in terms of "dynamic" gains.

a. The static gains arise, as a mere 10% trade increase leads to a 4% increase in per capita income. Reduced regulatory limitations to competition may lead to a 2% to 3% boost in per capita GDP in the OECD area. More effective customs procedures (i.e., trade facilitation) could strengthen global well-being by $100 billion. Total tariff liberalization in farming and industrial merchandise could boost global well-being by a further $100 billion. Greater profits would be anticipated because of the liberalization of services trade.

b. The dynamic gains, on the other hand, are related to trade-induced modifications to the long-run rate of production. Monetary and trade growth could well be greatly enhanced, offering an additional increase to global financial prosperity (OECD 2005).

3.3 RISK ASSESSMENT (RA) AND RISK MANAGEMENT (RM): HOW FAR CAN WE GO?

A ship in harbor is safe—
but that is NOT what ships are built for.

John A. Shedd

This section aims to evaluate the role of risk analysis in port management and the maritime industry and define its major components in a port's growth.

In 2010, IMO's Maritime Safety Committee 85 (MSC 85/26/Add.1) adopted, among other amendments to the ISM Code, a revision of clause 1.2.2.2, which stipulates the necessity for the maritime industry to evaluate the risks to ships, employees, and the environment deriving from shipboard operations (USCG NAVCEN 2010). Namely, enforced as of 2010, ISM's objective is to "assess all identified risks to its ships, personnel and the environment and establish appropriate safeguards" and its implementation, and define their role. The implementation of the risk assessment regulations is another significant landmark for the maritime industry, since comprehension of the risk element will enable port managers and the maritime industry as a whole to become safer and achieve sustainable growth.

Risk may be defined as the product of two elements: (a) the *probability* and (b) the *severity of consequences* (UN FAO 2013). Hence, the equation derives where

$$\text{Risk Assessment} = \text{Probability} \times \text{Single Hazard} \qquad (3.7)$$

Risk may derive from a single hazard, or with regard to a specific event–outcome combination, where multiple hazards (e.g., composite risks) are involved, for example, a Category 5 hurricane, and the multiple risks involved (loss of life, property, and the environment) within a specific radius in the vicinity of a port (NIST 2007). Figure 3.1 demonstrates the multiple risks that pertain to port management and the maritime industry as a whole.

A Composite Risk Index may be estimated as follows:

$$\text{Composite Risk Index} = \text{Probability} \times \text{Impact of Multiple Hazard Event} \qquad (3.8)$$

The US Coast Guard typically assesses the effect of the risk incident on a 1–5 scale: severity is measured from 1 to 5, where 1 denotes lack of severity and 5 is equal to a disaster. Likelihood of the implications is measured from 1 to 5, where 1 is equal to remote possibility, 3 offers a 50%/50% probability, and 5 foresees a high likelihood of occurrence (USCG 2013).

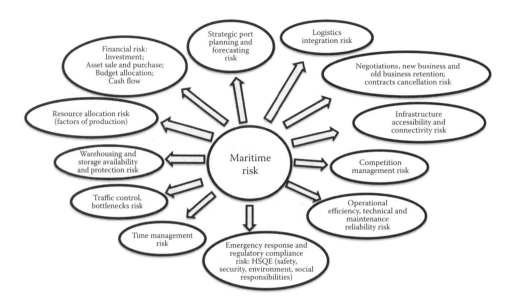

FIGURE 3.1   Maritime risk areas. (Courtesy of M.G. Burns.)

Risk analysis is rather difficult to measure, owing to the practical inability to measure the possible likelihood and impact of a complex risk. Elements such as time, resources, and human error make the risk management process rather complicated. In mathematical terms, risk magnitude equals the probability of incidence multiplied by the event's overall impact:

$$R_{\text{Magn}} = \sum_i \text{Lip(Li)} \qquad (3.9)$$

Risk analysis consists of three major components, that is, (1) risk management, (2) risk assessment, and (3) risk communication. At the same time, these components deal with the three principal elements of hazard: (i) probability, (ii) consequences, and (iii) impacts.

Over the past few years, the process of risk analysis has been increasingly upgraded and refined from segmented port operations and business evaluation to long-term corporate forecasting and planning. The typical sequence to be followed in risk analysis requires the risk managers, that is, the corporate decision makers, to form teams that will identify and assess possible risks.

Figure 3.2 depicts a sample of a port's risk assessment record sheet, which demonstrates how risk identification, mitigation and prevention, or recurrence is a part of the risk analysis and risk management process.

Port authorities need to commit to the hazard identification, monitoring, and mitigation, and commit to a continuous process improvement. The risk analysis process should encompass the port authority management and employees throughout all the levels of authority. Their expertise, qualifications, and observation can be a crucial component of risk assessment process. The port's risk assessment process will involve trustworthy internal

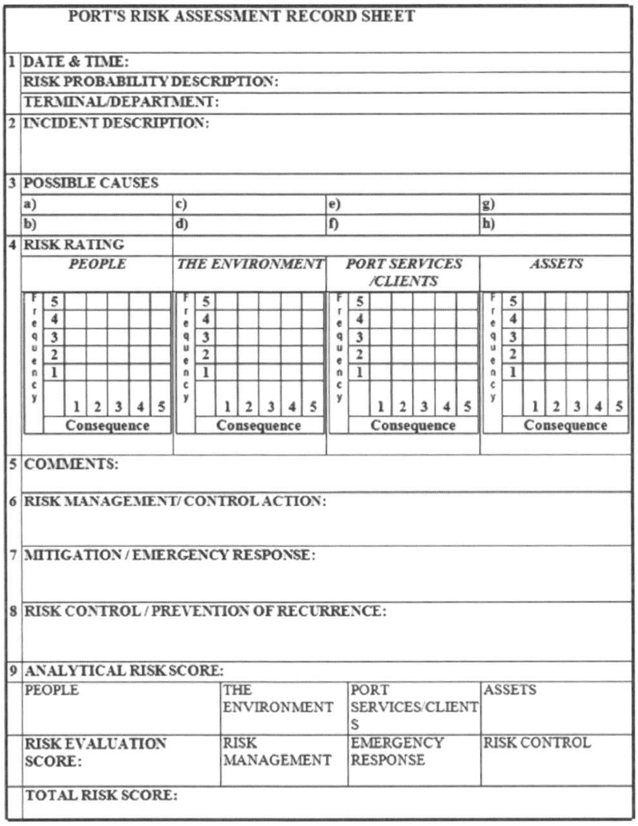

PORT'S RISK ASSESSMENT RECORD SHEET			
1 DATE & TIME:			
RISK PROBABILITY DESCRIPTION:			
TERMINAL/DEPARTMENT:			
2 INCIDENT DESCRIPTION:			
3 POSSIBLE CAUSES			
a)	c)	e)	g)
b)	d)	f)	h)
4 RISK RATING			

PEOPLE | THE ENVIRONMENT | PORT SERVICES /CLIENTS | ASSETS

Frequency 5 4 3 2 1 — Consequence 1 2 3 4 5 (four grids, one for each category)

5 COMMENTS:			
6 RISK MANAGEMENT/ CONTROL ACTION:			
7 MITIGATION / EMERGENCY RESPONSE:			
8 RISK CONTROL / PREVENTION OF RECURRENCE:			
9 ANALYTICAL RISK SCORE:			
PEOPLE	THE ENVIRONMENT	PORT SERVICES/CLIENTS	ASSETS
RISK EVALUATION SCORE:	RISK MANAGEMENT	EMERGENCY RESPONSE	RISK CONTROL
TOTAL RISK SCORE:			

FIGURE 3.2   Port's risk assessment record sheet.

and external communication networks of information exchange, such as port clients, the Coast Guard (Port State Control), supply chain partners, and so on.

Risk management measures, monitors, and controls the larger decision process: it brings together risk assessment methodologies with resolutions on how to tackle the risk. There are five elementary risk management approaches, which encompass several adaptations among the following options: (i) risk avoidance, (ii) risk mitigation, (iii) risk acceptance, (iv) risk outsourcing, and (v) risk sharing, that is, to third parties or allied organizations that are designated, available, and willing to support the port's risk mitigation, by providing financial, technical, operational, or other form of assistance. The entire risk management approach entails strategic and synchronized response or procedures that are developed to (i) eliminate the risk impact generated by an event–outcome combination; (ii) eliminate the likelihood of that event–outcome combination; (iii) reinforce, guide, and support emergency responders; (iv) expedite incident investigation, root cause analysis, and crisis management response; and finally (v) assist in recuperation. Risk mitigation strategies are categorized under three wide headings: financial, engineering, and managerial (NIST 2007).

The risk manager follows the risk assessment process, commencing with problem formulation and developing into five steps: (1) hazard identification, (2) hazard assessment

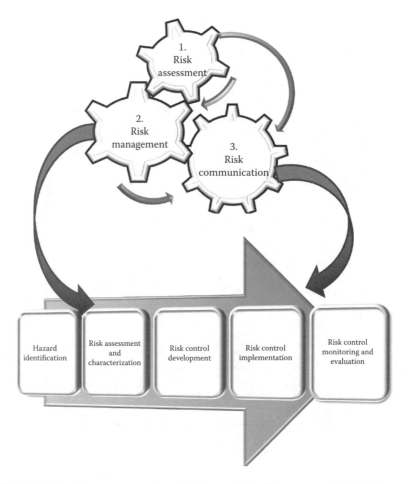

FIGURE 3.3   The components of risk analysis. (Courtesy of M.G. Burns.)

and characterization, (3) risk control development, (4) control implementation, and (5) supervision and evaluation (US Army 2006). Since the maritime industry is a high-risk, high-profit industry, it is imperative to implement risk communication as a nonstop, sustainable process.

Figure 3.3 demonstrates this holistic process that is applied to the entire business entity while enabling different company segments and divisions (e.g., terminals, ship types, etc.) to collaborate effectively.

Enterprise risk management alternatively defined as *strategic* or *holistic risk management* is a twenty-first century management discipline that demands global companies to identify all the possible risks they are likely to encounter, while centering on the upside, or increase of risk, as well as its downside, or decline. Risk managers have to decide which areas of risk should be evaluated and integrate these into the company's action plan, which must also be available to the stakeholders. ERM consists of accidental losses, as well as financial, strategic, operational, and other risks. Its success is determined by striking a balance between enhancing profits and managing risk.

Risk assessment is a methodical practice for identifying and assessing potential risks and measuring the impact or hindrance in achieving the port authorities' objectives, favorably or adversely. These incidents may occur externally (e.g., global financial developments, consumers, supply chains, trade routes, regulatory framework, and competition) or internally (e.g., personnel, process, and infrastructure). When these incidents intersect with an organization's targets, or can be forecasted to do so, they become risks (PWC 2012).

Typically, the maritime industry distinguishes risk into four major categories:

i. *Strategic*, pertaining to the port's values and long-term goals, which include company's reputation, innovation, differentiation, and internal and external competition, as well as the entrepreneurial aspects of negotiations, new business and old business retention, ability to meet contractual deadlines and obligations, customer loyalty, business cancellation, and so on.

ii. *HSQE, physical and regulatory compliance risk*, focusing on occupational health, safety, security, social responsibilities, and environmental risk. This risk segment encompasses the physical threat of natural disasters and environmental hazards, as well as the regulatory compliance, emergency planning, contingency planning, and response.

Port plans include physical risks, such as search and rescue, media handling, pollution (counterpollution; air, soil, and noise pollution), HazMat, HazWOper, chemical spillage, fire, grounding, collision, sinking, security (including bomb threat, terrorism, and piracy), emergency response and medical emergency, and so on. In addition, it covers operational, technical, and maintenance risks that are likely to derive pursuant to a physical or compliance hazard. Chapter 10 demonstrates the methodology for the risk assessment of physical risk. These econometric formulas address the asset's integrity, while incorporating the contribution of the human factor in the emergency response.

iii. *Operational*, focusing strictly on the port's operational efficiency, technical, and maintenance reliability risk, as well as encompassing the efficiency of the entire multimodal transportation and the supply chain, such as hazards related to infrastructure accessibility and connectivity, time management, traffic control, bottlenecks, warehousing and storage availability, logistics integration, and so on. Operational risk assessment may be estimated as follows:

$$OpSec_{base} = 100 - \frac{OpSec_{total}}{Scope + OpSec_{total}} \tag{3.10}$$

$$LC_{base} = Scope \times \frac{LC_{total} \times 0.1}{Scope + OpSec_{total}} \tag{3.11}$$

iv. *Financial*, which entails capital investment, banking and mortgage, currency, inflation, assets' sale and purchase, budget allocation and control, the economic aspects of the factors of production, cash flow, liquidity, taxation, and so on.

Financial-based assessment methods and strategies focus on three principal corporate areas:

1. A *conditional value at risk (CVaR)* is a portfolio risk measurement application that investigates financial risk exposure by employing a conventional algorithm formula that calculates the return on investment, over a specific time period. It is employed by risk managers, investors, and portfolio managers that wish to monitor and eliminate the likelihood of future financial losses.

   In a typical mean-variance versus mean-conditional value-at risk scenario characterized by a normal distribution of returns, both VaR and CVaR may be projected using just the initial two instances of the return distribution (see Rockafellar and Uryasev 2000a,b). Assuming that there is an estimate confidence level of 95%, the probability level for the VaR and CVaR is fixed at 5%, hence resulting in a portfolio with mean deviation μ_p = 10% and a standard deviation σ_p of 20% (Xiong et al. 2010). Consequently, the VaR and CVaR of the portfolio are VaR_p = 23% and $CVaR_p$ = 31.2% of the portfolio's commencing value, respectively, which would be expressed as

$$VaR_p = -\mu_p + 1.65\sigma_p \tag{3.12}$$

$$CVaR_p = -\mu_p + 2.06\sigma_p \tag{3.13}$$

2. *Loan-to-value (LTV) ratio* is a financial risk assessment ratio that is used by banks and investors in order to evaluate the risk of investment when purchasing an asset. Frequently, evaluations with higher LTV ratios are regarded as "higher risk," and this may be reflected in the contractual arrangements between the port authorities and investors. It is reflected as

$$\text{LTV ratio} = \frac{\text{Investment sum}}{\text{Evaluated asset price}} \tag{3.14}$$

3. *Credit analysis*, a common method used by investors. A risk assessment focused on a future financial partner's credit data will help the investor or the bank to conclude on an investment agreement and will likely assist in the stipulation of the repayment terms and conditions:

$$\text{Cash Returned on Capital Invested: } \frac{\text{EBITDA}}{\text{Equity}} \tag{3.15}$$

where EBITDA pertains to earnings before interest, taxes, depreciation, and amortization.

Capital employed equates to a corporation's long-term funds employed by the firm, that is, its equity plus noncurrent liabilities. Hence, it reveals the earnings and overall financial performance of a firm's capital investments.

Return on average capital employed (ROACE) estimates the average of opening and closing funds employed for a fiscal year or for a specified period to time. Depreciation and amortization need to be calculated separately.

$$\text{ROACE} = \text{EBIT}*(365/\text{time})/((\text{FP1[b][E]} + \text{FP1[b][NCL]} + \text{FP1[e][Equity]} + \text{FP1[e][NCL]})/2) \tag{3.16}$$

where FPb refers to the financial position at the beginning of the time period, FPe refers to the financial position at the end of the time period, E is the equity, and NCL denotes the noncurrent liabilities.

$$\text{Current Ratio:} \quad \frac{\text{Current Assets}}{\text{Current Liabilities}} \tag{3.17}$$

$$\text{Debt-Equity Ratio:} \quad \frac{\text{Total Liabilities}}{\text{Shareholders' Equity}} \tag{3.18}$$

In financial and strategic terms, risk is frequently considered as the prerequisite of profit: the "Risk-Reward Concept," alternatively quoted as "High Risk, High Gains," is one of the maritime industry's favorite mottos and guiding principles, suggesting the proven fact that the higher the risk of a specific venture, the higher the return.

Similar risk types are applicable to globalized, capital-intensive, technology-oriented, and regulatory-driven industries such as oil and gas, logistics, and manufacturing, to name a few.

Difficulties may arise when conducting risk assessment in global ports or trade and transport companies, as stakeholders may not be willing to share information as to past risks and future threats. For every hazard, a risk estimate is based on the likelihood of occurrence and the severity of the hazard. Difficulty also arises as a result of the difficulty to objectively assess a complex set of risk possibilities. A risk manager's perception on risk typically ranges between reluctance, prudence, and overconfidence.

3.3.1 The Risk Assessment Methodologies

The ability to grasp the extent and dimensions of risk is always dependent on internal and external factors affecting the port, the economy, and the market's volatility, which typically are unforeseen. Nevertheless, risk assessment offers sufficient feedback to the port management in order to pinpoint the elementary risks, prioritize the risk management resolutions, and formulate the port's strategy.

The methodologies to assess risk are categorized in three broad segments:

1. *Qualitative risk assessment*, which is based on statistical facts and figures. Risk is hereby classified as high, low, or medium. The limited sources of data gathered generate a rather biased or insufficient numeric and empirical basis. However, the accuracy of this primary data, coupled by specialized knowledge and recognition of possible areas of concern, may allow the risk identification and classification (PWC 2012).

2. *Semiquantitative risk assessment*, where a mathematical risk appraisal is calculated by using a combination of qualitative and quantitative data. An accurate outcome for this category of assessment will require a plethora of secondary data

that would be required in a full quantitative risk assessment, plus the selected, specific primary data used in a qualitative risk assessment.

3. *Quantitative risk assessment*, which needs calculations of two risk elements: the magnitude of the potential loss (*L*), and the probability (*p*) that the loss will occur. This assessment type offers numerical risk expressions and warnings of the potential risk hazards and uncertainties (WHO 1995).

The most commonly used quantitative technique of risk assessment is the *Annual Loss Expected (ALE) model*, which derives from the principle of expected loss. This is reflected by estimating the probability of negative risk and subsequent losses. It is expressed as follows:

$$\text{ALE} = (\text{Risk Probability}) \times (\text{Value of Loss}) \qquad (3.19)$$

and

$$\text{ALE} = \sum_{i=1}^{n} I(O_i)F_i \qquad (3.20)$$

where *I* reflects the potential loss caused by a future risk, *O* pertains to a number of unfavorable consequences, and *F* pertains to the event's frequency (Rot 2008).

In this quantitative econometric formula, the risk function variables have precise numerical values.

$$R = AV * E_{I,V,T} * P_{V,T} \qquad (3.21)$$

where risk (*R*) is evaluated in money terms, that is, loss of investment or loss owing to physical threat. *V* and *T* reflect the vulnerability and threat factors, which together with the exposure (*E*) and probability (*P*) factors, are initially measured in terms of levels 1–5 in the Risk Assessment Matrix, and consequently converted in a percentage. Time is expressed in years, months, or days.

3.3.2 Risk and Business Longevity

The element of risk is present in every business transaction. In fact, the maritime industry is considered as a high-risk, capital-intensive industry, together with all global trade and transport companies. Although port authorities are risk prone by nature, to some extent, they are capable of minimizing the repercussions, because of government support, that is, subsidies, and state, or regional backing, that is, European Union Funding.

The private sector, on the other hand, may be more exposed to risk. In particular, shipping companies that have penetrated multiple markets may be exposed to risk to a higher extent.

Successful companies in the early twentieth century had a life span of approximately 60 years, whereas this figure decreased to 25 years in the last decades of the twentieth century, 15 years in the first decade of the twenty-first century, and only 7 years in the second decade of our century. A company's longevity is the strongest possible marketing tool, as it verifies the company's successful strategies in risk assessment, forecasting, investment strategies, cash flow, alliances, and so on.

Here is a brief write-up on Odfjell Company, a 100+-year-old terminal and shipping major with a youthful and highly energetic attitude.

CASE STUDY: ODFJELL—CELEBRATING 100 YEARS OF ACHIEVEMENTS

While the risk factor is most common among seaports and maritime companies, some corporations prove their value and dynamics by longevity and sustainable achievements. Longevity proves safety, quality, and respect for people and the environment.

Odfjell has proven that successful longevity is a company's most powerful marketing message.

LESSONS ON HOW TO CREATE A 100-YEAR COMPANY, WITH HIGH-RISK, HIGH-GROWTH STRATEGIES

Odfjell was established in 1914 when Mr. Odfjell founded a shipping company with the purpose of buying, owning, and operating dry cargo ships, namely, carrying timber. Since then, the Odfjell family has taken great expansion leaps into a number of private companies and different market segments, through the establishment of Odfjell Tankers, Odfjell Management, Odfjell Terminals (Figure 3.4), and Odfjell SE, to name but a few. In 2014, that is, 100 years since the company's inception, Odfjell has achieved a leading global presence characterized by intelligent entrepreneurial strategies, penetration of the world's most profitable and productive market segments, but most of all a fine reputation of integrity, leadership, and dynamic vision for the future. The corporation is headquartered in Bergen, Norway, yet has offices in multiple strategic maritime and terminal centers in Europe, the Americas, Africa, and Asia.

Odfjell is a leading company in the global market for transportation and storage of bulk liquid chemicals, acids, edible oils, and other special products. The company owns and operates chemical tankers in international and regional trades, in

FIGURE 3.4 Odfjell Terminals, Houston (Texas, USA). (Courtesy of Odfjell Tankers.)

FIGURE 3.5   M/T Bow Sky, chemical tanker, DWT: 40,005 T. (Courtesy of Odfjell SE.)

addition to a network of tank terminals, which are traditionally named after trees, as a reference to the company's origins in the transport of timber (http://www.odfjell.com/Tankers/Pages/default.aspx).

Odfjell commenced the construction and commercial operations of the first small chemical tanker ships in 1937 and has gained a leading global market share in the industry's radical expansion since WWII. In the 1950s, Odfjell prevailed in the global chemical tanker trade (Figure 3.5), and in the 1960s, they innovated the first tanker with stainless steel tanks (*M/V Lind*).

Odfjell's strategy is to maintain and enhance their position as a leading logistics service provider for global customers. This will be achieved through efficient and safe operations of deep-sea and regional chemical tankers and tank terminals. Strategic trading patterns and economies of scale through volume and purchasing benefits are components of the corporate strategy leading to expansion and optimum fleet utilization.

Odfjell Tankers Ship Management undertakes the superintendence, crewing, technology management and development, risk and QHSSE management, planning and control, and training and procurement. With offices in Bergen, England, Houston, the Philippines (Training Academy), Sao Paulo, and Singapore, the company currently manages and oversees 54 vessels, while recruiting 2350 seafarers including cadets. The aim of Odfjell Ship Management is to persistently generate a work culture able to elevate their safety performance to a higher level. With this objective in mind, Odfjell has developed their organization through continuously delivering enhanced technical and marine superintendence capacities within the various groups (fleets), strengthening their training programs and implementing an efficient, well-defined key performance indicator (KPI) program (http://www.odfjell.com/Tankers/ShipManagement/Pages/ShipManagement.aspx).

ODFJELL TERMINALS

Odfjell Terminals constitutes a component of Odfjell SE, specializing in chemical tanker shipping and tank storage services. The company's head office for their global tank terminal activities is located in Rotterdam (The Netherlands), one of

FIGURE 3.6   Oiltanking Odfjell Terminals, Sohar, Sultanate of Oman. (Courtesy of Odfjell Company.)

the world's major clusters for tank terminal competence. The company's tank terminals are also located in Antwerp (Belgium), Sohar (Oman) (Figure 3.6), BIK (Iran), Singapore, Onsan (Korea), Dalian, Jiangyin, and Ningbo in China and Houston (USA). The latest acquisitions were developed in Charleston, USA, in Tianjin and in Quanzhou in China and will become operational in 2013, 2014, and 2016, respectively. Furthermore, the company also is part of a network of 11 terminals in South America and one in Canada, which are partly owned by related parties (http://www.odfjell.com/Terminals.aspx).

REFERENCES

Odfjell Tankers. 2013. Odfjell Tankers: For anything liquid. Available at http://www.odfjell.com/Tankers/Pages/default.aspx. Accessed August 25, 2013.

Odfjell Tankers. 2013. Ship Management. Available at http://www.odfjell.com/Tankers/ShipManagement/Pages/ShipManagement.aspx. Accessed August 25, 2013.

Odfjell Tankers. 2013. Odfjell Terminals. Available at http://www.odfjell.com/Terminals.aspx. Accessed August 18, 2013.

REFERENCES

American Association of Port Authorities (AAPA). 2012. Available at http://aapa.files.cms-plus.com/SeminarPresentations/2013Seminars/13MEDC/AgendaMEDC_13.pdf. Accessed July 4, 2013.

American Association of Port Authorities (AAPA). 2013. AAPA's port infrastructure investment survey. Available at http://www.aapa-ports.org. Accessed July 2, 2013.

A.P. Moller (APM). 2013. Sustainability Report, 2013. Editor in Chief: Louise Kjaergaard; Editor: Susanne Nielsen; Contributing writer: Eva Harpøth Skjoldborg. Published by Cool Gray A/S, Denmark.

Britannica. 2013. Encyclopedia Britannica Online. Encyclopedia Britannica Inc. Available at http://www.britannica.com. Accessed August 12, 2013.

CIA. 2013. The World Factbook, China. Available at http://www.cia.gov/library/publications/the-world-factbook/geos/ch.html. Accessed July 9, 2013.

Claycamp, H.J. and Massy, W.F. 1968. A theory of market segmentation. *Journal of Marketing Research* 5(4): 388–394.

Currier, J.P. (US Coast Guard Vice Admiral). 2013. Risk Management for the Proficient Operator. Senior Leadership Proceedings. Available at http://www.uscg.mil/senior leadership/DOCS/2013-08-01;%20USNI%20Proceedings.pdf. Accessed September 12, 2013.

Darwin, C. 1859. The origin of species, Chapter 4. *Natural Selection*, J. Murray (ed). London: Albemarle Street.

EPA. 2013a. Environmental Protection Agency Green Building. Available at http://www.epa.gov/greenbuilding. Accessed July 12, 2013.

EPA. 2013b. Environmental Protection Agency Green Power Partnership. Available at http://www.epa.gov/greenpower/communities/index.htm. Accessed July 12, 2013.

Forbes. 2012. China salaries now on par with US. Available at http://www.forbes.com/sites/kenrapoza/2012/06/14/for-some-jobs-china-salaries-now-on-par-with-u-s. Accessed June 20, 2013.

Geelong Port. 2013a. Geelong Port, Australia. Available at http://www.geelongport.com.au.

Geelong Port. 2013b. Land use strategy. Port of Geelong, Australia. Available at http://www.transport.vic.gov.au/__data/assets/pdf_file/0016/30841/PortofGeelongLandUseStrategy.pdf. Accessed July 12, 2013.

Global Innovation Index. 2013. Available at http://www.globalinnovationindex.org. Accessed July 2, 2013.

Indian Ports Association. 2013. Consolidated ports' development plan. Available at http://ipa.nic.in. Accessed July 12, 2013.

LOOP Port. 2013. LOOP port. Available at http://www.loopllc.com/About-Loop.aspx. Accessed July 8, 2013.

MARAD. US Maritime Administration. 2013. Port and terminal infrastructure investment roundtables. Available at http://www.marad.dot.gov/ports_landing_page/infra_dev_congestion_mitigation/port_finance/Port_Fin_Home.htm. Accessed July 8, 2013.

Marine Board Commission, NRC. 1985. *Dredging Coastal Ports, An Assessment of the Issues. Marine Board Commission on Engineering and Technical Systems*. Washington, DC: National Research Council. National Academy Press. Accessed July 8, 2013.

Maritime and Port Authority of Singapore. 2013. Available at http://www.mpa.gov.sg. Accessed July 8, 2013.

National Institute of Standards and Technology (NIST). 2007. Special Publication (SP) 800-30, NIST SP 800-60, Guide for Mapping Types of Information and Information Systems.

Nissan of Newport News. 2013. Available at http://www.nissanofnewportnews.com. Accessed July 2, 2013.

Northeast Diesel. 2008. EPA New England. Options for the marine ports sector: Green strategies for sustainable ports. Available at http://www.northeastdiesel.org/pdf/Green-Strategies-4-Sustainable-Ports.pdf. Accessed July 8, 2013.

Northern Economics, Inc. 2011. Planning for Alaska's regional ports and harbors. Prepared for U.S. Army Corps of Engineers Alaska District and Alaska Department of Transportation and Public Facilities. West Coast Corridors. Accessed July 2, 2013.

Noy, E. 2010. Niche strategy: Merging economic and marketing theories with population ecology arguments. *Journal of Strategic Marketing* 18(1): 77–86. Accessed April 16, 2013.

OECD. 2005. International trade. Free, fair and open? Available at http://www.oecd.org/trade/internationaltradefreefairandopen.htm. Accessed July 8, 2013.

Port of Helsinki. 2013. Available at http://www.portofhelsinki.fi/environment/environment_effects/energy_consumption. Accessed July 1, 2013.

Port of New York and New Jersey. 2013. Available at http://www.panynj.gov/about/energy.html. Accessed July 1, 2013.

Port of Portland. 2010. Environmental initiatives at seaports worldwide: A snapshot of best practices. Prepared for: Port of Portland, Prepared by: The International Institute for Sustainable Seaports. Global Environment and Technology Foundation. Available at http://ecbiz103.inmotionhosting.com/~getfor5/wp-content/uploads/2011/06/FINAL-Environmental-Initiatives-at-Seaports-Worldwide-April-2010.pdf. Accessed June 24, 2013.

Port of Rotterdam. 2006. Your chemical port of choice. Available at http://www.portofrotterdam.com/en/News/pressreleases-news/Documents/Your_chemical_port_of_choice-PDF_tcm26-20160.pdf. Accessed July 8, 2013.

Port of Virginia Authorities. 2013. Available at http://www.portofvirginia.com/facilities/nnmt.aspx. Accessed July 2, 2013.

Porter, M.E. 1980. *Competitive Strategy*. New York: Free Press.

Prokop, D., Kahumoku, D. and Kalugin, G. 2011. Alaska's lifeline. Cargo distribution patterns from the port of anchorage to Southcentral, Northern, Western and Southeast Alaska. University of Alaska Anchorage, College of Business and Public Policy. Department of Logistics and Port of Anchorage, Municipality of Anchorage. Available at http://www.portofalaska.com. Accessed July 2, 2013.

PWC. 2012. Risk in review. Rethinking risk management for new market realities. March 2012. PricewaterhouseCoopers LLP, a Delaware limited liability partnership. USA. Available at http://www.pwc.com/us/en/risk-assurance-services/publications/assets/risk-in-review-2012.pdf. Accessed May 12, 2013.

Ricardo, D. 1821. *On The Principles of Political Economy and Taxation*. 1973 reprint edited by Piero Sraffa with the Collaboration of M. H. Dobb. 8 volume 1, Section V, page 474. Liberty Fund, Indianapolis, USA.

Rockafellar, R.T. and Uryasev, S. 2000a. Conditional value-at-risk for general loss distributions. *Journal of Banking and Finance* 26(2002): 1443–1471. Elsevier.

Rockafellar, R.T. and Uryasev, S. 2000b. Optimization of conditional value-at-risk. *Journal of Risk* 2: 21–41.

Rot, A. 2008. IT Risk Assessment: Quantitative and Qualitative Approach. Proceedings of the World Congress on Engineering and Computer Science 2008. WCECS 2008, October 22–24, 2008, San Francisco.

Smith, A. 1776. *An Inquiry into the Nature and Causes of the Wealth of Nations*. London: Printed by W. Strahan and T. Cadell.

Smith, W.R. 1956. Product differentiation and market segmentation as alternative marketing strategies. *Journal of Marketing* 21(1): 3–8.

UN FAO. 2013. Food and Agriculture Organization of the United Nations. The State of Food and Agriculture. Rome, Italy. Available at http://www.fao.org/docrep/018/i3300e/i3300e.pdf. Accessed July 16, 2013.

UNCTAD. 2004a. Assessment of a seaport land interface: An analytical framework. Available at http://unctad.org/en/Docs/sdtetlbmisc20043_en.pdf. Accessed July 1, 2013.

UNCTAD. 2004b. World Investment Report. Available at http://unctad.org/en/Docs/ wir2004ch4_en.pdf. Accessed July 6, 2013.

Ungo, R. and Sabonge, R. 2012. A competitive analysis of Panama canal routes. *Maritime Policy and Management Journal* 39(6): 555–570.

US Army. 2006. Composite Risk Management. Field Manual No. 5-19 (100-14). Washington DC, July 2006. Available at http://www.cid.army.mil/documents/Safety/ Safety%20References/FM%205-19%20Composite%20Risk%20Management.pdf. Accessed May 25, 2013.

US Coast Guard Navigation Center (USCG NAVCEN). 2010. US Department of Homeland Security. Available at http://www.navcen.uscg.gov. Accessed August 5, 2013.

US Coast Guard Vice Admiral (USCG). 2013. Risk Management for the Proficient Operator. Senior Leadership Proceedings. Available at http://www.uscg.mil/seniorleadership/ DOCS/2013-08-01;%20USNI%20Proceedings.pdf. Accessed September 12, 2013.

US Green Building Council (USGBC). 2013. Leadership in Energy and Environmental Design (LEED). Available at http://www.usgbc.org/leed. Accessed July 8, 2013.

US Coast Guard Vice Admiral (USGS). 2008. US Geological Survey. 90 billion barrels of oil and 1,670 trillion cubic feet of natural gas assessed in the arctic. Available at http://www.usgs.gov/newsroom/article.asp?ID=1980. Accessed June 27, 2013.

Virginia International Terminals. 2013. Available at http://www.vit.org/vitinfo.aspx. Accessed July 2, 2013.

World Bank. 2013. World Bank, 2011 Census. Available at http://www.data.worldbank. org. Accessed July 2, 2013.

World Health Report (WHO). 1995. Bridging the Gaps. Report of the Director General. WHO, Geneva, Switzerland. Available at http://www.who.int/whr/1995/en/whr95_ en.pdf?ua=1. Accessed May 18, 2013.

Xiong, J.X., Ibbotson, R.G., Idzorek, T.M. and Chen, P. 2010. The Equal Importance of Asset Allocation and Active Management, Financial Analysts Journal, Volume 66, Number 2. CFA Institute.

Zabarenko W.R. 2013. Warmer climate to open new Arctic shipping routes by 2050: Study. Available at http://www.reuters.com/article/2013/03/08/us-climate-arctic-shipping-idUSBRE92718420130308. Accessed April 4, 2013.

CHAPTER 4

Port Operations

There is one rule for the industrialist and that is:
Make the best quality of goods possible
At the lowest cost possible,
Paying the highest wages possible.

Henry Ford

4.1 PORT MANAGEMENT SERVICES: TERMINAL OPERATORS; PROPERTY LEASING OPPORTUNITIES

Modern seaport facilities are technology-intensive platforms of carriage of goods, situated by a natural or man-made dredged port and containing multiple terminals, docks, and berths. Port management and terminal operations encompass a wide range of activities.

A port serves as a safety haven for ships and a cargo loading/unloading area; as a trade and transport link; as a hub center for sea, land, and air; as a bunkering supply station for fuel and diesel oil; and as a commercial, economic, and industrial zone, with banks, brokers, and agents of all kinds.

4.1.1 Port Management Services and Operations

There are approximately 9000 seaports globally, 3500 of which are medium to large in size. They all serve as the strategic transshipment links between inland and maritime transportation, and domestic and international trade. Because of their distinctive characteristics, market position, size and trade specialization, ports are classified into various categories. The general rule is that the larger the port's size, the more terminals a port has, and the more operational services and market segments it encompasses.

Seaports are complex entities, and their success formula lies in their ability to adapt in the modern business world. Prior to computerization, this was a laborious process that, because of communication and visibility hindrances, involved a wide margin of error. Modern ports now use efficient and interactive software that provide real-time operational organizing for all the port and terminal activities, including berthing, port operational schedule, allocation, and utilization of resources including loading/unloading equipment.

Ports typically function under a regulating system known as the port authority or port management. This is an agency—government or state, public or semipublic—legally established to operate ports and typically controlled by boards or commissions. The rapid evolution of globalization has brought about changes to the administrative role of ports within their national boundaries, their strategic role within the supply chain, and last but not least their operational role in terms of cargo handling.

Their navigational channel operations are in synchronicity with the superstructure and infrastructure that will enable the commodities to be effectively handled and transported in an intermodal or multimodal manner. Figure 4.1 demonstrates the different port functions and components of performance.

Twenty-first century ports are not immune to competition; hence, they are rapidly growing by utilizing their competitive advantages, which vary in terms of

- Economy
- Cargo volume and segments (dry bulk, wet, containers, Ro-Ro's, cruise ships, etc.)
- Culture, vision, and strategy
- Geographical location
- Legal and regulatory framework
- Layout, structure, and size
- Market and trade agreements
- Technology and innovation
- Working practices and corporate culture

In this manner, each port is unique.

The most noteworthy of a port's functions is its ability to adjust its cargo unloading/loading and carrying capacity to accommodate all of the market whims, fluctuations, and unexpected developments.

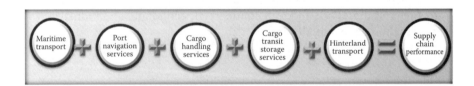

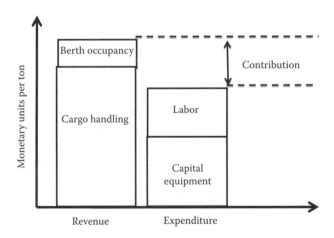

FIGURE 4.1   Port performance indicators.

4.1.2 The Harbormaster's Department and Functions

Harbormasters are designated civilians or naval officers, in charge of implementing the maritime rules of a specific port. Their overall duties include navigational safety including traffic and pilotage directions, security, marine environmental protection, and the operational and technical integrity of the port facilities.

In the United States, harbormasters are designated public servants, whose topographical zone of authority may vary, from a mega-port to a small harbor. Private dockmasters are an exception to the rule, as they usually supervise smaller docks or recreational boats. Regardless of the port's size, they supervise harbor patrol officers and port employees, such as dockworkers and maintenance workers.

Harbormasters also play the role of legal enforcement administrators and have the authority to investigate incidents and accidents pertaining to safety, security, and environmental pollution, as well as immigration and customs formalities and clearance issues. Harbormasters notify the authorities and hand over to them any persons suspected of committing a crime of any nature.

On the basis of the ship's size and type, harbormasters assign suited berths for each ship calling at the port. They provide the "Notice to Mariners," that is, all the safety navigational instructions to the ships' masters, covering each and every level of their passageway within the port, such as traffic obstructions, requirements for pilotage, locks if the port is tidal, berth shifting, mooring permitting and oversight, severe weather forecasts or other hazards, special aids to navigation instructions, lighthouse information, and so on. Ports provide navigation aids and vessel traffic services (VTSs) as parts of their traffic control management. The three main types of service that VTSs offer include (i) Information Service, (ii) Traffic Organization Service, and (iii) Navigational Assistance Service.

4.1.3 Terminal Manager

The terminal managers administer all aspects of port operations in a specific terminal, including the receipt, storage, and distribution of cargo by rail, sea, or road. Their activities encompass ship's loading and unloading, the operations of cargo handling equipment such as mobile cranes and other heavy plants, and the cargoes' ship–port warehouse distribution. They are in charge of stevedores (longshoremen) and union issues, as well as resource allocation management. They work with a team of vessel planners, supervisors, and port workers to ensure safe and efficient terminal operations and customers' satisfaction.

4.1.4 Vessels' Planning

Planning is carried out by the port's Senior Planner who supervises the Central Planners in assigning ships to berths aiming to avoid bottlenecks and promote safety, efficiency, and productivity. As different ships carry a wide range of cargoes, this dynamic team of experts ensures that all cargo shipments, including HAZMAT cargoes, are properly handled and certified and compliant to all mandatory regulations (safety, security, environment, quality, etc.). They acquire the ship's stowage plans and layout at an early stage, in order to coordinate the advance planning. For container ships, they consult the master plans to verify each container's slot, that is, the "bay, row, and tier," method which uses numerical coordinate systems where each container is numbered and its dimensions and precise location onboard the ship are registered. Hence, these records will enable liner companies to discharge the right container at the right port and truck its stowage space onboard the ship (and at the port), at any given time. These records remain with the port, the carriers, and the freight forwarders for future reference, in case of damage of the container box or the cargo, or any incident investigation.

The planners also verify that the cargoes' actual type and weight match with the data on the bills of lading. Finally, they supervise the berths' preparedness status and ensure that the cargo handling equipment is in an operating condition and geared up for the ships' operations.

4.1.5 The Four Stages of Port Management and Operations

There are four principal stages pertaining to the ships' stay at port, from the time of arrival at the port entrance, until the time of its departure. They entire protocol of port management and the breakdown of the time ships spend at port is defined in Figure 4.2 and is duly analyzed herewith.

The first stage gives the option to the ship to be served without having to enter the port, whereas the other three stages are analyzed in a chronological order.

Stage 1. Ports' Off-Port-Limits (OPL) Operations
   Seaports offer OPL operations services to the ships in transit that are not scheduled to visit the particular port of call for loading or discharging operations but are in need of specific port-related services. OPL services help ships achieve time efficiency and minimum deviation, low cost, and economies of scale.

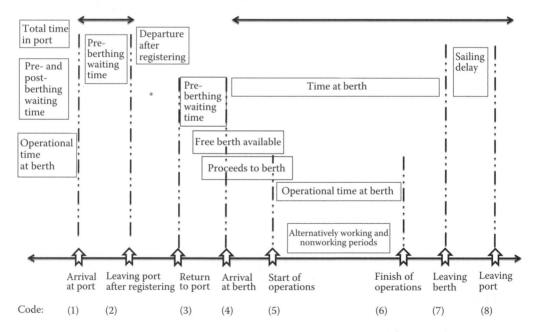

FIGURE 4.2 Breakdown of ship's time in port. (From UNCTAD, 1985. [UNCTAD/ ship/185] Nov. 1. Manual on a uniform system of port statistics and performance indicators. Available at http://r0.unctad.org/ttl/docs-un/unctad-ship-185-rev2/en/UNCTAD_ SHIP_185_Rev2e.Pdf [accessed on July 12, 2013].)

Since the ships' contractual obligations do not permit a deviation, ports can extend their time-efficient services to the OPL.

The OPL services are prearranged by the liner company's operations department and the ship's master, typically by VHF. The "rendezvous" position or meeting point is typically within a designated mileage, for example, 3–5 nautical miles or a specific latitude and longitude off port boundaries.

Designated service boats/launch boats or even helicopters for emergency services are employed by the port authorities. The services provided could be classified into the following categories:

a. Procurement, that is, the victualing of stores, spare parts, food supplies, and so on.
b. Passengers' drop-off, where each launch boat can accommodate about 20 persons, and charge per boat trip, instead of number of passengers, thus achieving economies of scale.
c. Crew changes.
d. Embarking and disembarking of vetting inspectors, marine surveyors, regulatory auditors, and repair teams.
e. In case of an accident or a medical emergency onboard, a helicopter can be employed to pick up the victims and deliver them to a designated hospital. This service can spare the shipowners a deviation claim on behalf of the charterers and prevent any navigational delays, while ensuring that the best possible medical and safety measures are taken.

Port OPL services are quite common, yet they are most frequent at ports that are situated in major navigation passages such as Singapore, Malta/Gibraltar, South Africa, and so on.

Stage 2. Port Operations and Berthing Management

This stage involves the ship's navigation and berthing at the port. From this stage onward, the port authorities will handle the ship's navigational, berthing, loading, and unloading operations, taking into consideration its commercial and contractual status, that is, depending on whether the ship is a *liner* or a *tramp*.

a. For *liner ships*, the port typically has a contract with the liner company, which is the ship's owner and manager. Large liner companies typically have a multiple year contract with the port, and their employees handle their fleet operations while being present at the port. Liner ships perform the same repeated and predetermined schedule at the same designated ports of call, in order to load or unload the cargoes of their numerous clients (charterers, freight forwarders). The liners' service network may include 8–12 ports of call per ship, while the port stay duration time may vary according to the cargo volumes per port, port traffic, and so on. Figure 4.3a and b demonstrate container ships operating liner services, namely, "Evergreen Lines" and "MSC" lines.

Since liner companies represent multiple charterers/freight forwarders, they assume a leader's position and thus can better control and manage port operations for their entire fleet. Their undisputed authority and procedural repetition simplify the port operations and standardize the paperwork exchanged between the owning company, the port, and the ship's master. For

FIGURE 4.3   Container ships operating in liner services. (a) "Evergreen Lines" container ship at the Port of Haifa, Israel. (b) "MSC" lines container ship at the Port of Haifa, Israel. (Courtesy of the Port of Haifa, Israel.)

example, there is a standardization of the navigational instructions e-mailed to the ship's master, the berthing instructions, port tariffs, and so on, which saves time and enhances the endeavor's effectiveness.

b. The *tramp ships*, on the other hand, trade in the spot market and conclude business usually with one or two charterers at a time. The charter party terms and conditions will stipulate the number of ports of call and the time allowed for loading or discharging. The agreement and decision-making process about each port of call will have to be decided between the shipowners and the charterers (freight forwarders), depending on the contractual duration and the level of authority that the owners wish to grant to the charterers. The communication protocol as to the port operations will depend on the charter party type that will be selected as a contractual agreement, that is, (i) a voyage C/P accommodates a single loading–unloading voyage; (ii) a time charter, for example, a 2-year contractual arrangement; and finally (iii) a bareboat charter, where the charterers have the commercial control of the ship and legally become the ship's disponent owners (see Section 4.4). Figure 4.4a and b show tramp ships, namely, *M/V Anemos*, a Supramax bulk carrier managed by "Aegean Bulk" of Greece, and *M/V Polska Walczaca*, a conventional Panamax bulk carrier of nine holds, managed by "Polsteam" of Poland.

FIGURE 4.4   Tramp ships at the Port of Haifa, Israel. (a) *M/V Anemos*, a Supramax bulk carrier managed by "Aegean Bulk" of Greece. (b) *M/V Polska Walczaca*, a Panamax bulk carrier managed by "Polsteam" of Poland. (Courtesy of the Port of Haifa, Israel.)

Services Prior to the Ship's Arrival

Berth Request

Prior to the ship's estimated arrival, its agents or managers (operations department) should submit an online request to the port, providing the following information:

- Vessel's name and previous names
- Ship's main particulars, for example, flag, ownership/management, charterers, deadweight (DWT), length overall (LOA), net tonnage (NT), gross tonnage (GT), draft, and so on
- Estimated date and time of arrival
- Estimated date and time of departure
- Cargo type and volume
- Cargo operations specified, that is, loading or discharging

Modern ports provide to the ship's registered and designated users—typically her agents or operators—online access to efficient, user-friendly berth request applications, enabling them to verify and monitor current status and amend or cancel their berth requests.

Notice of Arrival—72 Hours Prior to Ship's Arrival

At least 72 hours prior to the ship's arrival at port, the ship's operators, master, or agent should submit to the port, a Notice of Arrival, with particulars that pertain to the ship and its managers, cargo type and quantity, any operational information, and so on.

Special Provisions for "Cargoes of Particular Hazard"

Ships navigating in the United States and carrying cargoes of particular hazard should comply with the US 33 Code of Federal Regulations 126.3. The shipowners or operators or agents should forward an Advance Notice of Arrival to the US Coast Guard under copy to the port. As stipulated in 33 CFR 126.16, ports should install light alarms at designated areas, which, in case of an emergency, are activated by the responsible operator.

Cargo Manifest

The cargo manifest with all cargo information should be produced, prepared, and duly approved by the shipowners, masters, or agents, in line with the port requirements. It should be submitted to the port operators prior to the ship's arrival. A bank guarantee or other documentation pertaining to the payment of port expenses or disbursements should be readily available. After the completion of cargo operations, a statement reflecting any and all modifications of the manifest should be prepared and sent to the port.

Services During the Ship's Arrival

Pilotage Services

The use of pilots is mandatory in certain ports, terminals, canals, and inland waterways with tidal fluctuations, draft restrictions, heavy traffic, or hazardous waterbeds owing to reefs, and so on. The idea here is to avoid a ship's hydrodynamic problems (e.g., the suction effect caused by the ship's maneuvering in restricted water) or lack of visibility, which may lead to the ship's grounding, allision, or collision. Figure 4.5a–c illustrate the towing operations during a ship's entrance at port, mooring operations, and the ships' berthing arrangements.

Ports offer this service as an enhanced alternative to electronic navigation charts that provide weather and oceanographic information, yet do not have the local knowledge and experience that pilots deliver. The utility of pilots can be verified by the maritime accident statistics, where accidents, human error, claims, and legal cases have occurred even when pilots were not mandatory and were not used.

Berthing/Unberthing, Mooring/Unmooring, and Anchoring Operations

The ship's maneuvering for berthing/unberthing and its mooring/unmooring and anchoring entail ship-to-port collaboration and navigational proficiency by everyone involved, including tugs, mooring teams, and rope runners, among others. The ports' policies typically stipulate that the berthing operations are on a first come, first served sequence, yet special provisions can be made for different reasons. While ports aim for high berth occupancy rates, delays and traffic are common problems. The berthing process may vary depending on the weather conditions, as well as the docks, berths, or the terminals' characteristics. For safety reasons, appropriate berthing is mandatory in order for ships to commence their cargo operations, to be discussed in the next level.

(a)

(b)

(c)

FIGURE 4.5 Towing operations, mooring and ships at berth. (a) Towboat maneuvers container ship managed by "Zim London." (b) Mooring operations at berth: *M/V Norwegian Jade* managed by Norwegian Cruise Lines. (c) Ships at berth; (top left) passenger ferry *Nissos Rodos* managed by Hellenic Seaways, Greece; (bottom right) chemical tanker *Bentley I* managed by Bentley Marine, USA. (Courtesy of the Port of Haifa, Israel.)

Stage 3. Ship Operations at the Berth/Terminal

After the ship has met the navigational and documentation prerequisites, the operations stage involves the vessel's actual purpose of visiting the port, which may include loading, unloading, bunkering, victualing, and other operations. This is a critical point for the performance of the port, the ship, and the shippers. Teamwork is necessary in order to meet common goals such as holds' or tanks' cleaning, loading and discharging the cargo, cargo handling and stowage

(a)

(b)

(c)

FIGURE 4.6   Cargo handling operations. (a) Loading dry bulk cargo, Italy. (Courtesy of M.G. Burns.) (b) Loading timber cargo. (Courtesy of M.G. Burns.) (c) Rolling cars onboard Ro-Ro *Fides* managed by Grimaldi Lines of Naples, Italy. (Courtesy of the Port of Haifa, Israel.)

supervision, bunkering operations, and so on. All parties play an equally significant role in monitoring and measuring the results in terms of time, safety,
profitability, and efficiency. Figure 4.6a–e illustrate cargo handling operations,
namely, (a) grain loading operations, (b) ship loading at grain elevators, (c) cargo
hold cleaning (manually), (d) rolling cars onboard car carriers, and (e) container
handling operations.

Ship's Cargo Handling Equipment

To ensure operational safety, the ports' protocols typically require a permit from
the ship's master that will authorize them to examine and evaluate the operational efficiency of the ship's cargo handling equipment, while making available
for inspection, the respective certification.

Cargo Operations

Once the Notice of Arrival has been tendered and accepted, the vessel is officially
an arrived ship. The master is obliged to commence the freight operations within
a specified time while safeguarding the areas around the cargo operations.

The charter party agreement between the shipowners and the charterers and
eventually their instructions to the port will stipulate who will undertake the
freight operations; who will provide the cargo handling equipment; up to what
extent third parties, for example, stevedores (longshoremen), will be used; and

(d)

(e)

FIGURE 4.6   (Continued) Cargo handling operations. (d) Loading dry bulk cargo, Israel.
(Courtesy of the Port of Haifa, Israel.) (e) Container handling for CAI International, a
major container leasing company. (Courtesy of the Port of Haifa, Israel.)

so on. On the basis of written documentation, each party shares an agreed share of liability as to the safe and time-efficient operations of the ship. Modern ports have upgraded their cargo handling and surveillance technologies. At all times, the port monitors a database for cargo movements through sea and the hinterland. Its warehousing and cargo handling services include cargo transfer, cargo tallying, weighing, stuffing and unstuffing containers, palletization, repackaging, and labeling and marking hazardous and nonhazardous cargoes.

At this stage, the port is in charge of two separate elements: the ship's safe departure, and in case of cargo discharged at port, the cargo distribution through intermodal and multimodal services, within the supply chain.

Stage 4. Port and Terminal Operators' Logistics Networks

Ports' Positioning
In a matrix-like logic, a port's horizontal positioning entails its power over its competitors, while its vertical positioning pertains to its strategic alliances, such as private terminal operators, investors, long-term contractors and subcontractors, stevedores' companies, and so on. A port's market radius and the efficiency of their cargoes' spatial distribution are determined by its geographic location, its strategic alliances with terminal operators, and its supply chain integration and networking systems.

Logistics Networks and Partnerships
Competent ports actively pursue to attract skilled global partners such as terminal operators, to run a certain number of their containers, bulk and liquid cargoes, and multipurpose cargo terminals and become active members of the regional logistics infrastructure. In addition, ports seek for logistics operators in order to enhance the port's storage, cargo handling equipment, and warehousing services. At this stage, ports serve as distribution centers: they administer the cargo movements by utilizing their *warehouses, roads* (i.e., intrastate and interstate), and *rail* (i.e., on-dock and near-dock infrastructure). In the ports' centralized or satellite warehouses, which may be controlled by the port or its terminal operators, ports orchestrate all logistics activities including cargo forwarding, collecting, evaluation, and distribution. Hence, as terminal operators, investors and logistics' partners become stakeholders in the port, the port authorities undertake regulation and supervision of the entire span of activities in both the public and private sectors.

4.1.6 Marine Terminal Operator (MTO) Agreements and Leasing Opportunities

The profile of MTOs broadly expands from (i) maritime terminal facility owners, to (ii) shipowning (liners or tramps), to (iii) ship management firms, to (iv) oil or cargo majors, to (v) nonvessel operating common carriers, for example, freight forwarders that handle containerized or break bulk shipments and logistic companies. The services offered include warehousing, dockage, and wharfage, to name a few.

a. Public port authorities assume full ownership, monitoring, and controlling of the docks, berths, and other port facilities and frequently run the marine terminal in direct and full authority.
b. Private terminal operators are service providers that lease terminals, infrastructure, and superstructure from the landlord port authority and run the terminals as a private corporate entity.

At a global level, an identical application process is required, where MTOs file with the government's maritime authorities. In the United States for example, agreements conducted among MTOs or among ocean common carriers and MTOs, which pertain to ocean transport in the US foreign trade, must be filed with the Federal Maritime Commission and comply with the rules stipulated with the 46 United States Code (46 USC). Some of the activities that these legal entities are involved in, as permitted by such agreements, include (i) negotiating, fixing, or determining tariffs and rates; (ii) controlling various service-related issues; and (iii) participating in distinctive, exclusive, or cooperative operational and employment arrangements (Federal Maritime Commission 2013a, b).

A successful partnership among MTOs and port authorities can significantly affect the port's market power, by enhancing its activities, in areas such as terminal management, marketing and promotion, off-dock yard operations, rail yard operations, trucking, stevedoring (longshoremen), cargo handling equipment, maintenance, warehousing, container boxes logistics' distribution, cargo monitoring, safety and security surveillance, Hazmat and HazWOper response, emergency response, and so on.

The leasing agreements typically incorporate clauses, terms, and conditions that provide "for the maintenance and keeping in good condition and repair any marine terminal which is the subject of said agreement or agreements" (Statutes and Laws 2013).

4.1.7 Marine Terminal Operators and Leasing Opportunities: Case Studies

It has been verified that the corporate profile of the world's top ten terminal operators closely fits dedicated MTOs, shipowners, and investors.* Nevertheless, there is an interesting industrial diversification as to the type of companies that enter the market. It is worth noting that certain MTOs are manufacturers or cargo traders that entered into MTO agreements with ports in order to duly protect their own cargoes. Others, like *the Port of Miami Terminal Operating Company (POMTOC)*, represent an efficient stevedoring consortium and have penetrated the most competitive markets, through their expertise in cargo handling alone.

This section provides case studies of some of the most unconventional MTOs of the lot.

Cargill
> Cargill was founded in 1865 by the late William Wallace Cargill. From a small grain storage facility in Conover, Iowa, the business boomed as it followed the expansion of the railroad into northern Iowa after the Civil War. In its 150 years of operation, Cargill has developed into one of the greatest, privately owned corporations, producing agricultural, industrial, and food products.
>
> Cargill Marine Terminal Inc. is a respected terminal operator with numerous terminal projects and partnerships in Canada, the United States, Brazil, and the rest of the world. Every year, it operates 450 vessels in over 6000 ports globally and distributes at least 185 million tons of its cargoes to its global customers.

It generates 60% of its income outside the United States, thus contributing to the national balance of trade. It is the eleventh greatest Forbes 500 company and the leading privately held American corporation in terms of revenue. Its services pertain to terminal management, transportation, energy trading, pharmaceutical, steel, and financial and

---

* See Chapter 2 and Table 2.5.

risk management consulting offered to the global commodity markets. As of 2012, its revenue amounted to $133,958 billion, whereas its net income summed to $1175 billion dollars (Forbes 2013). Feeling tempted to do the math, it is good to know that Cargill's daily net income exceeds $3220 million, or $135,000 per hour!

Cargill owns and operates the greatest grain elevator complex in North America (Baie Comeau, Canada), which can hold 440,000 metric tons of grain. In the United States, it owns and operates 32 grain elevators through the Illinois, Minnesota, Ohio, and Mississippi rivers. These facilities ship approximately 8217 barges or 482.2 million bushels of grain annually to the US Gulf.

Cargill is a company of innovation, holding over 1800 global patents. It is increasingly dedicated to renewable energy, and it is actively involved in industrial and agricultural training, education and global charity, schooling, and nutrition projects (Cargill 2013).

Ports America, Inc.

Ports America, Inc., is the greatest American port and terminal operating enterprise, privately owned by AIG Highstar Capital and a combined entity of P&O Ports North America, MTC Holdings, and Amports, a marine cargo handler for automobiles.

They provide management, operations, and maintenance to 50 ports with 97 terminals in the United States, Mexico, and Chile. In 2012, their affiliate, Ports America International Holdings Cooperative, expanded their investment activities in the global leading port of Kaohsiung, Taiwan.

Their business has grown steadily, unaffected by the 2008 crisis. Each year, they handle over 2.7 million TEU (twenty feet equivalent units), 2.5 million vehicles, 8.9 million tons of general cargo, and 1.6 million cruise ship passengers, with an estimated total annual revenue of $2 billion. Their cargo operations include project cargo facilities, bulk, break bulk, world-class cruise terminals, intermodal facilities, and precision Ro-Ro handling (Ports America 2013; Solvere Market Intelligence 2013).

Highstar Capital is a private commercial infrastructure investment firm with an operationally centered, value-added financial investment strategy. Since 2000, the Highstar Team has controlled the investment of $7.6 billion in energy, services, and transportation infrastructure (Highstar Capital 2013).

The Port of Miami Terminal Operating Company

POMTOC, a longstanding terminal operator in Miami, was established in the early 1990s by four stevedore firms to operate a section of Dodge Island. It is the sole noncarrier owned terminal operator serving the maritime community since 1994 by offering an extensive combination of critical services to the global shipping industry. In their 16-year lease with the port, their annual dues exceed $33 million. Their customers, through a Terminal Operating System (TOS), have full visibility of containers moving through their gates, yard, and to/from vessels. TOS also allows importers, exporters, brokers, forwarders, and truckers to have immediate online access with real-time information on container availability, bookings, demurrage charges, online payments, truckers' interchanges, and gate activities by using the systems forecast feature. As it is rather unusual for a stevedore company to manage and operate an entire terminal in a major port such as Miami, POMTOC's market differentiation is exactly their unconventional background and their ability to secure quality business, thus competing terminals with a more "traditional" ship management background.

4.2 BERTHS, FACILITIES, AND EQUIPMENT

In a technology-intensive, capital-intensive industry such as maritime shipping, state-of-the-art designs and sustainable improvements are clearly driven by the market's demand in the most compelling manner. Port planners are focused on providing excellent operations increased efficiency of time, land, berthing space, and equipment. This is achieved by monitoring and controlling two major sectors:

i. Performance management, which focuses on efficiency
ii. Capacity management, which aims to optimize utilization and reduce costs in three core areas: (a) administration, (b) service, and (c) resources

Details of each are demonstrated in Figure 4.7.

4.2.1 Berth Performance versus Capacity

Port operations are improved through optimizing port capacity and overall performance.

A port's performance indicators are both financial and operational and are based on data collected from each terminal or berth. The principal productivity indicators pertain to its (i) output, (ii) utilization-to-capacity ratio, (iii) productivity, and (iv) service time.

a. A berth's output is measured in terms of cargo volume handled annually, whereas a ship's output estimates cargo handled per hour.
b. Utilization is measured in terms of berth occupancy, ashore equipment occupancy, and the occupancy of warehousing and storage areas.

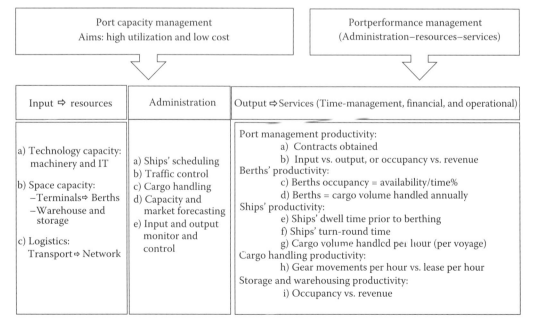

FIGURE 4.7   Port planning: performance and capacity management. (Courtesy of M.G. Burns.)

c. Productivity is measured by means of an estimation of traffic per year or through-put per year, or handling costs based on time efficiency, the factors of produc-tion, versus output and profitability. Namely, a ship's productivity is estimated in terms of cargo handled per hour; a berth's productivity is measured in terms of cargo handled per month or year; and the productivity of cargo handling equip-ment is measured in terms of movements per hour.

d. Since service indicators are associated with berthing, they are measured in terms of (i) berth and shore facilities availability/time (%), (ii) ship's dwell time prior to berthing, and (iii) ships' turnaround time, that is, until departure. Modern ports are integrated with the entire local logistics system; hence, their service indica-tors may entail cargo handling at port, as well as logistics turnaround time.

Port planning aims to properly schedule ships to berths in a manner that balances between congestion and underutilization, considering time allowance and the ship's prin-cipal characteristics, that is, ship's type, principal dimensions such as draft, LOA, DWT, and quantity of cargo to be loaded and discharged. Dwell time reflects the time cargo remains idle in a terminal's in-transit storage space or warehouse, in the process of dis-tribution and further carriage. Long cargo dwell times while at port is a vital concern of modern ports, as bottlenecks create slow process times, and may encourage the creation of new, competing trade routes.

In order to estimate in approximation the capacity of a berth, the following econo-metric formula is used:

$$BC = (O \times E \times EM \times H \times TEU) \qquad (4.1)$$

where BC is berth capacity; O is berth occupancy; E is the number of cargo handling equipment (e.g., two cranes); EM is cargo handling equipment per hour; H reflects a port's working hours a day, shifts per day, working days a week, and so on; and TEU denotes where the number of TEUs per crane move needs to be assessed (Port of Honolulu 2012).

4.2.2 Port Operations: The Place Where Capacity and Performance Meet

Docking Facilities
A docking facility is a port structure next to a pier, that is, between land and water, where ships can load and discharge cargo, and passengers can embark and disembark. It is a sturdy construction consisting of berths, where ships can secure themselves throughout their stay at the port. Docking allows for fixed infrastructure and mobile equipment, for example, cranes and derricks, port conveyor systems, port forklifts, forklift trucks, and other gear used for the cargo loading and unloading operations.

Berths
A berth is an allotted space at a dock or a wharf where a ship can dock or anchor, typi-cally designed alongside quays or jetties. Berths are classified according to the ship and cargo type they handle (e.g., tanker berth) and also according to their design and size (e.g., deepwater berth). Ships' berthing arrangements are subject to the berth's availabil-ity and suitability, that is, based on ship's size, type, and requirements for cargo handling shore facilities.

A terminal's characteristics that affect layout and performance include topology characteristics; dock and berth construction on coastline, that is, how linear and sheltered it is; berthing capacity and ability for simultaneous cargo operations; cargo handling capacity; cargo handling equipment; land filling and availability for additional storage of cargo and equipment; warehousing availability; and intermodal connectivity (Maine DOT 2007).

A key concern on behalf of the port authorities pertains to the time restrictions and the berth usage, in a manner that an optimum berth occupancy ratio is achieved, with eliminated waiting time prior to berthing. From a ship's perspective, time lost is the accumulated waiting time including waiting time for pilot, for tugs, and for berth. Once the ship berths, additional time may be lost while waiting for the cargo or the ashore cargo handling equipment. Process and technology-based improvements are intended to maximize berth occupancy, while minimizing dwell time for ships and turnaround times for the entire multimodal transportation process.

Berth designs have been developed over time in order to increase safety and efficiency, while reducing vessel operation time. As global trade grows, berths need to accommodate more and larger vessels, and this imposes significant pressure to keep abreast with the technological and operational standards. The strength of the supporting wharves must be increased in order to accommodate faster dockside cranes with better antisway load control. The ship-to-shore connectivity must be ensured as container vessels become larger and wider. In the traditional berths, dockside container cranes will need to extend their workable outreach by longer boom designs. On the other hand, indented berths allow for two to nine ship-to-shore gantry cranes (quay cranes) to simultaneously load and unload containers from both sides of the vessel.

Figure 4.8a–d demonstrate modern port berths and cargo handling equipment, as analyzed in this section.

Deepwater Berths
As their name suggests, these are the deepest berths presently available to accommodate the largest ships' size. In view of the new post-Panamax generation ships, modern ports prioritize investments and work on improving the technical capacity and commercial aspects of the harbors, in order to attract contracts for the largest ships:

Tankers:
- Ultra large crude carriers with an LOA of 415 m (1361.55 ft) and a draft of 35 m (114.82 ft)
- Very large crude carriers with an LOA of 330 m (1082.68 ft) and a draft of 28 m (91.86 ft)

Containers:
- New generation: Maersk Triple E-Class container ships, with an LOA of 400 m, that is, almost a quarter of a mile long, a draft of 14.5 m (48 ft), and a carrying capacity of 18,000 TEUs
- Post-Panamax containers with an LOA of 366 m (1200 ft), a draft of 15.2 m (49.9 ft), and a carrying capacity of over 12,000 TEUs

To accommodate these larger ship sizes, global ports invest in deepwater berths. For example, the Port of Southampton, UK, has a 500 m long capacity berth, a deepwater quay area of 1.87 km, with 16 m depth alongside and 16 quayside gantry cranes with super post-Panamax capacity (DP World Southampton 2013).

(a)

(b)

FIGURE 4.8 Terminals, berths, piers, and cargo handling equipment. (a) Carmel Terminal. Port of Haifa, Israel (2013). (b) An oil terminal on the right side. Haifa's historic Crying Pier on the left.

(c)

(d)

FIGURE 4.8 (Continued) (c) Twin Spreaders, handling 20 TEU containers. Carmel Terminal, Port of Haifa. (d) Grain uploader, Dagon Terminal, Port of Haifa. (Courtesy of the Haifa Port Authorities, Israel.)

National trade agreements and the potential of concluding large-scale contracts are another reason that nations and ports develop deepwater berth investment plans. Vale of Brazil, the world's leading iron ore producer and mining corporation, has invested billions of dollars to construct an unrivaled fleet of very large ore carriers to carry their steel making commodity to China and other global clients (China Shipowners' Association 2011; Reuters 2011; Bloomberg Businessweek 2012; Vale 2013). Since Chinese ports are not ready to accommodate this large-scale fleet traffic, it is committed to building 440 deepwater berths by 2015. These examples demonstrate how fast the maritime industry reflects the global trade patterns and logistics requirements.

Designated berths per ship type

a. *Container berths*: Containerization and the opportunity for fast and efficient transshipment resolved many of the problems that break-bulk shipping entailed, and made intermodal and multimodal transportation possible. A container terminal's designs include the traditional one-sided marginal berths and the indented berths. Modern container ports have large-capacity container yards, or a storage area alongside the quay, to stack container boxes. Accessibility and hinterland connectivity are important, that is, the existence of on-dock or near-dock rail yard facilities.

Furthermore, cargo handling equipment should be carefully selected to match their expected volume of cargoes: mobile harbor cranes, ship-to-shore container cranes, straddle carriers, dockyard cranes, fixed and rail mounted cargo cranes, rail mounted stacking cranes, rubber tire gantry cranes, crawler cranes, and so on.

The cargo handling equipment's technical specifications and cargo carrying capacity (measured in tonnage, i.e., metric tons [MT]) should cover and, if possible, exceed their clients' required services. Cranes can be particularly designed to a single container box size (i.e., 20 TEUs or 40 TEUs).

b. *Wet and dry bulk terminals* are frequently related to the cargo side of business, that is, oil majors and commodity key players. In the cases where third-party private operators can enter this market, their earnings equally derive from storage and land space charges, and cargo handling. On the other hand, container terminals' earnings are not as affected by cargo handling. It is also worth noting that national trade agreements mostly affect raw material and energy sources; hence it is the wet and dry bulk terminals that are mostly affected.

India is an example where increasing volumes of coal and various other commodity imports are powered by the government's investment of over $11 billion, in the development of 50 new seaports by 2015 (Port Strategy 2009).

Dry bulk terminals specialize in handling bulk products, like minerals, grains, woodchip, cottonseed, clinker, coal, cement, and so on. Automation and efficient use of technology enable direct transshipment and logistics agility. Products arrive at the port by sea, rail, or trucks, and are discharged from hopper cars at the terminals' purpose-designed discharge stations, which are linked with railway tracks. Depending on the cargoes, ships are loaded by employing excavators and conveyor belts or pipelines, mobile cranes, vessel loaders, bucket wheel dischargers, loading spouts, grabs, and so on. Conveyor systems are utilized to directly move the freight from a regional industrial zone or silo to the ship. Silos and storage facilities are usually found alongside the berth and have modern conveyor systems transfer the commodities to the storage areas or to the ships. Improved storage facilities require large, automated warehouses and the capacity to separate different cargo types. Dry bulk berths seek to improve operating efficiency and loading volume capacity by reducing loading times, thus meeting the industry's goals.

Vessels are loaded using either excavators and conveyor belts or pipelines. The equipment used for loading or unloading cargo onto or from the vessel depends on the characteristics of both the vessel and the cargo. Ships typically have their own cranes, while ports also have their own mobile cranes to accommodate their clients.

c. *Tanker berths for crude and refined oil—bulk oil jetties*: With growing market expectations, the efficient terminal management and operations are critical for liquid bulk terminals. Berths for crude oil tanker ships are typically located alongside the refinery's maritime terminal. Typically, the crude oil cargo is unloaded via pipelines to storage tanks in the refinery. The shipping terminal usually has berths to load refined oil products. Refined products are ready to be distributed locally and overseas, mostly by sea transport (Chevron 2013).

Liquid shipments are loaded alongside a terminal by means of pipelines, pumps, and hoses. Since time efficiency is crucial, the industry strives for high loading and unloading rates through pumping ability and pipelines' size. Once the tanker is ready and all required terminal and tanker valves in the loading system are open, the loading operations begin. Shipowners and cargo owners are occupied with ships' overall performance and limited port stay. In tanker ships, this is measured in terms of their pumping capacity and loading/unloading performance. Detailed logs of the cargo operations and pumping capacity should include pump discharge, suction pressures, and RPM rates, which will help the shipowners in providing evidence for their vessels' performance and efficiency.

The tanker vessels' stability and stress factors will determine the tanks' loading or unloading sequence. Typically, loading commences at a slower rate, which gradually increases to the highest levels. Technology enables the vessels' remote checking of performance and temperature conditions throughout loading and unloading operations. Monitoring and controlling of the loading and unloading rates, with frequent ullage measurements, are logged in the deck log book at least every hour (UK PANDI 2003).

d. *Product cargoes, LNG/LPG berths*: These handle oil and gas-related products, usually in liquid form. Vessels are loaded via loading arms containing the pipelines. Storage facilities for the products are usually some distance away from the berth and connected by several pipes to ensure fast loading.

e. *Chemicals and fertilizers' berths*: Specialized terminals are designated for the carriage of chemical products and fertilizers such as phosphate, urea, and so on. Terminals are typically designed with designated areas for storage and operations.

For chemical facilities located in the port's vicinity, direct handling is achieved in accordance with the factory-ship plan. Contemporary ports are equipped with technological solutions that handle chemicals and fertilizers. To reduce spillage, completely closed conveyor systems and large storage sheds are used for transporting chemicals and fertilizers through the port. In addition, specialized wastewater management systems are built. Dedicated terminals, berths, and warehouses are connected to railway and designated tank trailers for road network systems.

f. *Cruise ship berths*: These berths and terminals are designed to fit the requirements of the tourist industry. In addition to increased safety compliance, cruise terminals are designed in a manner of luxury and convenience focused on (a) terminal's accessibility from nearby parking, bus stops, and passenger drop-off areas; (b) efficient passengers' boarding through boardway bridges and gangway and jet-way systems; (c) protection from weather, through, for example, covered walkways; and (d) spacious terminal and superstructures for passenger reception and baggage handling spaces, provisions, and warehouse areas. The entire terminal area should be designed in a manner that comfortably accommodates thousands of passengers for the large cruise ships.

Floating docks offer a consistent level for cruise ship passengers as they embark and disembark ships, providing added safety. Floating docks are designed in a manner that provides increased marine security, and special designs are featured in areas of large tidal fluctuations. Other docks are designed to include bus staging, pedestrian covered floats, passenger boarding systems, and pedestrial/vehicular transfer bridges (PND Engineers 2013).

g. *Ro-Ro berths* are designed either (a) for *passenger/Ro-Ro carriers*, which combine the luxury and safety features of a cruise terminal, or (b) as parts of designated car carrier terminals, which are typically leased by car manufacturers and are especially designed to fit a large capacity of cars, without the passengers' extra requirements.

Typically, Ro-Ro berths feature spacious car park areas in their vicinity, enabling the grouping and storing of freight before distribution. Technology should allow simultaneous, independent loading of two or more decks of large car parks, or side-loaders for loading lorries on the upper car decks.

Pontoon technology makes it possible to load or unload three ships at the same time on the same terminal. Both sides of the pontoon are ballasted independently by high-capacity pumps using seawater, for stability reasons, allowing for different-sized vessels to be handled quickly and efficiently.

h. *Lay-up berths or layberths*: A berth used for idle (lay-up status) vessels. A berth where no loading or unloading takes place. Lay berth and lay-by berth (below) may be used somewhat interchangeably for intermediate (two- to seven-day) periods. One factor to consider is the mooring arrangements offered, which usually relate to the size of the lay-up area. In a very large open space, such as found in Southeast Asia, vessels can be parked far apart and allowed to weathervane around their anchors. In smaller bays, as found in Europe, ships are usually nested together and held fast in special mooring arrangements. Ships are also sometimes packed side by side along the coast. Lay-up strategies are generally described as hot, warm, or cold, depending on the period of repose involved—a hot lay-up can be likened. In a "hot" lay-up, the entire crew is kept on board and the machinery is kept running—the ship is, basically, parked in anticipation that it will get work soon. Ships are generally kept in this state for a period of one to six months, although extensions of up to a year are not unknown. It is reported that maintaining the hot lay-up condition can require up to 70% of the ship's normal running costs. A "warm" lay-up takes the sleep a little deeper, typically lasting six months to a year, although extensions of up to three years are not unknown. In the warm lay-up state, there is a skeleton crew on board, with some systems deactivated but still a fair amount of maintenance activity. Users report that this level of dormancy costs up to 40% of the vessel's running costs. A "cold" lay-up generally means inactivity for up to five years. The normal crew is dismissed and replaced by a crew of watchmen or engineers, whose job is to do only maintenance work necessary to forestall deterioration of the hull structure and machinery as long as possible (Burns-Kokkinaki, ABS 2009).

4.3 THE PORT AND CHARTER PARTY TERMS

4.3.1 Charter Party Types

One key difference between the liner and the tramp trade pertains to the types of contracts of affreightment they use. While the liner trade uses simple and concise documents such as bills of lading and booking notes, the tramp trade uses a more detailed, structured, and legally complex document: The Charter Party. Table 4.1 demonstrates the principal differences between the liner trade and the tramp trade.

A charter party is a legal contract of affreightment common in the tramp trade, by which a shipowner designates one or more specified vessels to carry a charterer's specified cargo quality and quantity between designated ports, berths, or area ranges. Charter parties stipulate the legal system governing the agreement, which is usually under the US or British law. The incorporation of the C/P Arbitration clause provides that any dispute arising between the shipowners and the charterers can be resolved through arbitrators.

For centuries, charter parties have been the prevailing contractual documents in the tramp shipping trade. In fact, the term charter party derives from the Latin "Carta Partita," which means "paper divided in two." Its name signifies the customary act found in numerous centuries old contracts of writing the contract on a piece of paper and subsequently tearing it into two irregular portions. Each contractual party would obtain half a piece of paper. The authenticity of the documents and their identical contents would be verified if the two torn pieces of paper fitted together.

TABLE 4.1   Differences between Liner Trade and Tramp Trade

	Liner Shipping	Tramp Shipping
Contract of affreightment	Bill of lading Booking note Lump sum charter	Charter party and bills of lading
Carrier type	Common carrier	Private carrier
Standardized contract of affreightment clauses	Yes	No
Charge and liabilities	Freight charge only The carrier undertakes all charges and liabilities in case of delays, accidents, or third-party disputes, e.g., port authorities	As the charterers designate the ports/berths of load and discharge, they share liabilities and charges with the owners, as stipulated in the C/P
Voyage route	Scheduled and regular	Flexible, agreed between the two parties
Option for multiple charterers and parcel cargoes on a single carrier	Typical	Less likely
Option for multiple ports of call on a single voyage	Typical	Less likely

Source: M.G. Burns.

Charter parties are classified into three categories: (i) voyage C/P, (ii) time C/P, and (iii) bareboat or demise C/P. Three key elements will determine which charter party type is most appropriate for a fixture:

a. The element of time, that is, time duration and liabilities owing to delays
b. The element of place

In voyage C/Ps, the nomination of loading and discharging berths or ports, will determine when the ship is an "arrived ship." A berth C/P or a port C/P will determine each party's liabilities in terms of place and time.

In time C/Ps and bareboat C/Ps, the charterers are free to trade in a wide range of global ports, with only a list of territorial exclusions and cargo exclusions as stipulated in the C/P. The owners receive the daily hire regardless of whether the port is at sea or at berth.

c. The element of control

In voyage C/Ps, the owners have a greater control over the ship as well as its commercial, financial, navigational, operational, and technical decisions, related to place, time, money, and cargo.

The element of control and decision making increases in time C/Ps, whereas in bareboat C/Ps, the charterers are considered as the ship's disponent owners and therefore control most of the ship's functions.

In detail, the three charter party types include the following:

i. A *voyage charter party* pertains to a contract for a particular voyage, where a predetermined freight payment is paid to the shipowners per metric ton of cargo (i.e., $30 PMT). In this short-term contract, the shipowners assume the greatest possible commercial, financial, and operational control over the vessel. The ship must be redelivered to its owners within a certain time frame, after which the charterers should indemnify the shipowners with special reimbursement, the payment of which is calculated as per the C/P clause stipulations. Because of its short duration, this contractual arrangement is suitable (i) when carrying irregular cargo volumes, and the future cargo flow cannot be predicted, or (ii) in a volatile market that lacks visibility necessary for long-term planning.

ii. A *time charter party (TCP)* stipulates the terms and conditions in which the charterers will hire the vessel for an agreed period—anywhere from a few months to multiple years. In this arrangement, the charterers assume a greater control of the ship's operations and directly give operational and commercial orders to the master under copy to the owners. Under a TCP, the charterers typically undertake the voyage costs:

Voyage costs = ship's bunkers (i.e., fuel and diesel oil) + port charges (i.e., port, light, and canal expenses; tugs and pilots; cargo handling; agency fees; etc.)

Meanwhile, the shipowners cover the running—or fixed ship's costs:

Running or fixed costs = overhead costs + crew wages, navigational + insurance + maintenance, spare parts, and repairs + bonded stores + lubricants

Needless to say, the shipowners' administration and insurance costs are still covered by the company.

Payment in time C/Ps is reimbursed by multiplying the daily hire by the months or years of the contract. Typically, the charterers pay a monthly hire to the owners. Because of its longer duration, this charter party is suitable (i) to carry large and regular volumes of cargoes, for example, within a trade agreement between governments, or import/export agreements and the subsequent distribution of raw materials, or manufactured goods. In this case, the charterers already have multiple year contracts with cargo buyers (importers) and are looking for a regular, reliable means of transport.

iii. A *bareboat or demise charter party* enables the charterers to become the vessel's disponent owners for an agreed time—typically multiple years.

In this agreement, the shipowner agrees for the ship to be administered, recruited, technically maintained, and run by its new disponent owners. The shipowners will not be informed about the ship's operational or commercial activities and will not be informed about the ship's itinerary and ports of call. The charter party allows for certain exclusions, that is, (i) geographical areas where its actual owners wish to be excluded from the ship's trade routes, for example, war zones, piracy zones, or areas of political or trade conflict; and (ii) cargo types to be excluded, either because of its potential safety hazards (i.e., HAZMAT cargoes) or because of the difficulty in properly cleaning the ship's holds, and the future commercial complications (e.g., cement in bulk, whose residues are difficult to clean, and would disable the ship to carry edible bulk cargoes in the future, such as sugar, grains, etc.). Except for these two exceptions in cargo and ports, its new owners will be free to select any cargo type, berth, and port in the world.

The bareboat C/P is frequently associated with ship management agreements, new building contracts, and, more frequently, the sale and purchase option, during or at the end of the bareboat C/P contract. The C/P should stipulate which party, that is, the shipowners or the disponent owners, is in charge of the ship's hull and machinery, protection and indemnity, crew, war, and piracy insurance. Moreover, bareboat charterers frequently decide to undertake the ship's safety (ISM), security (ISPS) and environmental (ISO 14001, VGP, BWST, etc.) management as a means of controlling the ship's performance.

This contractual arrangement is suitable for large-scale charterers, frequently terminal operators that wish to fully control the quality and performance of their BB fleet, with minimum external intervention or conflict. Other BB owners pursue this type of contracts in order to retain privacy as to their clients and cargoes loaded and discharged. A BB contract will enable them to trade with their own global clients in their own global terminals. Finally, the third reason why charterers would pursue to bareboat a ship entails their potential to subhire the ship to other charterers, at a profitable hire.

The maritime industry can demonstrate a plethora of different contracts of affreightment for voyage, time charter and bareboat C/Ps, as well as forms used for new building contracts, repairs, ship lay-ups, demolitions, repairs, and so on. Table 4.2 shows the prevailing standard charter party forms, contracts of affreightment, and miscellaneous maritime contracts.

TABLE 4.2   List of Standard Charter Party Forms, Contracts of Affreightment, and
Miscellaneous Maritime Contracts

Voyage C/Ps	Time C/Ps	BS/L, Waybills, CGO Receipts
Asbatankvoy	Gentime	Bimchemvoybill 2008
Gencon 94	NYPE 93	Cementvoybill 2006
Graincon	Asbatime/NYPE 81	Coalorevoybill
BHP Billiton Voy 2003	BHP Billiton Time 2003	Combiconbill
Shellvoy 6	Shelltime4	Congenbill 2007
Amwelsh 93	NYPE 1946	Conlinebill 2000
BPVoy4	Shell LNGTime 1	Ferticonbill 2007
Bimchemvoy 2008	Baltime 1939 (rev. 2001)	Genwaybill
Cementvoy 2006	Bimchemtime 2005	Grainconbill
Norgrain 89	Boxtime 2004	Heavyconbill 2007
Ferticon 2007	BPTime 3	Heavyconreceipt 2007
Synacomex 2000	Gastime	Heavyliftvoybill
Rio Doce Ore	Supplytime 2005	Hydrobill
Austwheat 1990		Linewaybill
Cruisevoy	Bareboat	Multidoc 95
Gasvoy 2005	Barecon 2001 (revised)	Norgrain Bill
Heavycon 2007	Barecon 1989 (past edition)	Yarabill
Heavyliftvoy		Yarawaybill
Hydrocharter	Newbuilding	
Nipponcoal	Newbuildcon	Sale and Purchase
Nipponore		Saleform 2012
Projectcon	Ship Repair	Bimcosale
Worldfood 99	Repaircon	Nipponsale 1999
Yaracharter		Saleform 1987 (past edition)
Coalorevoy	Lay Up	Saleform 1993 (past edition)
Gencon 76 (past edition)	Layupman	
Heavycon (past edition)		Demolition
		Demolishcon

Source: M.G. Burns.

4.3.2 Charter Party Clauses and Areas of Dispute

Successful negotiations entail not only the direct financial benefits deriving from the hire, and not only in the long-term opportunities that the specific fixture may bring. Each and every clause among the standard forms and rider clauses of the C/P contract may be potentially ambiguous, with areas of potential dispute. It is important to remember that in case of a dispute, arbitrators and the court of law interpret the C/P clauses as a whole and never each clause separately.

Negotiating the amendment of a single clause may not be a commercial victory, if other clauses are conflicting. Adding a special rider clause may not be a commercial victory; in fact, rider clauses are frequently ambiguous and conflicting with the main form stipulations. A single word may alter the significance of a clause, and an oversight is possible during the long hours of intensive negotiations.

The most common areas of potential dispute focus on safety, time, and money:

a. *Safety*: Port-related claims related to the port's safety. The interpretation varies between a "port" charter party and a "berth" charter party. Frequently, safety issues are related to time delays and, consequently, money.

b. *Time*: The time of the ship's delivery to her charterers and redelivery to her owners is critical.

   Furthermore, *voyage charter parties*, in particular "berth" C/Ps, may increase the shipowners' liability pertaining to timely arrival at berth, and so on. In a charter party, the element of time is crucial (a) in determining the time that a ship is an "arrived ship," (b) in estimating when laytime commences, and (c) in determining whether time delays are due to port congestion, ship's mechanical failure, navigational or operational issues, severe weather conditions, act of war, etc.

   On the other hand, in *time C/Ps and bareboat C/Ps*, the charterers are in charge of the time, schedule, and delays from the contractual time of the ship's delivery, until its redelivery.

   The *off-hire clause* in TCPs signifies the period when the vessel is unable to perform (and time is lost) due to reasons pertaining to the ship's side, such as equipment breakdown, or time lost because of deviation from the course of the voyage. In this case, the vessel is officially "off-hire," and time stops to count until the problem is rectified and the vessel is "on-hire" again. The hire payment is not estimated or reimbursed for the "off-hire period."

   Piracy and "off-hire" have been a critical area of concern, with recent legal disputes that seek to answer whether a ship that is seized by pirates is under "off-hire" or not.

c. *Money*: Safety and time disputes always have an impact on money, that is, the payment of hire, off-hire issues, demurrage claims, accidents and negligence liabilities, and withdrawal and payment deductions.

The two focal points of a charter party agreement are as follows: (i) the ship's delivery and redelivery, that is, the time when a vessel is delivered to her charterers and is redelivered to her owners. All three charter party types are focused on the relevant clause. (ii) In voyage charter parties, the ship's arrival at her designated loading and discharging areas (berths or ports), where time commences for the ship's loading and discharging operations.

4.3.3 The Port and Charter Party Terms

Generally speaking, port-related claims are the most common type in the maritime industry, as the port is an area where most maritime accidents occur—because of congestion, stress, and the simultaneous operations taking place. Hence, a port is an area where time counts for the cargo to be handled with utmost efficiency, and an area where financial claims can arise, because of time or safety issues.

Among the three different C/P types, it is the voyage charter and the time charter parties that mostly relate to port clauses and claims. The bareboat charter party is the type where the charterers have undertaken the operational, financial, and commercial management of the ship. Since the owners' control over the ship is minimum, the number of disputes between charterers and shipowners would be eliminated. Disputes would arise because of (i) hire payment disruptions, (ii) the ship's deterioration owing to lack of maintenance or crew inefficiency, and finally (iii) breach of the cargo exemption or area exemption clauses.

This subsection examines the most critical clauses and wording pertaining to (a) the ship being an "arrived ship," (b) the distinction between a port and berth charter party, and (c) the elements of ship safety while at port.

Arrived Ship, Port C/P, and Berth C/P

In voyage charter parties, the charterer designates ports, sets itineraries and furnishes cargoes, and is allowed to load and discharge the cargo within a specific period, that is, laytime. Demurrage is paid once laytime is exceeded. In an ideal scenario, the ship arrives at her port(s) with no delays. In real maritime business, delays and misunderstandings as to the C/P wording are quite common. In voyage charter parties, the ship's master needs to ascertain at which point the ship has become an "arrived ship" as defined in the contract's terms. This is determined from the C/P stipulations, and the distinction between a "port C/P" and a "berth C/P."

In case of a "berth C/P," the vessel is an "arrived ship," and laytime commences once it arrives at its designated berth.

Characteristics of a Port or Berth Charter Party

In a port C/P, a named or unnamed port is cited as the vessel's stipulated loading or discharging area. In a *port C/P*, the ship can only be considered as "arrived" and consequently eligible to submit notice of readiness (NOR) and commence her cargo operations once she has arrived at the port while awaiting for her berthing or instructions from charterers.

On the other hand, in a *berth C/P*, a named or unnamed berth is cited as the vessel's loading or discharging area. Hence, for the ship to be considered as "arrived," and hand the NOR, she must arrive at the designated berth. This suggests that the vessel may have arrived at its port and be subjected to tremendous delays and congestion until it reaches its designated berth. Any time delays are incurred due to congestion while waiting for berthing; they are for the owners' account.

It is favorable for owners to recognize that most voyage C/Ps are "port" contracts, and yet the slightest clause amendment can change this. Although the wording in the newly amended clause might be straightforward, problems arise when conflicting clauses create ambiguity.

The elements of ship safety while at port

The element of ship's safety while at port encompasses the moment the ship has officially entered the port's limits, throughout the ship's stay at port, until the ship's departure. Numerous claim cases have arisen among shipowners and charterers on the grounds of breach of contract, determined by the ship being an "arrived ship" at a "safe berth" or a "safe port." During the charter party negotiations, the charterers need to nominate a safe loading berth or port and a safe discharging berth or port. Shipping practice requires the ship to safely get into the port, commence its cargo operations, and depart, without its structural integrity being jeopardized.

An accurate characterization of port safety was provided in the "Eastern City" case of 1958 (Leeds Shipping Co v Société Française Bunge. 2 Lloyd's Report 127–131) as formulated by LJSellers:

> A port or berth will not be safe unless, in the relevant period of time, a particular ship can reach it, use it and return from it without, in the absence of some abnormal occurrence, being exposed to danger which cannot be avoided by good navigation and seamanship.

From a legal and arbitration standpoint, there is a significant distinction between naming designated ports (e.g., Houston) and providing a wide geographic region (e.g., US Gulf). The latter option of providing a safe port warranty is most favorable for the owners, since the charterers are obliged to load and discharge at safe ports or berths among all the ports in the region, while offering multiple alternatives in case their nomination at a later date is not accepted as safe by the owners. In the first option, once both parties consent to specific named ports or berths, the shipowners are obliged to carry out their contractual obligation and proceed to these ports, even if safety issues arise at a later date.

The general safety rule in most cases (other than war, etc.) is that a port may be perfectly safe for most ships, but unsafe for one particular ship. Hence, each charter party contract should reflect the conditions pertaining to the one particular ship that is designated within the contract.

Typical issues of safety pertain to the ship's larger size in proportion to the nominated port, including the following:

i. *Draft restrictions, maneuverability, and overall ship's size* while at port are combined with port, canal, or berth restrictions. Many claim that cases have been associated with the ship's particulars or technical specifications that are perfectly safe, legal, and suitable for most global ports, yet could be unsuitable for a particular port.

ii. *Tidal ports* may be temporarily unsafe, or to be precise, increased safety measures are required.

iii. *Negligence* gives rise to safety liabilities, that is, in cases where the port's safety issues pertain to navigational negligence and equipment failure. Furthermore, safety may include metal objects left at the seabed by other ships, such as anchors, and so on.

iv. *The act of God*, that is, extreme weather conditions beyond the master's control. Hence, safety issues may impose ships' delays or deviations, and the shipowners need to prove that time lost and pertinent deviation are closely associated to the extreme weather conditions.

v. *Ship security* is frequently associated to safety, when events of war risk and ship seizure, political turmoil, social unrest and strikes, and port and canal closures occur. Numerous port closures have occurred during wartimes, invasions, guerilla warfare, and even protests. The more strategic the position of a port or canal, the more serious its impact is. A renowned canal closure is the Suez Canal closure during the Suez Crisis of 1956, as well as its numerous closures during the 1960s and the 1970s.

  Legally speaking, *ship security* frequently involves lack of safety. Maritime security is distinguished into (a) piracy, whose motive is predominantly financial, and (b) terrorism, which intends to generate fear and is based on political, religious, or ideological motives. Pursuant to maritime legislation and the international law, actions taken against ships by pirates and terrorists are regarded as criminal acts; however, the political and social consequences of the latter are more severe. In the event of piracy, the target(s) selection is determined by monetary gains (e.g., seeking ransom for a large ship carrying high-value cargo), combined with low-security measures. In the case of a terrorist attack, the aim is to create damage, that is, loss of human life and destruction of property, business, and natural resources (Burns 2013).

The safety, arrival, security, and negligence issues discussed herewith are duly demonstrated in Section 4.5 through the examination of relevant legal cases.

Situations where a charter party stipulates the port or berth of loading or unloading, however, include no warranty with regard to the safety of these ports or berths and there is no intended safety warranty. Whereby, as soon as the shipowners have consented to the C/P wording, they are held legally liable for the ship's safety while at port.

Politically related risks, a ship's seizure, piracy, or other security issues may also arise within a charter party. In fact, commercial ships are frequently confined in war zones and piracy-prone areas. Arbitration in these cases aims to verify whether the port was safe at the time of the charter.

As a conclusion, a port's or berth's safety requires that a ship enters, loads/unloads, and exits in full safety. The wording of pertinent charter party clauses will determine elements of authority and liability among the shipowners and the charterers:

A. Voyage Charter Parties
   I. Asbatankvoy, issued by the Association of Shipbrokers and Agents (ASBA) (for wet bulk/tanker ships)
      Clause 4 of this C/P form pertains to the naming of loading and discharge ports after the C/P has been signed, but at least 24 hours prior to the ship's readiness to sail (ASBA 2013):

      (a) The Charterer shall name the loading port or ports at least twenty-four (24) hours prior to the Vessel's readiness to sail from the last previous port of discharge, or from bunkering port for the voyage, or upon signing this Charter if the Vessel has already sailed. However, Charterer shall have the option of ordering the Vessel to the following destinations for wireless orders: on a voyage to a port or ports in: (…) loading and discharging port(s).

      Clause 9 of this C/P form pertains to the ship's safe berthing and shifting, which will eventually determine the laytime:

      The vessel shall load and discharge at any safe place or wharf, or alongside vessels or lighters reachable on her arrival, which shall be designated and procured by the Charterer, provided the Vessel can proceed thereto, lie at, and depart therefrom always safely afloat, any lighterage being at the expense, risk and peril of the Charterer. The Charterer shall have the right of shifting the Vessel at ports of loading and/or discharge from one safe berth to another on payment of all towage and pilotage shifting to next berth, charges for running lines on arrival at and leaving that berth, additional agency charges and expense, customs overtime and fees, and any other extra port charges or port expenses incurred by reason of using more than one berth. Time consumed on account of shifting shall count as used laytime except as otherwise provided in Clause 15.

Pursuant to numerous legal disputes, the interpretation of this clause entails the charterers' obligation to appoint a safe place (any place, e.g., port, terminal, wharf, etc.) for the cargo operations to take place. By this clause, the charterers designate places or wharfs that are ports or places that are "reachable on the ship's arrival" as well as safe for the vessel. The clause stipulates the "ship's arrival" from a logical or commercial perspective of a port entering a harbor, and not from the strictly legal perspective of being an arrived ship when the master hands the NOR over to the charterers. Consequently, the charterers undertake the responsibility of any time delays, obstructions, port traffic, and act of God.

II. Gencon, issued by BIMCO (for dry bulk ships)

Laytime in a Gencon C/P form is stipulated in clause 6, where items (a) and (b) in particular state that the cargo will be loaded in a specific number of days/hours as stipulated in Box 16 (BIMCO Charterparty 2013).

Section (c) clarifies the time that laytime counts: in case NOR is provided up to an including 12:00 hours, it will commence at 13:00 hours, whereas if it is provided later than 1:00 hours, it will commence at 06:00 hours the following working day. Once the master deems that the vessel is in all respects ready, the NOR will be handed even prior to customs' clearance, even while the vessel is in free pratique, and time counts as if the vessel was in her berth. The WIBON term (Whether in Berth or Not) would hereby shift the liability of delays arising due to port congestion to the charterers:

(a) Separate laytime for loading and discharging

The cargo shall be loaded within the number of running days/hours as indicated in Box 16, weather permitting, Sundays and holidays excepted, unless used, in which event time used shall count. The cargo shall be discharged within the number of running days/hours as indicated in Box 16, weather permitting, Sundays and holidays excepted, unless used, in which event time used shall count.

(b) Total laytime for loading and discharging

The cargo shall be loaded and discharged within the number of total running days/hours as indicated in Box 16, weather permitting, Sundays and holidays excepted, unless used, in which event time used shall count.

(c) Commencement of laytime (loading and discharging). Laytime for loading and discharging shall commence at 13.00 hours, if NOR is given up to and including 12.00 hours, and at 06.00 hours next working day if notice given during office hours after 12.00 hours. NOR at loading port to be given to the Shippers named in Box 17 or if not named, to the Charterers or their agents named in Box 18. NOR at the discharging port to be given to the Receivers or, if not known, to the Charterers or their agents named in Box 19. If the loading/discharging berth is not available on the Vessel's arrival at or off the port of loading/discharging, the Vessel shall be entitled to give NOR within ordinary office hours on arrival there, whether in free pratique or not, whether customs cleared or not. Laytime or time on demurrage shall then count as if she were in berth and in all respects ready for loading/discharging provided that the Master warrants that she is in fact ready in all respects. Time used in moving from the place of waiting to the loading/discharging berth shall not count as laytime. If, after inspection, the Vessel is found not to be ready in all respects to load/discharge time lost after the discovery thereof until the Vessel is again ready to load/discharge shall not count as laytime. Time used before commencement of laytime shall count. Indicate alternative (a) or (b) as agreed, in Box 16.

III. Shellvoy (for wet bulk/tanker ships)
      The vessel's safety while at port or at berth is stipulated in both Shellvoy 5
      and Shellvoy 6, Clause 4, where it is clearly expressed that the safety at port,
      berth or transshipment operations, or ship-to-ship transfer is not the char-
      terers' responsibility, as long as they can prove that they have exercised due
      diligence:
      The clause indicates that

      Charterers shall exercise due diligence to order the vessel only to ports and
      berths which are safe for the vessel and to ensure that transhipment opera-
      tions conform to standards not less than those set out in the latest edition
      of ICS/OCIMF Ship-to-Ship Transfer Guide (Petroleum). Notwithstanding
      anything contained in this charter, Charterers do not warrant the safety of
      any port, berth or transhipment operation and Charterers shall not be liable
      for loss or damage arising from any unsafety if they can prove that due dili-
      gence was exercised in the giving of the order, or if such loss or damage was
      caused by an act of war or civil commotion within the trading areas defined
      in Part 1 (D/E).

IV.   TankerVoy 87, issued by BIMCO (for wet bulk/tanker ships)
      Clause 8 of TankerVoy 87 specifies under which circumstances the NOR
      may be tendered to charterers. Clause 9 is interrelated as it provides the
      prerequisites for laytime to commence, demurrage, and time lost (BIMCO
      Charterparty 1987).

      Clause 8: Notice of Readiness
         When the vessel has arrived at a customary anchorage or waiting place for
         each loading and discharging port or place and is ready to load or discharge,
         notice of readiness (which may be tendered at any time on any day) shall
         be given to Charterers or their agents by letter, telegraph, telex, radio or
         telephone, berth or no berth. An oral notice shall be confirmed promptly in
         writing.
      Clause 9: Laytime
      (a) The laytime specified in Part I (I) shall be allowed to Charterers for loading
         and discharging of cargo and other Charterers' purposes. Other than when
         the vessel loads or discharges cargo by transhipment at sea, laytime shall
         commence at the first loading and at the first discharging port or place six
         hours after the tender of notice or upon arrival in berth if that occurs ear-
         lier, and at any subsequent port or place laytime shall resume when notice
         is tendered. Time shall run until hoses have been disconnected, which shall
         be effected promptly, but if the vessel is delayed after disconnection of hoses
         for more than two hours awaiting bills of lading or for other Charterers'
         purposes, time shall continue to run from disconnection of hoses until the
         termination of such delay.
      (b) Time lost owing to any of the following causes shall not count as laytime
         or for demurrage if the vessel is on demurrage: (i) awaiting next high tide or
         daylight to proceed on the inward passage (...); (ii) actually moving from a
         waiting place on an inward passage to a loading or discharging berth (...);
         (iii) in handling ballast unless carried out concurrently with cargo operations
         such that no time is lost thereby; (iv) stoppages on the vessel's orders, break-
         down or inefficiency of the vessel, negligence or breach of duty on the part of
         the Owners or their servants or agents or strike, lockout, or other restraint of

labour of the vessel's crew; (v) strike, lockout or other restraint of labour of pilot or tug personnel.

(c) Subject only to Clause 9(b)(iv), if the vessel loads or discharges cargo by transhipment at sea, all time from the vessel's arrival at the transhipment place until final unmooring of the lightening vessel at the end of transhipment operations shall count as laytime or for demurrage if the vessel is on demurrage.

V. BPVoy4, issued by British Petroleum Shipping Ltd. (for wet bulk/tanker ships) Clause 5 of this charter party entails the term *port*, which may include port, berth, dock, loading, or discharging anchorage of offshore location, submarine line, single point, or single buoy mooring facility, alongside vessels or lighters, or any other place whatsoever as the context requires (British Petroleum Shipping Ltd. 2013).

Clause 6 stipulates certain terms and conditions under which the NOR shall be accepted under two conditions, namely: (a) if the vessel is proceeding directly to her loading or discharging place, is securely moored, and if appropriate, her gangway is in place; or (b) (i) if the vessel has been instructed to anchor and wait in the designated area where similar vessel types usually wait; and (ii) free pratique has been granted or will be granted within 6 hours after the NOR has been tendered, and (iii) if in the United States, a US Coast Guard Tanker Vessel Examination Letter has been granted, or if in non-US port, a similar letter has been obtained.

B. Time Charter Parties

I. NYPE 93 (New York Produce Exchange), issued by the Federation of National Associations of Ship Brokers (FONASBA 1993)

In Clause 5 pertaining to Trading Limits, this charter party states that *"the Vessel shall be employed in such lawful trades between safe ports and safe places within excluding as the Charterers shall direct,"* whereas it provides the option of exclusions. Once both sides agree to the ports' safety as stipulated in this clause, the ship's navigation through unsafe ports and places will signify a breach of contract and liabilities for any damages incurred.

II. Shelltime 4 (amended in 1993)

Clause 4 entails the agreed Time Charter Period, Trading Limits, and Safe Places. Namely, in section (a), charterers have the right to order the ship to *"ice-bound waters or to any part of the world outside such limits provided that Owner's consent thereto,"* as long as the consent is not unreasonably withheld. Charterers will undertake to pay for the insurance premium related to their order. 4(c) refers to the due diligence exercised by the charterers in order to "ensure that the vessel is only employed between and at safe places (which expression when used in this charter shall include ports, berths, wharves, docks, anchorages, submarine lines, alongside vessels or lighters, and other locations including locations at sea) where she can safely lie always afloat." Charterers' liability is expressly related to any loss or damage incurred "by their failure to exercise due diligence."

Due diligence is a condition frequently stipulated in time charters. Consequently, for any dispute arising therefrom, the arbitration or the court of law will have to determine the extent to which due diligence was exercised by the charterers.

4.4 SHIPYARDS

Building ships and navigating them utilizes vast capital at home;
it employs thousands of workmen in their construction and manning;
it creates a home market for the products of the farm and the shop;
it diminishes the balance of trade against us precisely to the extent of
freights and passage money paid to American vessels, and
gives us a supremacy upon the seas of inestimable value in case of foreign wars.
My opinion is that in addition to subsidizing very desirable lines of ocean traffic,
a general assistance should be given in an effective way.

US President Ulysses S. Grant, Message to the Congress, March 23, 1870

4.4.1 Introduction: The Global Shipbuilding Market

A port's financial and commercial activities are closely intertwined with the shipbuilding industry for two reasons: First, most shipyards are managed by port authorities and lease space from ports or are located in a port's vicinity. Second, ports' revenue increases as the global fleet grows: the more ships are built, ports grow and enjoy increased demand for their services.

According to the IMO, the global oceangoing merchant fleet exceeds 55,138 ships with a carrying capacity of 991,173,697 GT and 1,483,121,493 DWT and an average age of 19 years. 85%. In the global merchant fleet order book, dry bulk carriers prevail, comprising one-third of the new orders, followed by offshore, containers, LNG carriers, and tanker ships (IMO 2012).

Customarily, ports and shipyards are the nodal points where civilizations and technologies meet. Globalization has enabled the exchange of technologies and innovations, either through ships calling at their global ports or at the shipyards during ship repairs. As the environmental regulations of the most advanced countries did not encourage ship demolitions and the development of scrapyards, a plethora of developing countries undertook an increasing market share of ship scrapping and recycling. As modest and unprofitable as it may sound, the art and science of scrapping enabled many less developed economies to enhance their knowledge on the technology patents of new, original ship designs that they could now reproduce, combine, and use as prototype where new ideas were based upon. Hence, shipyards are the focal points of innovation, production, economic stimulus, and trade cycles.

The art of shipbuilding is the foundation of the shipping industry, as it enables vessels to be seaworthy, utilizes the latest technologies to ensure ships' performance and safety, and allows shipping conglomerates to gain a competitive edge through innovation. EcoDesigns and EcoShips are the latest and most desired features that ensure environmental protection, energy efficiency, and clever naval architecture and engineering, all in one attractive package. The shipbuilding industry embraces the economic and commercial aspects of the maritime industry and is an active ingredient and impetus of the shipping market fluctuations.

And while shipyards crave for the shipowners' interest confirmed through a growing order book, it is the very same oversupply that will affect the shipyards' future business and thus define the cyclical manner in which shipping cycles work.

4.4.2 The Utility of Shipbuilding

According to the US Maritime Administration (MARAD) (2013), shipbuilding is pivotal to a nation's growth, as its economic significance affects the nation's GDP, employment, and labor income.

The industry's economic multipliers can be distinguished into three economic levels, that is, direct, indirect, and induced:

a. *Direct impact* is assessed on the basis of the country's GDP, labor revenue, and employment opportunities.
b. *Indirect impact* is calculated in a broader manner, as it includes the estimated benefits throughout the entire supply chain, that is, its growth, labor revenue, and employment opportunities generated. The term *supply chain* encompasses commodity traders, for example, for raw materials; manufacturers and suppliers of spare parts; and the entire logistics network.
c. *Induced impact* focuses on the financial benefits deriving from the national and regional sales, employment opportunities, revenue, GDP and PPP, government spending patterns, and so on.

4.4.3 The Components of Shipbuilding

Traditionally, shipyards undertake three major functions (shipbuilding, ship repairs, and ship scrapping) throughout their commercial life and are demonstrated in Figure 4.9. The economic life expectancy of a ship is typically 20–25 years, depending on its structural fatigue, the quality of construction materials, and its history of accidents.

Naval architecture and naval engineering encompass the first two functions, that is,

i. *Shipbuilding*, which entails the construction of a ship's hull and the installment of machinery and systems. Larger, modern shipyards are built with sizeable drydocks and slipways. They contain a wide range of cranes (e.g., goliath gantry cranes, floating dock cranes, plate handling cranes, etc.), and other workstation lifting gear, for all the phases of their shipbuilding operations. In addition, they host a number of drydocks, dust-free storage areas and warehouses, extremely sizeable locations for the shipbuilding process, painting facilities, slipways, and so on.

Each shipyard, based on their technology, internal process, and logistics infrastructure, may follow different construction patterns. Upon finalization of the shipbuilding agreement, shipyards implement a program that depends on technical and other parameters, that is, financing and payment installments, process times and date of delivery, logistics pertaining to materials and equipment, weather conditions, availability of prefabrication spaces, landing spaces, drydocks, order of shipbuilding as per owners' agreements and contracts, and so on.

FIGURE 4.9   Shipyards and the three main stages of a ship's commercial life. (Courtesy of M.G. Burns.)

A shipyard's efficiency is measured in terms of quality, quantity, innovation, and variety. Namely:

- *Quality* is measured in terms of technical, engineering, and architectural competence.
- *Quantity* is measured in terms of their shipbuilding capacity/ship tonnage.
- *Innovation and variety* refer to the implementation of advanced technologies and diverse ship designs, types, and sizes.

The shipbuilding process is the most demanding in terms of skills, capital, and technology. The process of ship design and construction is highlighted in Figure 4.10.

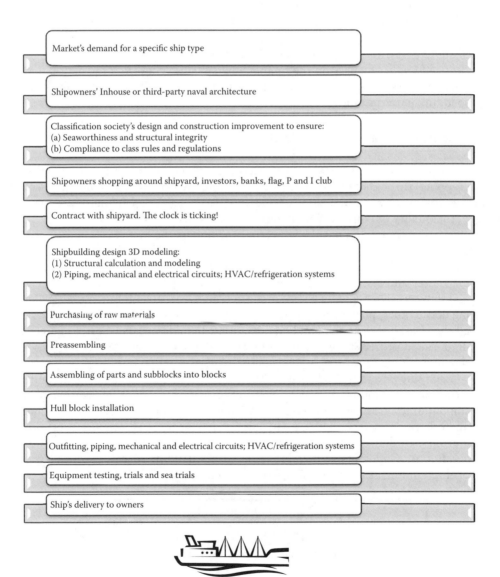

FIGURE 4.10   Ship design and construction process. (Courtesy of M.G. Burns.)

ii. *Ship repairs*, which pertain to ship alterations, retrofitting programs, and engineering service; floating dock repairs; repairs alongside or at anchor; installation and repairs/replacement of pipelines, pumps, and valves; planned maintenance, painting, and sandblasting; and so on.

iii. *Ship recycling*, scrapping, or breaking pertains to a vessel's demolition, dismantling of its useful machinery, and the utilization of its recycled materials commencing from the ship's superstructure and upper decks, and proceeding to the removal of the ship's sections and larger machinery through the use of heavy lift cranes. In the past, ship demolition was performed in the port's vicinity, in designated drydocks. Because of the increasing emerging environmental rules, the scrapping activities have shifted to developing or less developed regions.

The industry needs to scale down and consolidate in sectors such as the following:

- Time management issues and logistics disruptions create delays and breach of the delivery clauses in the shipbuilding contract.
- Delays in receiving manufactured goods and raw materials.
- Owners' or class requirements for improved designs and redesigns.
- Oversupply of spares and inventories, in an effort achieve economies of scale. The opposite is achieved: diseconomies of scale.
- Pilferage and damage of shipbuilding materials and equipment owing to negligence.
- Inefficient time management and utilization of human resources. Gap between working hours and true production.
- Frequently, third-party agreements between shipyard, repair teams, manufacturers, suppliers, and so on, are not honored. If the agreement is not back to back, it is a problem.

4.4.4 Intellectual Property Rights

Among the most significant challenges in the shipbuilding industry, ownership of intellectual property rights seems to be on the top of the list. In a process where multiple companies, frequently have conflicting interests, collaborate in order to improve and materialize a ship design, a common question that arises pertains to the ownership of the design and its pertinent patents. Typically, the shipowners employ their own in-house naval architects, yet the initial design may be distributed to the future ship's classification society, shipyards, and third-party consultants, and extended to the bank or shareholders, the owners' Protection and Indemnity Club, engineers, and so on.

Modern technology has enabled shipyards to copy ship designs and patents, even as a ship arrives for repairs and drydock and routine maintenance. Problems arise as most shipbuilding contracts do not stipulate for property rights. For this reason, the IMO and key players have raised the industry's awareness and have promoted the development of efficient IT security in the shipyards' everyday business. Pursuant to industrial investigation, leakage related to intellectual property and innovation is frequently due to carelessness in the professionals' daily interactions with liaising companies in the broad business cycle.

Considerable amounts of sensitive information are forwarded through e-mail and intercompany software and hardware. Because of its complexity, the shipbuilding process requires the exchange of data between multiple entities and stages during a ship's construction (e.g., from a ship's design development, the manufacturing and assembly process throughout the shipyard's logistics chain, and regulatory compliance processes).

Therefore, shipbuilding companies find it increasingly difficult to implement intellectual property rights' applications that can be used in most other industries. Modern methods of protecting original ideas and patents must encompass both corporate policies and targeted IT security actions in order to monitor IT activities and prevent any future leakage (CESA 2011).

4.4.5 The History of Shipbuilding

England has a long history in shipbuilding; as it led the Industrial Revolution, that is, from 1760 to 1840, it was established as the primary shipbuilding power globally, for many decades to come, when London was established as the international capital of finance. The collapse of the shipbuilding industry in the twentieth century occurred during two periods related to fiscal downturns. First, from 1909 to 1933, centered during World War I (1914–1918), and the consecutive years when the US economy gained global leadership. Second, between 1960 and 1993, when shipyards were affected by economic policies that aimed to reduce inflation to the detriment of economic growth. The mass privatization and deregulation brought about radical changes in the British shipbuilding industry, and at the time, Japan was ready to take over.

From the 1950s until the 1990s, Japan and Europe dominated 90% of the global shipbuilding industry, but progressively Japan became the leader. Japan drove the global shipbuilding industry from the 1950s, for at least four decades, followed by the United States, England, and Germany as key global players. The dynamic Japanese economic growth reached a peak in the 1980s but collapsed during the country's historic "Lost Decade" (失われた10年). Fueled by an asset value bubble that affected banks, investors, and companies, a wave of consolidation severely affected the Japanese shipyards. At the time, South Korea was in a position to assume leadership.

In the 1970s, South Korea initiated its new shipbuilding endeavors, having been influenced by Japan's success and taking advantage of its low-cost labor. The 2000s saw Asia's miracle, led by China, and its fast-paced evolution, which overtook Japan and Korea's lead in the shipbuilding industry. Emerging shipbuilding nations are now India, the Philippines, Brazil, Turkey, and Vietnam, their competitive advantage being low-cost factors of production, in particular labor. As the new stage for shipyards is now introduced, nations are looking into alternative growth strategies.

Meanwhile, the shipbuilding industry in the Western Hemisphere is debilitated despite its strong innovation and effective production control, unable to compete with Asia's lower salaries, weaker and thus more attractive currencies, less stringent environmental regulations, and minimal occupational liability laws.

As of 2013, China, South Korea, and Japan are the leading shipbuilding nations. Emerging shipbuilding countries with increasingly large shipyard activities include the Philippines, Brazil, India, Australia, Germany, and Turkey, among others.

4.4.6 Reasons for Shipyards Losing Market Share

In observing the manner in which key shipbuilding nations succeed each other in leadership, there seems to be a pattern in the ways that competitive advantages are achieved and certain key points that maritime nations should be aware of. In examining the global history of shipbuilding, the following observations are made:

    i. Lack of innovation and modernization. Ship designs must be in line with the new market demand.

   ii. Leading shipbuilding nations in Asia tend to invest in the development of fewer, larger shipyards, as opposed to the numerous, scarce shipyards in the Western Hemisphere.

  iii. Poor production management methods.

  iv. Need for low cost. High labor cost, poor resource allocation, and utilization.

   v. Domestic steel industry—low steel cost. Interestingly enough, the major steel producers are the world's leading maritime nations. According to the World Steel Association (2013), the leading crude steel producers are China with 716.5 million tons annually, the European Union with 169.4 million tons, Japan with 107.2 million tons, and the United States with 88.6 million tons, followed by South Korea, Brazil, the Philippines, India, Germany, Taiwan, and Vietnam. Domestic shipyards must be attractive from a global perspective, that is, must have a competitive edge based on either low cost (labor, raw materials) or technological innovation, or both.

  vi. Cheap currency is needed. A nation's currency determines its global attractiveness. A depreciated currency is most attractive to the global market. Most traditional shipbuilding nations lost their competitive edge once their currency was appreciated.

 vii. Require upstream and downstream linkages to industries, for example, marine equipment, steel and ship-related services, steel mills, engine manufacturers.

viii. Need for government support, subsidies, and private investment.

  ix. Regulatory framework: strict/rigid environmental and OSHA regulations. Environmental restrictions, energy consumption, and emissions limits. OSHA and employees' liabilities are an additional cost, which is not the same for all nations. Many nations with a competitive edge are compromising strict regulations and employees' liability laws, for the sake of production. Historically, the nations that lost power have very strict regulations, which seem to reduce their ability to offer low-cost services.

4.4.7 Contemporary Shipbuilding Trends

In the shipbuilding industry, this meant that long-established banks and shipping finance houses could no longer grant their capital for new or secondhand ships. The economic regression halted the shipbuilding industry, and ignited a fleet oversupply owing to the dispersion of global trade.

> *Oversupply simply reveals that when too many entrepreneurs have the same great idea, it is no longer a great idea.*

The global economic crisis of 2008 brought about a reality check for shipyards, the shipping industry, and nations alike, as well as the opportunity for a market reshuffling based on their true strengths and weaknesses. During this time, shipyards collapsed, the exorbitant ships' prices and freight rates came back to their normal, preaugmented rates.

The postcrisis years have brought a market clearing and a price adjustment that requested the repositioning and reinventing of the shipyards, the shipping industry, and the global economy. In the same manner that many shipyards' business declined, other nations assumed a leadership role and dynamically pursued to revive the dream of national shipbuilding.

Overall, the leading shipbuilding nations seem to enjoy government support in terms of subsidies, regulations, and financial incentives, whereas economies that are incapable of achieving competitive advantages through cost efficiency, experience a decrease in their market share. With an apparent shift from the West toward the East, western ship-yards manage to survive mainly due to naval/marine shipbuilding and less due to their commercial shipbuilding activities.

1. Chinese Shipbuilding

   China, with 45% of the global market share, runs at least 1647 shipyards, with large-scale production concentrated in 212 major shipyards. China's advantages entail low labor costs coupled with strong government support. The swift expansion of Chinese shipbuilding is directly related to the government's macroeconomic plans. By means of its "five-year plans," the Chinese administration structures its recommendations for the expansion of diverse industrial sectors to assure their future viability and market potential.

   China's 11th National 5-year Economic Plan (2006–2010) was particularly focused in the maritime industry. The nation's 12th plan (2011–2015) is focused in three core areas: the establishment of Hong Kong as a global finance center, the spread of development from the established coastal urban center to the inland region, and the implementation of stricter environmental regulations.

   What can be read between the lines is that China intends to equally support smaller shipyards and industrial zones. China's distinguished regime allows the central control of all of the nation's industries, including all of its shipyards.

   Most important, we can foresee a significant growth of the Hong Kong stock exchange and banking systems. This strategic movement is likely to draw more global investors and further establish China's power.

   The miracle of China depends to a large extent on the nation's strategic ability to utilize its factors of production in the most efficient way. In 2012 and 2013, shipyard profits have been crumbling as national production diminished by 50% over the past two years. Their future depends on the government subsidies they are due to receive in this critical era.

   In order to forecast the future of Chinese shipyards, one has to predict whether the central government plans will focus on retaining the smaller shipyards despite their losses, or, the most likely scenario, to carefully plan their differentiation in a multiproduct scheme across China's coastline.

2. Korean Shipbuilding

   South Korea has a 29% market share of the global shipbuilding market. Korea's shipbuilding industry is mostly within the private sector, whereas its production concentrated in 24 major shipyards. South Korea guides the production of larger oceangoing ships including large containers and tanker ships, drill ships, luxury cruise ships, and LNG ships. The nation has some of the world's largest and most productive shipyards globally, as they deliver one large new building every four days. The leading global shipyards belong to South Korea and include the world's largest, that is, Ulsan's Hyundai Heavy Industries shipyard, closely followed by Samsung Heavy Industries and Daewoo Shipbuilding and Marine Engineering. It is worth noting that Maersk's largest container ships are being built in the Daewoo shipyard. This contract is of strategic significance to Korean shipbuilding.

3. Japanese Shipbuilding

   Japan currently has an 18% market share of the global shipbuilding market. It operates over 1000 shipyards, predominantly privately owned. Among them, six Japanese shipbuilders are among the top 30 global frontrunners, as measured by order books (OECD 2013). The Japanese shipbuilding industry is concentrated mostly within the private sector, whereas its production concentrated in 51 major shipyards. The country's recent currency (yen) depreciation has been most promising for Japanese shipyards, especially to shipowners that are interested in fuel-efficient vessels.

4. The European Union

   The European shipbuilding industry consists of Belgium, Bulgaria, Croatia, Denmark, Finland, France, Germany, Greece, Italy, Lithuania, The Netherlands, Norway, Poland, Portugal, Romania, Spain, and the United Kingdom. The EU currently represents 3% of the market. Over 300 shipyards are available to build, retrofit, convert, and maintain merchant and naval ships. Each year, the European shipyards receive roughly €30 billion revenues, whereas about 75% of the vessels they build are actually for export markets (CESA 2013).

5. The United States

   In the United States, the Merchant Marine Act of 1920, also known as the "Jones Act," is a US federal statute that regulates coastal shipping. Under the Jones Act in the United States, any vessel navigating among two US ports and thus moving domestic cargoes shall be a US flagged ship, constructed in the United States, owned by US citizens, and crewed by US citizens and US permanent residents. The Jones Act and the US government's recent announcement on the subsidies of numerous American shipyards are intended to be an effort to revive American shipbuilding. The vast oil and gas reserves along with the nation's commitment, that by 2016 America will be a major oil and gas exporter, are significant enough reasons to evolve the domestic shipbuilding industry. In this commitment to growth, it is necessary for the country to become globally competitive in terms of cost, as a strong US dollar combined with high labor and high cost in all factors of production challenge the United States to develop its competitive edge. For this reason, there is a compelling need for process improvement, cost efficiency, and an economic and currency reform in order for American shipyards to be in a position to offer globally competitive prices.

6. Advanced economies in the Western Hemisphere: United States and EU

   In observing economies like the United States and the European Union, it becomes apparent that their current strong currencies and high labor costs make them uncompetitive for the shipbuilding industry.

   At this stage, shipbuilding prices are falling, currently reaching the pre-2002 levels. The world shipbuilding market is characterized by oversupply, slashed prices, and limited profit.

   Based on this market, the key element to keep in mind is that nations need to pick their battles and specialize in the areas that are most useful and profitable.

7. Emerging shipbuilding powers: India, Brazil, and the Philippines

   In the same manner, in a comparison between the three emerging shipbuilding powers, it appears that *India* is the nation with the lowest currency and the lowest labor costs; hence, it has a greater chance of achieving rapid growth, which might propel to become one of the top three Asian shipbuilding nations.

8. The Philippines

   The *Philippines* also has an attractive combination of low currency and labor costs, plus the geographical advantage of being in the vicinity of the mega-economies of Asia. This argument can be verified by the first shipyard outsources from Korea and Japan that the Philippines has already secured, that is, the Korean Hanjin H.I. Subic and the Japanese Tsuneishi-Cebu.

9. Brazil

Last but not least, there is *Brazil*, a strong economy with significant maritime achievements and state-of-the-art infrastructure investments. One might think that the nation's significant oil, gas, and ethanol energy production would boost its shipbuilding activities. In reality, the country's strong currency, high taxes, and restrictive "local content" laws for vessels used in Brazilian waters will challenge the country's ability to compete with key global shipbuilding players.

Upon critically evaluating the national steel production levels of the nations in question, at this stage, there does not seem to be a close connection between steel production and growth in shipbuilding. This becomes apparent in countries like the United States and the EU, which are advanced, yet their high steel production does not seem to boost their shipbuilding marketing edge.

In the end, what does become apparent is that currency, labor costs, and government regulations are interrelated in influencing a country's industrial production capabilities, with shipbuilding being a prominent example. Further consideration of how a country's currency may affect its ability to succeed in capital-intensive industries such as shipbuilding is provided in Chapter 12, which discusses potential "Currency Wars."

4.4.8 Shipbuilding and Oil Market Analysis

The shipbuilding industry is a technology-intensive industry that is dependent upon the energy market fluctuations for the following reasons:

a. Ship construction requires energy to produce ships; hence, the price of oil determines the industry's profit margin.
b. Shipbuilding is a derived demand, directly dependent on the global demand for ships and cargoes. Higher oil prices will reduce global trade and the need for new ships.

The developed and rapidly developed economies have been increasing oil consumption for a number of years. The recent discoveries of oil and gas reserves in the United States and Brazil are likely to boost global trade in the next decade or so. The patterns of the oil distribution are typically defined by the most advanced economies, who are also the price setters. The key factors in influencing exports demand are currencies and rates of exchange, GDP, oil production salaries and income, and relative export price. Also, the oil imports demand is influenced by currency and rate of exchange, GDP, PPP, and import prices. In examining the global and regional configurations of imports and exports, one can estimate the demand for shipping and consequently the demand for shipbuilding.

4.4.9 Global Market Analysis

The conclusions pertaining to the current and future trends of the shipping industry are shown in Table 4.3. Figures 4.11 through 4.14 indicate that the main factors that determine a nation's shipbuilding growth potential are its currency and its labor costs.

Among the world's leading economies, the ones that thrive in global shipbuilding are the ones with the lowest currency and the ones with the lowest labor cost. On the other hand,

TABLE 4.3 Top Shipbuilders and Steel Producers

Timeline	Nation	Shipbuilding Order Book CGT	% Market Share per CGT	Currency	Currency against US Dollar	Steel Production, Million Tons	Labor: per Capita PPP ($)	Current Status
2010 to present	China	29,361	33.1	Yuan	0.16	716.5	9300	First shipbuilding nation. Leading economy. Controls trade growth Low currency. Low labor cost First in steel production
1990s to present	South Korea	24,164	27.3	Won	0.00090	69.3	32,800	Second shipbuilding nation. Strong economy. Very low currency. Moderate labor cost Moderate steel production
1950s to present	Japan	12,223	13.8	Yen	0.010	107.2	36,900	Third shipbuilding nation. Strong economy Very low currency. High labor cost Moderate steel production
1860s to present	EU	1114	2.9	Euro	1.33	169.4	35,100	1% of global shipbuilding capacity. Strong economy Very strong currency. High labor cost High steel production
2000s to present	Brazil	738	1.9	Real	0.44	34.7	12,100	Emerging shipbuilding nation. Strong economy High currency. Low labor cost
2000s to present	Philippines	405	1.1	Peso	0.023	0	4500	Emerging shipbuilding nation. Very low currency. Very low labor cost Moderate steel production

(continued)

TABLE 4.3 (Continued) Top Shipbuilders and Steel Producers

Timeline	Nation	Shipbuilding Order Book CGT	% Market Share per CGT	Currency	Currency against US Dollar	Steel Production, Million Tons	Labor: per Capita PPP ($)	Current Status
2000s to present	India	144	0.4	Rupee	0.017	76.7	3900	Emerging shipbuilding nation. Strong economy Very low currency. Very low labor cost Moderate steel production
1900s to present	United States	257	0.7	Dollar	1	88.6	50,700	World's leading economy. Controls trade growth Strong currency. Moderate steel production High labor cost

Source: M.G. Burns, based on data from CIA World Factbook, 2013. Available at http://www.cia.gov (accessed on July 10, 2013); World Steel Association, 2013. Available at http://www.worldsteel.org (accessed on July 12, 2013); Bloomberg News, 2013; OECD Korea, 2012; OECD, 2008. OECD Council Working Party on Shipbuilding (WP6). The Shipbuilding Industry in China. June 2008. Available at http://www.oecd.org (accessed on July 1, 2013); OECD, 2013. OECD Council Working Party on Shipbuilding (WP6). The Shipbuilding Industry in Japan. April 2013. Available at http://www.oecd.org (accessed on July 1, 2013); KOSHIPA 2013. Korean Shipbuilding Association. Available at http://www.koshipa.org (accessed on July 5, 2013); SAJ, 2013. The Shipbuilders' Association of Japan. Shipbuilding Statistics, March 2013. Available at http://www.sajn.or.jp (accessed on July 1, 2013); CESA, 2013. Community of European Shipyards Associations. Available at http://www.cesa.eu (accessed on July 5, 2013).

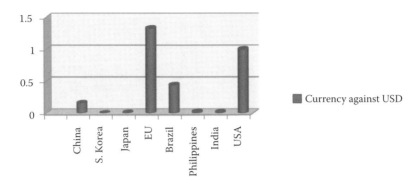

FIGURE 4.11 Top shipbuilders: currency. (Courtesy of M.G. Burns, based on data from CIA World Factbook, 2013. Available at http://www.cia.gov [accessed on July 10, 2013]; World Steel Association, 2013. Available at http://www.worldsteel.org [accessed on July 12, 2013]; Bloomberg News, 2013; OECD Korea, 2012; OECD, 2008. OECD Council Working Party on Shipbuilding [WP6]. The Shipbuilding Industry in China. June 2008. Available at http://www.oecd.org [accessed on July 1, 2013]; OECD, 2013. OECD Council Working Party on Shipbuilding [WP6]. The Shipbuilding Industry in Japan. April 2013. Available at http://www.oecd.org [accessed on July 1, 2013]; KOSHIPA 2013. Korean Shipbuilding Association. Available at http://www.koshipa.org [accessed on July 5, 2013]; SAJ, 2013. The Shipbuilders' Association of Japan. Shipbuilding Statistics, March 2013. Available at http://www.sajn.or.jp [accessed on July 1, 2013]; CESA, 2013. Community of European Shipyards Associations. Available at http://www.cesa.eu [accessed on July 5, 2013].)

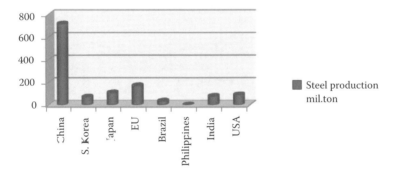

FIGURE 4.12 Top shipbuilders: steel production. (Courtesy of M.G. Burns, based on data from CIA World Factbook, 2013. Available at http://www.cia.gov [accessed on July 10, 2013]; World Steel Association, 2013. Available at http://www.worldsteel.org [accessed on July 12, 2013]; Bloomberg News, 2013; OECD Korea, 2012; OECD, 2008. OECD Council Working Party on Shipbuilding [WP6]. The Shipbuilding Industry in China. June 2008. Available at http://www.oecd.org [accessed on July 1, 2013]; OECD, 2013. OECD Council Working Party on Shipbuilding [WP6]. The Shipbuilding Industry in Japan. April 2013. Available at http://www.oecd.org [accessed on July 1, 2013]; KOSHIPA 2013. Korean Shipbuilding Association. Available at http://www.koshipa.org [accessed on July 5, 2013]; SAJ, 2013. The Shipbuilders' Association of Japan. Shipbuilding Statistics, March 2013. Available at http://www.sajn.or.jp [accessed on July 1, 2013]; CESA, 2013. Community of European Shipyards Associations. Available at http://www.cesa.eu [accessed on July 5, 2013].)

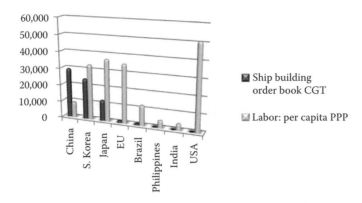

FIGURE 4.13   Top shipbuilders: labor. (Courtesy of M.G. Burns, based on data from CIA World Factbook, 2013. Available at http://www.cia.gov [accessed on July 10, 2013]; World Steel Association, 2013. Available at http://www.worldsteel.org [accessed on July 12, 2013]; Bloomberg News, 2013; OECD Korea, 2012; OECD, 2008. OECD Council Working Party on Shipbuilding [WP6]. The Shipbuilding Industry in China. June 2008. Available at http://www.oecd.org [accessed on July 1, 2013]; OECD, 2013. OECD Council Working Party on Shipbuilding [WP6]. The Shipbuilding Industry in Japan. April 2013. Available at http://www.oecd.org [accessed on July 1, 2013]; KOSHIPA 2013. Korean Shipbuilding Association. Available at http://www.koshipa.org [accessed on July 5, 2013]; SAJ, 2013. The Shipbuilders' Association of Japan. Shipbuilding Statistics, March 2013. Available at http://www.sajn.or.jp [accessed on July 1, 2013]; CESA, 2013. Community of European Shipyards Associations. Available at http://www.cesa.eu [accessed on July 5, 2013].)

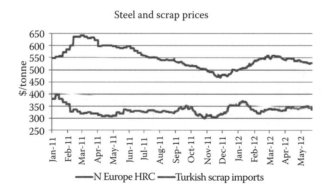

FIGURE 4.14   Top shipbuilders: Currency, labor, and steel production. Note: Top price index: N. Europe HRC; bottom price index: Turkish scrap imports. (From French International Ship Register [RIF, Registre International Francais], 2012. Available at http://www.rif.mer.developpement-durable.gouv.fr/en/about-the-tonnage-r40.html [accessed July 12, 2013].)

strong economies need to build their own ships at a navy and commercial level. The three questions that one needs to ask when observing the top shipbuilding countries are as follows:

1. *How does shipbuilding contribute to a nation's economy?* In the current over-supply of shipyards, and low ship prices, a cost–benefit analysis would not be very encouraging, as the profit margin would be diminished.

2. *Based on the findings of a cost–benefit analysis, is shipbuilding justified as a sound and feasible investment?* A comparison between shipbuilding and other investment projects will evaluate the degree of usefulness and profitability.
3. *Does the country build ships for domestic or global trade purposes?* Domestic-purpose shipbuilding has a strategic significance, which exceeds the commercial benefits. Through the course of history, shipbuilding and the ability of a nation to extend its marine fleet to protect national sovereignty have been critical (e.g., American Revolution, US Civil War, WWII and Liberty ships, etc.).

4.4.10 Conclusions

Global trade growth has reached unparalleled levels over the past decades. However, the global economy and consequently the shipbuilding industry have not fully recovered since the 2008 global financial crisis. An oversupply of global shipyards, in combination with lower freight rates for shipowners, has resulted in lower shipbuilding earnings and slightly reduced volumes as verified by the annual shipbuilding order books. This suggests that shipyards will continue to build and deliver similar volumes of ships, although likely at a lower profit margin.

Banks have restricted their financing to shipbuilding activities, imposing additional strain on the shipping and shipbuilding industries that both suffer from oversupply. This additional pressure may not affect the world's mega-shipyards but is likely to strike the smaller players.

While focusing on the top three shipbuilding nations, that is, China, South Korea and Japan, a restrictive credit policy in Chinese banks and investors, coupled with sluggish Japanese and South Korean currencies, is apparent. These trends have the potential to result in cutthroat competition among the global shipbuilding leaders. As a result, the newbuilding prices tend to drop, thus minimizing profit margins in the Chinese shipyards.

The past decade has been marked as one of the most lucrative decades in the shipbuilding industry since World War II. To be precise, the world order book from 2002 to 2012 has grown at an unprecedented pace. Over these years, literally hundreds of new Eastern Hemisphere shipyards have been established, while in the Western Hemisphere, numerous traditional shipyards of historical significance have collapsed.

As much as the early 2000s fostered years of growth and abundance, the 2008 global crisis emerged to highlight the previously established trade imbalances. Systemic failures as well as financial and credit crisis led to unsustainable thresholds for external debt, both private and public. The dawn of a currency crisis and the legendary currency wars also emerged, yet their true impact and dimensions are to be felt after 2014 (for a detailed market forecast, see Chapter 12).

4.5 PORT AGENTS: LINER SERVICES, TRAMP TRADE, AND OFFSHORE SUPPORT AGENTS

The concept of *ship agents* or *port agents* arose centuries ago, as shipowners and cargo owners needed to protect and control their ships and cargoes in every single port of call, in every part of the world.

In global port operations where time efficiency is money, it is necessary for independent and reliable business entities to be present during the ship's arrival, cargo, and bunkering operations. Agents would ensure that every single action that occurs at port and registered in the ship's "statement of facts" and log books has been duly preplanned, compliant and in line with the principals' instructions.

In the early days where no cell phones, Internet, or satellite systems existed, an agent was the eyes and the ears of their principals. As technology evolved, the industry's complexity requires ship agents more than ever, their roles being literally to witness all port operations, represent their principals in transactions with third parties, and verify that proper shipping practices were followed. A successful agent literally needs to act on behalf of his or her principals, with utmost zeal and professionalism. During the past centuries, an agent was an independent individual, familiar with a specific port's requirements, whereas from the 1960s to date, large agency firms increasingly tend to form an efficient geographic network that expands over an entire country or continent.

4.5.1 FONASBA

The international "Federation of National Associations of Ship Brokers and Agents" (FONASBA) and the US Association of Ship Brokers and Agents (ASBA) are independent membership trade associations that guide ship agents through the issuance of regular standard agreements, contracts, and advice. Since its inception in 1952, FONASBA has established a set of standards aiming to establish a holistic support system for port agents and shipbrokers. As of 2010, FONASBA had 36 full members, 12 associate and candidate members, and five club members (FONASBA 2010).

Its General Agency Agreement for liner services is signed among liner principals and general agents. The agreement was modified and implemented in 1993 and has been strongly recommended by BIMCO.

FONASBA's "Quality Standard for Shipbrokers and Agents" has been approved by INTERTANKO, which has also endorsed FONASBA'S Port Tramp Agency Working Group. FONASBA has also established an ambitious plan of quality standardization for port agents and brokers. The Quality Standard will provide shipowners and operators with a best practice selection system, in order to evaluate the best industry professionals in terms of economic status, expertise, and industrial know how (FONASBA Annual Report 2010).

4.5.2 Agency Selection and Practices

The agents are selected by the principals' operations department, initially by "shopping around" and comparing services, prices, company reputation, feedback from the ship, recommendations from business partners, and so on.

Although there is no formal contract to be signed between the agents and their principals in the tramp trade, the appointment confirmation that the principals e-mail to the agents contains a concise confirmation of the appointment, the agreed agency fees, the provisional services required with the agreed tariffs, and so on. This confirmation, along with the e-mails exchanged by both parties, has legally binding consequences, and if needed can be used for claims, arbitration, or the court of law. The principals are legally liable for the consequences of all written instructions and subject to evidence and circumstances, the tacit authority, and implied guidelines given to the agents. The key elements here include what instructions were given to the agents, what services have been undertaken by the agents *prior to the ship's departure*, and at what price.

The initial agents' appointment letter is accompanied by funds remitted to the agents prior to the ship's arrival at the specific port of call, reflecting agency fees and specified tasks. The day that the vessel departs from the port, the agents produce the disbursement account (D/A). This is a package of numerous pro forma invoices, payment receipts, and supporting documents pertaining to the agency's services offered for the specific ship at the specific port.

In case the agents are the appointed entity as stipulated by the C/P, the D/A is forwarded to both the shipowners and the charterers, as the services rendered jointly involve instructions received by both parties. In the case of protective agents, the D/A is submitted to the agents' exclusive principals.

4.5.3 Port Agency Responsibilities

A port agent's responsibilities literally encompass and oversee all ship's operations while at port. There are three principal areas of agency service, which can be carried out by the same agency companies:

a. *Agency for the tramp trade*, which covers an extensive collaboration of a shorter duration, yet lacking a contractual agreement and deep involvement in the principals' financial and commercial transactions.
b. *Agency for the liner trade*, which entails a long-term contractual agreement and the agents' authority to support their principals' marketing, financial, logistics networking, customer support, and operational endeavors.
c. *Agency for offshore support and logistics.* Modern port agents become increasingly accustomed to the industry's new trends and expand to the oil and gas offshore and logistics agency services. The industry increasingly requires the support of ship and offshore agencies, which have managed to bridge the maritime, inland, and oil and gas industry (see Chapter 11).

Despite their functional differences, many of the agents' duties are overlapping. An analysis of the agency types and functions is covered herewith.

a. *Liner Services Agency Duties*
The liner trade differs with the tramp trade as it offers scheduled, consistent services to predetermined trade routes and carries cargoes from multiple charterers and complex logistics networks. These special characteristics necessitate more complex, all-encompassing agency services, which cover the duties of a booking agent, an operations coordinator, a shipbroker, a commodity broker, and, depending on the agreement, sometimes reaching the authorities of a ship manager. Their monthly fees depend on the fleet size, schedule regularity, and duties.

A FONASBA "Standard Liner and General Agency Agreement" is signed between the principal, that is, the liner owners, and the port agents, and stipulates terms related to the services to be provided and the payment terms (FONASBA 2013).

Figure 4.15 demonstrates a FONASBA Agency Agreement.

FONASBA   The Federation of National Associations of Ship Brokers and Agents   FONASBA

STANDARD LINER AND GENERAL AGENCY AGREEMENT

Revised and adopted 2001

Approved by BIMCO 2001

It is hereby agreed between:

....of...(hereinafter referred to as the Principal)

and

....of...(hereinafter referred to as the Agent)

on the ....day of ....20.........

that:

1.00   The Principal hereby appoints the Agent as its Liner Agent for all its owned and/or chartered vessels including any space or slot charter agreement serving the trade between ....
and ....

1.01   This Agreement shall come into effect on ....and shall continue until.....................
Thereafter it shall continue until terminated by either party giving to the other notice in writing, in which event the Agreement shall terminate upon the expiration of a period of .....months from the date upon which such notice was given.

1.02   The territory in which the Agent shall perform its duties under the Agreement shall be...............
hereinafter referred to as the "Territory".

1.03   This Agreement covers the activities described in section 3...............................

1.04   The Agent undertakes not to accept the representation of other shipping companies nor to engage in NVOCC or such freight forwarding activities in the Territory, which are in direct competition to any of the Principal's transportation activities, without prior written consent, which shall not unreasonably be withheld.

1.05   The Principal undertakes not to appoint any other party in the Agent's Territory for the services defined in this Agreement.

1.06   The established custom of the trade and/or port shall apply and form part of this Agreement.

1.07   In countries where the position of the agent is in any way legally protected or regulated, the Agent shall have the benefit of such protection or regulation.

1.08   All aspects of the Principal's business are to be treated confidentially and all files and records pertaining to this business are the property of the Principal.

2.0   **Duties of the Agent**

2.01   To represent the Principal in the Territory, using his best endeavours to comply at all times with any reasonable specific instructions which the Principal may give, including the use of Principal's documentation, terms and conditions.

2.02   In consultation with the Principal to recommend and/or appoint on the Principal's behalf and account, Sub-Agents.

2.03   In consultation with the Principal to recommend and/or to appoint on the Principal's behalf and account, Stevedores, Watchmen, Tallymen, Terminal Operators, Hauliers and all kinds of suppliers.

2.04   The Agent will not be responsible for the negligent acts or defaults of the Sub-Agent or Sub-Contractor unless the Agent fails to exercise due care in the appointment and supervision of such Sub-Agent or Sub- Contractor. Notwithstanding the foregoing the Agent shall be responsible for the acts of his subsidiary companies appointed within the context of this Clause.

2.05   The Agent will always strictly observe the shipping laws and regulations of the country and will indemnify the Principal for fines, penalties, expenses or restrictions that may arise due to the failure of the Agent to comply herewith.

1

FIGURE 4.15   FONASBA agency agreement.

3.0   **Activities of Agent** (Delete those which do not apply)

3.1   **Marketing and Sales**

3.11   To provide marketing and sales activities in the Territory, in accordance with general guidelines laid down by the Principal, to canvass and book cargo, to publicise the services and to maintain contact with Shippers, Consignees, Forwarding Agents, Port and other Authorities and Trade Organisations.

3.12   To provide statistics and information and to report on cargo bookings and use of space allotments. To announce sailing and/or arrivals, and to quote freight rates and announce freight tariffs and amendments.

3.13   To arrange for public relations work (including advertising, press releases, sailing schedules and general promotional material) in accordance with the budget agreed with the Principal and for his account.

3.14   To attend to conference, consortia and /or alliance matters on behalf of the Principal and for the Principal's account.

3.15   To issue on behalf of the Principal Bills of Landing and Manifests, delivery orders, certificates and such other documents.

3.2   **Port Agency**

3.21   To arrange for berthing of vessels, loading and discharging of the cargo, in accordance with the local custom and conditions.

3.22   To arrange and co-ordinate all activities of the Terminal Operators, Stevedores, Tallymen and all other Contractors, in the interest of obtaining the best possible operation and despatch of the Principal's vessel.

3.23   To arrange for calling forward, reception and loading of outward cargo and discharge and release of inward cargo and to attend to the transhipment of through cargo.

3.24   To arrange for bunkering, repairs, husbandry, crew changes, passengers, ship's stores, spare parts, technical and nautical assistance and medical assistance.

3.25   To carry out the Principal's requirements concerning claims handling, P & I matters, General Average and/or insurance, and the appointment of Surveyors.

3.26   To attend to all necessary documentation and to attend to consular requirements.

3.27   To arrange for and attend to the clearance of the vessel and to arrange for all other services appertaining to the vessel's movements through the port.

3.28   To report to the Principal the vessel's position and to prepare a statement of facts of the call and/or a port log.

3.29   To keep the Principal regularly and timely informed on Port and working conditions likely to affect the despatch of the Principal's vessels.

3.3   **Container and Ro/Ro Traffic**

Where "equipment" is referred to in the following section it shall comprise container, flat racks, trailers or similar cargo carrying devices, owned, leased or otherwise controlled by the Principal.

3.31   To arrange for the booking of equipment on the vessel.

3.32   To arrange for the stuffing and unstuffing of LCL cargo at the port and to arrange for the provision of inland LCL terminals.

3.33   To provide and administer a proper system, or to comply with the principal's system for the control and registration of equipment. To organise equipment stock within the Territory and make provision for storage, positioning and repositioning of the equipment.

3.34   To comply with Customs requirements and arrange for equipment interchange documents in respect of the movements for which the Agent is responsible and to control the supply and use of locks, seals and labels.

3.35   To make equipment available and to arrange inland haulage.

3.36   To undertake the leasing of equipment into and re-delivery out of the system.

3.37   To operate an adequate equipment damage control system in compliance with the Principal's instructions. To arrange for equipment repairs and maintenance, when and where necessary and to report on the condition of equipment under the Agent's control.

2

FIGURE 4.15   (Continued) FONASBA agency agreement.

3.4   General Agency

3.41   To supervise, activities and co-ordinate all marketing and sales activities of Port, Inland Agents and/or Sub-agents in the Territory, in accordance with general guidelines laid down by the Principal and to use every effort to obtain business from prospective clients and to consolidate the flow of statistics and information.

3.42   To supervise and co-ordinate all activities of Port, Inland Agents and/or Sub-agents as set forth in the agreement, in order to ensure the proper performance of all customary requirements for the best possible operation of the Principal's vessel in the G.A.'s Territory.

3.43   In consultation with the Principal to recommend and/or appoint on the Principal's behalf and account Port, Inland Agents, and/or Sub-Agents if required.

3.44   To provide Port, Inland Agents and/or Sub-agents with space allocations in accordance with the Principal's requirements.

3.45   To arrange for an efficient rotation of vessels within the Territory, in compliance with the Principal's instructions and to arrange for the most economical despatch in the ports of its area within the scope of the sailing schedule.

3.46   To liaise with Port Agents and/or Sub-agents if and where required, in the Territory in arranging for such matters as bunkering, repairs, crew changes, ship's stores, spare parts, technical, nautical, medical assistance and consular requirements.

3.47   To instruct and supervise Port, Inland Agents and/or Sub-Agents regarding the Principal's requirements concerning claims handling. P & I matters and/or insurance, and the appointment of Surveyors. All expenses involved with claims handling other than routine claims are for Principal's account.

3.5   Accounting and Finance

3.51   To provide for appropriate records of the Principal's financial position to be maintained in the Agent's books, which shall be available for inspection and to prepare periodic financial statements.

3.52   To check all vouchers received for services rendered and to prepare a proper disbursement account in respect of each voyage or accounting period.

3.53   To advise the Principal of all amendments to port tariffs and other charges as they become known.

3.54   To calculate freight and other charges according to Tariffs supplied by the Principal and exercise every care and diligence in applying all terms and conditions of such Tariffs or other freight agreements. If the Principal organises or employs an organisation for checking freight calculations and documentation the costs for such checking to be entirely for the Principal's account.

3.55   To collect freight and related accounts and remit to the Principal all freights and other monies belonging to the Principal at such periodic intervals as the Principal may require. All bank charges to be for the Principal's account. The Agent shall advise the Principal of the customary credit terms and arrangements. If the Agent is required to grant credit to customers due to commercial reasons, the risk in respect of outstanding collections is for the Principal's account unless the Agent has granted credit without the knowledge and prior consent of the Principal.

3.56   The Agent shall have authority to retain money from the freight collected to cover all past and current disbursements, subject to providing regular cash position statements to the Principal.

3.57   The Agent in carrying out his duties under this Agreement shall not be responsible to the Principal for loss or damage caused by any Banker, Broker or other person, instructed by the Agent in good faith unless the same happens by or through the wilful neglect or default of the Agent. The burden of proving the wilful neglect of the Agent shall be on the Principal.

4.0   Principal's Duties

4.01   To provide all documentation, necessary to fulfil the Agent's task together with any stationery specifically required by the Principal.

4.02   To give full and timely information regarding the vessel's schedules, ports of call and line policy insofar as it affects the port and sales agency activities.

4.03   To provide the Agents immediately upon request with all necessary funds to cover advance disbursements unless the Agent shall have sufficient funds from the freights collected.

4.04   The Principal shall at all times indemnify the Agent against all claims, charges, losses, damages and expenses which the Agent may incur in connection with the fulfilment of his duties under this Agreement. Such indemnity shall extend to all acts, matters and things done, suffered or incurred by the Agent during the duration of this Agreement, notwithstanding any termination thereof, provided always, that this indemnity shall not extend to matters arising by reason of the wilful misconduct or negligence of the Agent.

3

FIGURE 4.15   (Continued) FONASBA agency agreement.

4.05   Where the Agent provides bonds, guarantees and any other forms of security to Customs or other statutory authorities then the Principal shall indemnify and reimburse the Agent immediately such claims are made, provided they do not arise by reason of the wilful misconduct or the negligence of the Agent.

4.06   If mutually agreed the Principal shall take over the conduct of any dispute which may arise between the Agent and any third party as a result of the performance of the Agent's duties.

5.0   Remuneration

5.01   The Principal agrees to pay the Agent and the Agent accepts, as consideration for the services rendered, the commissions and fees set forth on the schedule attached to this Agreement. Any fees specified in monetary units in the attached schedule shall be reviewed every 12 months and if necessary adjusted in accordance with such recognised cost of living index as is published in the country of the Agent.

5.02   Should the Principal require the Agent to undertake full processing and settlement of claims, then the Agent is entitled to a separate remuneration as agreed with the Principal and commensurate with the work involved.

5.03   The remuneration specified in the schedule attached is in respect of the ordinary and anticipated duties of the Agent within the scope of this Agreement. Should the Agent be required to perform duties beyond the scope of this Agreement then the terms on which the Agent may agree to perform such duties will be subject to express agreement between the parties. Without prejudice to the generality of the foregoing such duties may include e.g. participating in conference activities on behalf of the Principal, booking fare-paying passengers, sending out general average notices and making collections under average bonds insofar as these duties are not performed by the average adjuster.

5.04   If the Tariff currency varies in value against the local currency by more than 10% after consideration of any currency adjustment factor existing in the trade the basis for calculation of remuneration shall be adjusted accordingly.

5.05   Any extra expenses occasioned by specific additional requirements of the Principal in the use of computer equipment and systems for the performance of the Agent's duties to the Principal shall be borne by the Principal.

5.06   The Principal is responsible for all additional expenses incurred by the Agent in connecting its computers to any national or local port community system.

6.0   Duration

6.01   This agreement shall remain in force as specified in clause 1.01 of this Agreement. Any notice of termination shall be sent by registered or recorded mail.

6.02   If the Agreement for any reason other than negligence or wilful misconduct of the Agent should by cancelled at an earlier date than on the expiry of the notice given under clause 1.01 hereof, the Principal shall compensate the Agent. The compensation payable by the Principal to the agent shall be determined in accordance with clause 6.04 below.

6.03   If for any reason the Principal withdraws or suspends the service, the Agent may withdraw from this agreement forthwith, without prejudice to its claim for compensation.

6.04   The basis of compensation shall be the monthly average of the commission and fees earned during the previous 12 months or if less than 12 months have passed then a reasonable estimate of the same, multiplied by the number of months from the date of cancellation until the contract would have been terminated in accordance with clause 1.01 above. Furthermore the gross redundancy payments, which the Agent and/or Sub-Agent(s) is compelled to make to employees made redundant by reason of the withdrawal or suspension of the Principal's service, or termination of this Agreement, shall also be taken into account.

6.05   The Agent shall have a general lien on amounts payable to the Principal in respect of any undisputed sums due and owing to the Agent including but not limited to commissions, disbursements and duties.

7.0   Jurisdiction

7.01   a) This Agreement shall be governed by and construed in accordance with the laws of the country in which the Agent has its principle place of business and any dispute arising out of or in connection with this Agreement shall be referred to arbitration in that country subject to the procedures applicable there.

b) This Agreement shall be governed by and construed in accordance with the laws of .... and any dispute arising out of or in connection with this Agreement shall be referred to arbitration at ...., subject to the procedures applicable there.

c) Any dispute arising out of this Agreement shall be referred to arbitration at....................subject to the law and procedures applicable there.

(subclauses [a] [b] & [c] are options. If [b] or [c] are not filled in then [a] shall apply.)

4

FIGURE 4.15   (Continued) FONASBA agency agreement.

REMUNERATION SCHEDULE BELONGING TO STANDARD LINER AND GENERAL AGENCY AGREEMENT

Between..and..date...........................
       (As Principal)   (As Agent)

The Agent is entitled to the following remuneration based <u>on all total freight</u> earnings (including any surcharges,(eg BAF, CAF) handling charges (e.g. THC) and freight additionals including inland transport which may be agreed) of the Principal's liner service to and from the Territory to be paid in Agent's local currency. The total remuneration per call shall not in any case be lower than the local fee applicable

I   A.   Where the Agent provides all the services enumerated in this Agreement the Commission shall be:

             Services outward......................%   [Min .... per cent or tonne/cbm]   } MIN
                                                                  }
                   inward......................%   [Min .... per cent or tonne/cbm]   } LUMP SUM
                                                                     }

      2.   ....% for cargo when only booking is involved. [Min .... per cent] } PER

      3.   ....% for cargo when only handling is involved. [Min........ per cent] } CALL ....

             ("only handling" in the remuneration schedule is so defined that the duties of an Agent are to call forward and otherwise arrange for the cargo to be loaded on board, where the specific booking has been made elsewhere and acknowledged as such by the shipper as nominated for the Principal's service.

      4.   In respect of movements of cargo outside the Agent's Territory....................% of the gross total freight is payable in cases where only collection of freight is involved.

      5.   An additional fee for containers and/or units entering or leaving the inventory control system of the Agent a fee of...................... per unit.

II   A.   ....% for cargo loaded on board in bulk. [Min .... per tonne/cbm]

      2.   ....% for cargo discharged in bulk.   [Min .... per tonne/cbm]

III   Where the Agent provides only the services as non-port agent the remuneration shall be:

      When actually booked/originating from this area:

      1.   Services outward ....%   [Min .... per cent or tonne/cbm]

                   inward ....%   [Min .... per cent or tonne/cbm]

      2.   An additional fee for containers and/or units entering or leaving the inventory control system of the Agent a fee of .... per unit.

IV   Where the Agent provides only the services as non-port agent the remuneration shall be:

      1.   ....% for cargo loaded on board in bulk. [Min .... per tonne/cbm]

      2.   ....% for cargo discharged in bulk.   [Min .... per tonne/cbm]

V   Clearance and ship's husbandry fee shall be as agreed.

VI   A Commission of .... % shall be paid on all ancillary charges collected by the Agent on behalf of the Principal such as Depot Charges, Container Demurrage, Container Damage etc.

VII   Communications: The Principal will either pay actual communication expenses on a cost plus basis or pay a lumpsum monthly on an average cost plus basis, to be review able.

VIII Travelling expenses: When the Agent is requested by the Principal to undertake journeys of any significant distance and/or duration, all travel expenses including accommodation and other expenses will be for the Principal's account.

IX   Documentary and Administrative Charges: Such charges to be levied as appropriate by the Agent to cargo interests and to remain with the Agent even if related to the trade of the principal.

X   In case of Transhipment Cargo, a transhipment fee of .... per cent/tonne/cbm is charged by the Agent.

....
        PRINCIPAL   **AGENT**

FIGURE 4.15   (Continued) FONASBA agency agreement.

A liner agent's areas of responsibility encompass the duties of a commodity broker and a shipping line's operations coordinator on one hand. On the other hand, their traditional ship agent's duties resemble the tramp trade agent's responsibilities.

In terms of cargo handling and operations, a liner agent's work may entail the following principal stages:

Offering freight quotations, promotion, and marketing services
  – Distribution of the fleet's sailing list. Networking and pursuing commercial negotiations with manufacturers, importers, exporters, cargo forwarders, and so on. Circulating the liner company's newsletter, obtaining feedback, and reporting their findings to their principals.

Handling inward and outward freight
  – Handling inward freight and providing the shippers with information on the carriage, tracking, and discharge of goods. Booking outward freight and providing the shippers with information on the carriage and tracking and loading of goods. Producing and monitoring inward and outward booking lists.

Administering freight dispatch
  – Gathering documentation related to the loading and discharging of freight, such as bills of lading, cargo manifests, delivery orders, shipping permits, and so on. Liaising with the charterers or cargo receivers as to the bills of lading (Bs/L), dispatch and delivery of the cargo. Ensuring the freight integrity, and that data are in line with the Bs/L information. Administering freight collection prior to the ship's departure and before the Bs/Ls have been delivered. Supporting customs' formalities and documentation process. Attending cargo delivery.

Reporting to their principals
  – Reporting market trends and actual cargo booking agreements and current status of freight orders. Freight estimation and dispatch arrangements based on cargo dimensions and weight. Estimated ship's schedule revisions and port stay, based on cargo loading and discharging volumes.

b. *Tramp Trade Agency Duties*

A charter party agreement signed between the shipowners and the charterers stipulates in great detail the terms and conditions of carriage of goods by sea. Since the port is a focal point of this contract, the Agency Clause is established to verify whether the ship's agents at both the loading and discharging port(s) are formally nominated by the charterers (i.e., charterers' agents) or the shipowners (owners' agents).

During C/P negotiations, both sides pursue to nominate the agents, as this will ensure that a trusted, loyal entity has been appointed to look after their own interests. The final decision of who appoints the agents depends on many factors, such as (a) each company's market power to impose their own terms and condition during negotiations, (b) the willingness of both companies to compromise and establish a long-term partnership, but most important, (c) the charter party's type and duration: a charter party's type and consequently its duration will determine the party responsible for the agents' appointment.

In voyage charter parties, the short duration of the agreement entails a single visit at the loading and discharging port(s) of call. Each party can divide the agency appointments, for example, the owners may nominate the agents at the loading port(s), and the charterers may nominate them at the discharging port(s).

In time charters and bareboat charter parties, the charterers become the ship's disponent owners, a fact that suggests that although they do not own the ship, they assume full commercial control over its operations. The time charterers assume full control of the ship's trade routes, which over the course of years may involve literally dozens of seaports and agency nominations. Generally, in a T/C agreement, the shipowners pay the ship's running costs (fixed costs) and the charterers are in charge of the operational costs (variable), which depend on the selection of ports of call, which entirely depend on the charterers. Agency nomination and agency fees are therefore for charterers' account. Under this arrangement, the nominated agents consider the charterers as their principals, yet they must equally support the shipowners and the ship's master and provide them with the services required.

Husbandry or protective agents

Whatever the agreement, the entity that has not nominated the C/P agents can still appoint their own *Protective Agents*, which will serve exclusively the interests of this entity, while still working close with the formal appointed agents. A protective agent is a different agency company to the one stipulated in the C/P agreement. Since the agents nominated by the charterers may be in charge of all the cargo handling-related services, the husbandry services include different or overlapping services, such as the following:

- Crew handling; in case of crew changes, agents in collaboration with the company's operations department can arrange for crew visas, launches and transportation, hotel reservations, and so on.
- Crew medical and various services, that is, visits to doctor, hospitalization, crew transportation, correspondence, and communication arrangements.
- Spare parts, purchase, customs' clearance (inward and outward), dispatch and delivery onboard the ship.
- Third-party liaison with contractors, that is, repair teams, shipyards, drydocks, and so on.
- Bunkering operations (IFO, MDO, lubricants).
- Cargo sampling and bunkers' sampling and dispatch to laboratories.
- Verification of actual port activities and events, which are already handled by the appointed agents, that is, verification of strikes, weather conditions, cargo handling, cargo damage.
- Ship-to-shore communication and contact with local authorities.

c. *Agency for Offshore Support and Logistics*

Port agency services now involve an active partnership with offshore specialists, and their new role is highly supportive and essential for the effective execution of oilfield services. While their daily duties very much resemble the duties of a protective agent, the difference lies in the offshore industry's geographical concentration, limited mobility, and continuous, long-term collaborations with drilling units and contractors that remain in a designated area for months, even years.

Modern offshore and shipping agencies now require increased efficacy, as their collaborations and agency attendance spread out in the critical areas of offshore salvage and emergency response, logistics, procurement, transportation, repairs, and so on.

Their new clientele now expands to the following:
- Drilling contractors
- Seismic survey ships
- Offshore exploration, production units, and support vessels
- Floating production systems
- Transportation, installation, and commissioning contractors
- Construction and engineering contractors
- Subsea construction contractors
- Dredging companies
- Accommodation barges
- Pipe laying and cable-laying companies

4.5.4 General Agency Duties, for Tramp, Liner, and Logistics Services

An agent's duties in terms of operations in the maritime and logistics industries overlap and are hereby analyzed.

A) *Prearrival Preparations*
Typically, the charter party stipulates that the ship's master needs to provide the ship's estimated time of arrival (ETA) notices for 4 days, 3 days, 2 days, and 24 hours prior to the ship's arrival; hence, the port authorities, the charterers, the shipowners, and the agents are duly notified. Once the master has informed the agents of the ship's ETA, the agents commence the preparations for the ship's stay at port. Hence, prearrival is a synonym to preplanning.

At this stage, the agents need to report to their principals the port's overview in terms of the following:

a. Socioeconomic trends such as trade agreements, investment, leasing opportunities, and so on
b. Political events such as strikes
c. The port's strategy, such as expansion or construction plans
d. Discrete, yet accurate information about companies in the same supply chain, potential partners, or potential rivals

Indeed, the agents should be the eyes and the ears of their principals.

The master forwards to the port authorities, through the agents, a copy of the cargo manifest, the bills of lading. If the ship is already loaded, the stowage plan is required, that is, a diagram demonstrating the stowage placement of cargoes. In bulk carriers, it shows cargo loaded in every hold. In general cargo ships, it also reveals cargo loaded on above-deck or between-deck spaces. For container ships, the stowage plan reveals the container boxes' positioning onboard.

Modern tanker ships use hydrostatic calculation software that shows the liquid cargo contained in each tank. Generally, shipowning companies use stowage software to utilize the ship's full loading capacity in the most efficient manner. Copies of the stowage plans, Bs/L, and cargo manifest are dispatched to the agents at the discharge port (D/P), to facilitate preplanning, that is, reserve the berth, shore facilities, storage space, stevedores, and so on, as needed.

Before the ship arrives at the port, the agents will receive from the ship's master copies of the ship's documentation, for further processing with the port authorities, upon the ship's arrival:

For ship's quarantine/free pratique:
- Vessel's deratting certificate
- Crew and passengers' list
- Embarkation and disembarkation list for crew and passengers
- World Health Organization vaccination list
- Shipping declaration of health statement
- Past ports of call list
- Drugs list (ship's medical chest)

For port's customs:
- Advance notice of arrival
- Crew and passengers' list
- Embarkation and disembarkation list for crew and passengers
- Drugs list (ship's medical chest) and firearms declaration
- Bonded stores list
- Crew declaration form

The pilots are notified if and where necessary tugs are reserved for the ship's berthing and the berthing process has been scheduled. Once the berthing time is known, ashore arrangements can be made, including cargo handling, and booked services such as storage, as well as stevedores, can all be scheduled in full synchronicity for the ship's berthing operations. Other operational activities that are secondary to the cargo may include bunkering operations, ship chandlers, spare parts delivery, and so on. Last but not least, other components of the supply chain, such as cargo forwarders and multimodal transportation providers (rail, road), are all in a standby mode for the ship's arrival.

B) *Vessel's Arrival at Port*

The agents declare the ship's arrival to the port authorities. The agents support the master in *Customs Clearance and Free Pratique*. Meanwhile, the master or the agent can tender a NOR to the charterers, notifying them that the ship is in all respects ready to load or discharge. In case during or en route to cargo operations an adversity occurred beyond the master's control (e.g., act of God, a hurricane, adverse currents, etc.), the agent will help the master in submitting a *Letter of Protest*. Although this document does not have a legal standing, in the court of law or arbitration, in case of a claim, it will be treated as admissible evidence.

The agent organizes any surveys, inspections, audits, or repairs that are scheduled to take place onboard the ship, in liaison with the respective authority, that is, Port State Control/Coast Guard, ship's flag/registry, vetting inspections, cargo inspections/preloading inspections, shipyard and repair team, and so on.

C) *Agent's Services During the Ship's Stay at Port*

During a ship's stay at the port, the agents are duly responsible for providing specific services pertaining to port expenses; ship, cargo, and crew expenses; and charges pertaining to inspectors, auditors, shipyards, repairs, and so on, as shown in Table 4.4.

TABLE 4.4   Port D/A—Extended

Principals:	Owners/Charterers		
Ship Agents:			
Vessel's Name:	Previous Names:		
Voyage Number:			
Ship's Particulars:	IMO Number:		CALL SIGN:
	DRAFT:	GT:	NT:
	LOA	TDW:	
Current Port:	ARR: (Date and Time)	B:	SLD:
Previous Port:			
D/A Currency:	Rate of exchange:		

Port Expenses (including canal or channel expenses)
Port expenses:
Lightering expenses:
Pilotage charges:
Towage charges:
Mooring/Unmooring:
Customs:
Vessel shifting:

Port Transportation
Launch:
Car hire:

Cargo Expenses
Stevedoring (Longshoremen) fees:
Cargo handling equipment: cranes, derricks, winches, etc.
Tally expenses:
Overtime expenses:

Vessel's Expenses
Cash to master:
Bank charges:
Stores/provisions:
Fresh water:
Ship chandlers/victualing:
Fresh food provisions (vegetables, fruit):
Spare parts:
Cell phones leasing expenses:

Shipyard expenses:
Drydock expenses:
Ship repairs' expenses:

Transportation
Launch:
Car hire:

(continued)

TABLE 4.4   (Continued) Port D/A—Extended

Crew and Passengers' Expenses

Medical and hospital expenses:
Hotel accommodation:
Subsistence expenses (meals, etc.):
Cell phones leasing expenses:
Crew changes' traveling expenses:

Transportation
Launch:
Car hire:

Inspection/survey/audit expenses
Hotel accommodation:
Subsistence expenses (meals, etc.):
Cell phones leasing expenses:
Crew changes' traveling expenses:

Transportation
Launch:
Car Hire:

Various Expenses
Representation and entertainment expenses:

Communication expenses (telephone, Internet, fax, courier, postage):

Agency fees:
Overtime:

Total Expenses:

Cash advance received: From: Owners/Charterers   Date:

Total Amount Due:

Source: The author's extended version based on BIMCO's Standard Disbursement Account.

D) *Loading, discharging, and cargo handling*

   i. Cargo verification: Prior to the ship's berthing, the agents have verified that the freight has arrived at a port's designated storage area and that its quantity, type, and condition are in accordance with the bills of lading and the cargo manifest.
  ii. HAZMAT/dangerous goods: In case the ship's cargo entails hazardous materials as full cargo or parcel cargo, the agent ensures the proper handling, marking, labeling, and storage conditions, in accordance with its documentation, UN Number and material safety data sheet, safety data sheet, or product safety data sheet.
 iii. Cargo and bunkers' sampling and laboratory analysis: Another area of support is the case where a laboratory analysis is required for the cargo or the ship's bunkers (IFO, MDO). In this case, the agents will liaise with laboratories for the sampling analysis.
  iv. Machinery damage and leasing: In case the ship's cargo handling gear breaks down, the agents in coordination with the shipowning company undertake the leasing of dock equipment.

v. Claims, cargo damage: In case of any claims, for example, cargo damage, machinery breakdown, or other delay that would result into an off-hire, the agents administer off-hire and on-hire formalities, and the respective surveys.

vi. General average: Furthermore, the agents will support the ship and its cargo in case of general average. In this legal maritime principle, in case of an emergency, if freight needs to be jettisoned or any charges arise, the financial damage is shared by those involved with the ship and its freight, in proportion to the value of their contribution exposed to the common hazard.

vii. Logistics: While complying with the ship's charter party agreement pertaining to the cargo transportation, as well as the cargo documents, the agent ensures the cargo follows the preplanned logistics route with the designated carrier.

viii. Cargo manifest: In the event of errors in freight estimations, cargo manifest corrections are made. Finally, once the agent has confirmed the cargo's integrity and quality, plans are made for the subsequent port of call.

E) *Vessel's Sailing*
The agents arrange the vessel's clearance and sailing preparations with the master. Where necessary, pilots are arranged to escort the vessel to the outer port limits. Any outcomes pertaining to ship's surveys, inspections, repairs, claims, cargo damage, and certificate renewals are forwarded to the ship's principals, that is, the shipowners or charterers. Pursuant to the ship's sailing, the agents gather all the original invoices, vouchers, and supporting documents, and produce the D/A.

F) *Ships' Husbandry, Ship's Protective Agency, and Ship Management*
In maritime law, a protective agent provides ships' husbandry services and is thereby referred to as "ship's husband." As their duties and the duration of their contractual agreements with their principals tends to extend in time, modern shipping agents tend to expand their activities by offering ship management services, thus increasing their involvement and control over the ship's performance and commercial, technical, and financial operations. An interesting observation of the modern shipping market is the fact that owing to the ship agents' deep and varied knowledge of shipping, more agencies become involved in relative services, such as crew agencies, ship management, and so on.

G) *Ships' D/As and Incident Investigation*
As a conclusion, the examination of a ship's D/As serves as a most detailed log book that offers ample evidence as to the ship's overall performance and activities, encompassing its seaworthiness, repairs, inspections, crew changes and health issues, cargo handling operations, and so on. Ideally, a company's incident investigation should also encompass the company's accounts.

4.6 PORT-RELATED CLAIMS AND LEGAL LIABILITIES

4.6.1 Conflict Resolution: Arbitration versus Court of Law

In theory, an examination of the charter party clauses may provide the reader with elementary information on the required action and performance. However, shipping practice over the past few decades can prove that an accurate legal interpretation of the charter party clauses is not an easy task. Throughout a charter party's life, both the charterers and the shipowners need to exercise due diligence in the performance of their assigned tasks.

Along the way, disputes may arise. In this case, typically the owners' or charterers' department that has identified a potential claim closely liaises with their company's in-house legal and claims department. These professionals will evaluate the situation, usually with the support of their Protection and Indemnity Club. In the best-case scenario, an amicable resolution is reached, where both parties seek for a solution that seems fair to both and continue their agreed partnership with minimum interruptions. In case an amicable resolution cannot be reached, the case will have to move through the legal path, that is, arbitration, or the court of law.

Arbitration has been the maritime industry's most popular path for claims resolution, as opposed to most other industries that seek a remedy in a court of law. A charter party's arbitration clause will specify which legal system will apply for this agreement. During the twentieth century, the British and US legal systems prevailed within the charter parties. Over the past few years, the emerging economies and the industry's developments have introduced a number of other national judicial systems, that is, China, Canada, the Netherlands, and so on, as alternatives. In addition, the clause will stipulate in which ways the nonbreaching party may seek for remedy, that is, through arbitration or a court hearing. The principal difference between arbitration and a court hearing pertains to maritime expertise, cost, formality, and simplicity. Most important, both parties involved would prefer arbitration rather than a court of law as a milder dispute resolution that will ideally not burn the commercial bridges among their companies. Upon determining the merits of the case by an arbitration tribunal, an arbitration panel of maritime specialists will typically issue the arbitration decision and order the enforcement of an arbitration award. The most preferred arbitration seats are London, Paris, New York, Geneva, Zurich, Singapore, Stockholm, Tokyo, Hong Kong, and China, to name a few.

4.6.2 Port-Related Claims and Charter Party Clause Interpretation

In order to avoid unnecessary claims and legal disputes, contracts of carriage must be interpreted very strictly in order to conform to the required navigational, commercial, technical, operational, financial, and other aspects entailed.

Among the three major C/P types, it is the voyage charter and the time charter parties that require a high level of collaboration and unity in navigational, operational, and financial issues, in order to eliminate port-related clauses and claims.

The bareboat charter party is the type where the charterers have undertaken the operational, financial, and commercial management of the ship. As the ship's control in these critical areas shifts from the shipowners to the demise charterers, disputes arising because of the operational decision making would usually be eliminated. On the other hand, bareboat disputes would arise owing to (i) hire payment disruptions, (ii) the ship's deterioration caused by lack of maintenance or crew inefficiency, and finally (iii) breach of the cargo exemption or area exemption clauses.

Clause stipulation should be worded so as to avoid implications. However, the consistently high volume of claims, disputes, and arbitration instances suggests that even single-word amendments have the power to entirely alter the context of a charter party, with legal consequences arising therefrom. This section demonstrates a number of characteristic port-related cases, in order to examine the pitfalls of contractual interpretation and expand the readers' perception to port claims.

4.6.3 Port or Berth Charter Party

A common issue for shipowners in voyage charter parties pertains to the false perception that by amending one clause, they can convert a berth charter party into a port charter party. Based on this assumption, conflicting laytime clauses are stipulated. Frequently, the owners agree to the nomination of busy ports of call, where the ship's waiting time for berthing may range from a few days to over a month. As an outcome, not only are the owners burdened with the ship's fixed and running costs, but charterers do not accept the owners' demurrage claims. The general principle here is that even if a port is nominated, the presence of other clauses where a "safe berth" or "good and safe berth" is referred strongly implies that this is considered as a berth charter party in the court of law.

In the case of Novo Logistics SARL v Five Ocean Corporation (The Merida) (EWHC 2012b), M/V Merida ("the vessel") was "fixed" under a voyage C/P to load steel plates from Xingang, China, to Cadiz and Bilbao, Spain. When the ship arrived at her loading port, she had to wait for 20 days in order to proceed to her berth. Shipowners would seek for a remedy through a demurrage exceeding half a million US dollars, under the supposition that this was a port charter party. When the written agreement among the two parties was examined, it was observed that the recap did not make any reference to a particular charter party form. It stipulated that the loading and discharge should take place between: "One good and safe charterers' berth terminal 4 stevedores Xingang to one good and safe berth Cadiz and one good and safe berth Bilbao." Clause 2 of the charter party pertaining to laytime contained both a designated berth and a port, that is, "one good and safe port/one good and safe berth," and concluded that all time including shifting from anchorage toward the berth will count as laytime. When reading the charter party as a whole, the court determined that this was a berth charter party.

4.6.4 Nominating a "Safe Port"

Numerous claims and legal cases have determined that a port can be a perfectly safe haven for most ships and yet may be unsafe for certain ships. Therefore, charterers should be cautious when nominating a port as "safe" and consider the particularities of each port, that is, its weather conditions, draft restrictions, berths, and so on, in combination with a ship's type, design, particulars, and dimensions, among others.

Partederiet "Primo" versus "Crispin Co Ltd."
M/T Primo is a 60,000 MT double-hull crude oil tanker ship that was fixed under Asba 11 C/P form to load at 1/2 safe ports in Argentina, with an option to discharge at numerous global ports. The charterers instructed the master to load a portion of the cargo at La Plata anchorage and the remaining at Loading Zone Charlie.

The Rio de la Plata is a shallow, wide estuary formed by the Parana and Uruguay rivers. It is broken down into an inner tidal river and an exterior estuarine area. Because of its landlocked shape, every time there are powerful winds from the southeast, the water level rises. In these conditions, low-lying areas along the right bank are affected by floods and heavy rainfall, regionally referred to as "sudestadas." Damages in some cases have exceeded $250 million.

Because of the ship's size, that is, beam and draft restrictions, the shipowner was reluctant to allow the ship to enter La Plata, anticipating maneuvering and navigational challenges. The ship's draft limitations and the expected underwater impediment were a serious area of concern, as the ship's variable pitch propeller type was not as robust and sturdy as a fixed propeller.

As the ship arrived at La Plata, the master anticipated the charterers to provide solutions pertaining to the ship's navigation through the channel by the assistance of tugs. When the charterers announced they would cancel the charter, the ship's master agreed to enter by a letter of protest and is accompanied by three tugs. Upon completion of loading operations, a further 26-hour delay was incurred, as the master had to wait for the high tides in order to proceed for lightering. The shipowners' demurrage claims exceeded $200,000 for time lost, whereas the charterers'counter-claims exceeded $2 million.

To solve this case, the arbitrators had to determine if La Plata was a safe port for M/T Primo. Initially, the master's reservations in entering La Plata and navigating with a note of protest were considered reasonable. Subsequently, it was verified that M/T Primo was a larger vessel compared to the typical ship sizes that navigate in La Plata. It was decided that the charterers had to reimburse the shipowners with the claimed amount (Maritime Advocate 2013; Society of Maritime Arbitrators 2013).

The lesson learned from this case study is that safety at a specific port or berth should be examined for each ship separately, depending on its size, design, technology, and so on. An area may be perfectly safe for certain ships but not for others.

In the case of the M/V Eastern Eagle, a port may be considered unsafe in case the master has carefully calculated the sea route, yet its ranging water draft is not accurately mapped and the port's navigational aids, including buoys, pilotage, and so on, are insufficient. In this case, there is a breach of contractual condition on behalf of the charterers, related to their safe port obligation.

Based on the case of the M/V Adamastos SMA 3416 (SMANY 1988), the claims and legal process following a ship's accident while at port and owners' claims that the port is unsafe should focus on the following key points:

1. Based on the existing risk of hazard, it must be established whether the accident could have been preventable, pursuant to optimum shipboard navigation and safety practices.
2. The communication channels are examined, in particular between (a) the charterers and the owners, (b) the port authorities and the charterers, and (c) the port authorities and the owners. In case of an accident, it is elementary for all parties concerned to communicate effectively without depriving any party of any development or physical access to the scene of the incident.
3. Objective evidence examined includes log books and official documentation signed and exchanged between these three entities. The authorities duly evaluate the consistency, quality, and reliability of documented information. In case of discrepancies or inconsistencies, the court of law or the arbitration will consider this information as unreliable, which will be to the party's disfavor.

4.6.5 When a "Safe Port" Becomes "Temporarily Unsafe"

Under certain conditions, a perfectly safe port may temporarily become unsafe, owing to unanticipated hazards, for example, as a result of an act of God, which may subsequently lead to equipment failure and impose hazards to the ships and their cargoes. In the best-case scenario, an unsafe port may cause delays in the cargo operations and prolong the ship's stay at port. In worst-case scenarios, crew and port employee accidents, as well as damage to ships, their cargoes, and to the port assets, may occur. In these cases, the instructions given to the master throughout the ship's navigation and entry toward the port will be considered. In essence, the charterers need to prove that the port's unsafe characteristics did not preexist at the time they nominated the port.

In the *M/V Hermine* case (*Unitramp vs. Garnac Grain Co Inc.*) (Lloyd's Rep. 1979) and the *M/V Count* case *(Independent Petroleum Group Limited vs. Seacarriers "Count" Pte Limited [2006], EWHC 3222 Comm)*, the ports were unsafe for a long period, which commenced before the charterers' port nomination.

In the case of *M/V Count*, the ship was delayed in reaching her loading port, as the channel was obstructed because of an inbound ship, the *M/V British Enterprise*, which grounded, and a few days after the port's emergency response, the ship grounded again. The *M/V Count* moved toward its berth of discharge a few days later, yet its outbound sailing was obstructed because of *M/V Pongola*'s grounding in a position close to the previous vessel's grounding. The *M/V Count* was again delayed, and the shipowner filed for a claim on the grounds of a breach of contract on behalf of the charterer, pertaining to the harbor's safety. The judge decided that the port's unsafe characteristics preexisted at the time of nomination. It is these characteristics that imposed a sustainable risk to the ships. In this case, both the arbitration and the court of appeals rejected the charterers' claims on the port's temporary lack of safety owing to bad weather (EWHC 2006).

4.6.6 When Communication Becomes a Prerequisite to Port Safety

The *Dagmar* case (*Tage vs. Montoro SS*) (Lloyd's Rep. 1968) verifies that if a port produces and makes available weather reports and relevant statements are circulated in a consistent manner, it is necessary for the ships' side to obtain them. A Baltime charter party contract stipulated that *M/V Dagmar* would be trading among safe ports. The ship navigated to her loading port, Cape Chat, Quebec, and two days after her loading operations, extreme weather conditions, that is, strong winds and high seas, caused the ship's grounding. The court discovered that the port's weather forecasting process was satisfactory, and yet the port became unsafe for the shipowners and the ship, as they did not receive the weather broadcast from their agents. This case study verifies that a port's weather conditions can be mitigated if ships receive correct weather forecasting through the appropriate communication channels. It was the charterers' responsibility to provide weather forecasts, and failure to do so rendered the port unsafe for the ship, at charterers' fault.

In the case of *Slebent Shipping vs. Associated Transport* (SMANY 2003), the Star B by SMANY, *M/V Star B*, is a general cargo ship "fixed," under an NYPE time charter party of 1999, to load her cargo from Brazil to her three discharging ports, namely, San Juan (Puerto Rico), Rio Haina (Santo Domingo), and Kingston (Jamaica). Because of congestion at the ship's second discharging port, the charterers nominated an alternative port to Rio Haina, that is, the proximate port of Boca Chica. As the ship navigated into the channel to Boca Chica in the escort of pilots, the ship grounded and suffered extensive damage as a result. The shipowners initiated an arbitration hearing process against the charterers, on the grounds that their instructions for the ship to enter the port of Boca Chica violated the "safe port warranty" stipulation of their charter party. The owners' objective evidence provided support that the port was unsafe even during pilotage, owing to "deficiencies in the entrance buoys, the range markers, the charts, the navigation guides." The arbitrators' tribunal focused on the issue of liability and on postponing the issue of damages, if relevant, to a later hearing. The arbitrators established that the liability for the grounding should be distributed among the parties, with no reference as to the parties' percentage of liability.

The lessons learned pertain to the two contractual parties: the first major issue in the case lies with the owners and the ship's master, and his right to either refuse to enter an unsafe port or enter with a note of protest, holding the charterers liable for any and all damages incurred. The second major issue pertains to the charterers and their capacity to prove that the accident was not due to the port's lack of safety but due to the master's poor seamanship.

4.6.7 Stevedore Damage and Bills of Lading Stipulations

Rabaul Stevedores Ltd vs. Seeto [1984] PGNC43; [1984] PNGLR248; N483 (October 5, 1984) is a characteristic case pertaining to the B/L exemption clause, referring to the protection of stevedores while performing services of contract, and their exclusion from any liability referring to damage or pilferage (PACLII 1984).

During loading and unloading operations in different parts of the world, incidents of stevedore damage or pilferage are not uncommon. Correspondingly, other parties such as subcontractors, agents, ship chandlers and other "servants" may be involved in a claim case. Regardless of whether they are union or nonunion members, the contract of affreightment, that is, the charter party or a bill of lading, may stipulate their level of liability or totally exempt them from any liability whatsoever.

In most voyage charter parties, the stevedores are the charterers' servants, yet they are supervised by the ship's master and officers. The stevedore damage clauses are typically formulated in two options.

In the first option, in case of stevedores' damage, the charterers assist the owners in settling the damage claim or dispute. A shipowner initiating a claim or dispute against stevedores in an overseas legal system may not be an appealing option. Hence, in case that the stevedores cause cargo damage or fail to comply with the instructions of the master, the latter has the right to hand over to the charterers a notice of protest, seeking for remedy for damages suffered by their servants, that is, the stevedores. This complex communication process of uncertain outcome requires a thorough drafting of the stevedore damage clause both for voyage and for time charter parties.

In the second option, any such damage is for the charterers' account; this clause is most desirable for shipowners who shift the responsibility to the charterers. This shift of liability to the charterers eventually forces them to closely monitor and control the stevedores' performance. In the world's most established ports, stevedores will cover any and all expenses for damages, even after the ship's departure from the specific port of call. However, in certain developing or less developed markets, stevedores are unlikely to cover the damages after the ship's departure, the reason for this being a generous cash refund on behalf of their insurance company.

In time charter parties, the charterers are legally liable to supervise the stevedores, and therefore their performance is more efficient.

In order to avoid stevedore damage claims and complications, both parties, that is, the owners and the charterers, should ensure that the C/P clause stipulation is in line with the bill of lading, with a duly reference to the C/P.

Back-to-back liability should be ensured when a chain of C/Ps and agreements exists between numerous key players, that is, owners, charterers, the port, subcontractors, and so on. In these complex legal contracts, a claim case may need to be shifted from one legal entity to another, and it is the C/P clauses stipulating the liability, time delays, and financial issues that will determine the case's outcome.

Inconsistent contracts and "sub silentio" clauses, that is, implied but not expressly stated clauses, may shift the liability toward the weakest link across the chain. On the other hand, a back-to-back contract is highly desirable as the afflicted party will recover its damages through the linear formation of the contracts.

4.6.8 Safe Port, Security, and Loss of Time

A safe port is not only related to the ship's type, size, and draft, but in addition has to provide safe navigation practices and provide good seamanship on behalf of the ship's master. When a port's tides, currents, and weather conditions can be foreseen and harnessed, a port can still be safe. On the other hand, a port's safety is not restricted to its weather conditions. A port may be unsafe because of disruptions caused by political and social instability, warfare, invasions, and similar conditions. In these cases, ports may be captured or isolated in the vicinity of the port or within port limits, and may be unable to conclude its commercial purpose, that is, loading or discharging, for reasons beyond its control.

Another aspect of port safety pertains to its permanent or temporary nature. Ports may impose a permanent or temporary danger, for example, traffic, a hurricane, depth changes, swells, tides, currents, and so on. In this case, arbitration and the court of law will investigate whether the port's condition preexisted the charter party date or the charterers' port nomination. Also, in case of a port's temporary hazard, the elements to be examined will focus on the master's, owners', and charterers' ability to assess the level and potential duration of the hazard; that is, it may be easy to examine the tide table in a tidal port, and special tidal prediction programs can monitor the low tides and high tides. At the same time, it may be practically impossible to forecast the duration of announced extreme weather phenomena. Finally, any navigational disruptions on behalf of the port, such as equipment failure or misalignment of buoys, will be evaluated either as temporarily unsafe, outside the port's liability and fault (i.e., weather), or as a permanent lack of safety, with inadequate safety measures and supervision on behalf of the port.

In certain claims cases, while the vessels are en route to their destination, they get hijacked and held by pirates to ransom for a number of weeks or even months. In most of these cases, the charterers count this loss of time as off-hire, while the shipowners' defense is based on the argument that piracy is a ship's involuntary seizure, unrelated to a ship's deficiency or fault whatsoever, yet directly related to the charterers' nomination of port(s). For example, ships transiting the Suez Canal have the risk of encountering Somalian pirates.

Arbitrators and courts of law try to resolve such cases by reading and evaluating the entire charter party as a unitary exercise, that is, as a whole, as opposed to isolating the piracy clauses alone. Particular attention is paid to all references pertinent to risk allocation and issues of performance, liabilities, off-hire, and time lost.

A critical evaluation of the recent claims cases related to piracy suggests that shipowners lose the case mostly because of the piracy and off-hire-related wording. The vast majority of clauses are ambiguous and contradicting, while at the same time there is no stipulation explicitly referring to piracy. Most clauses referring to the ships' arrest, involuntary detention, or seizure do not mention the risk of piracy; hence, the cases are lost (M/V Stefanos, EWHC 2012a).

4.6.9 Ship's Off-Hire, Time Lost, and Piracy Clauses

While a ship is under a time charter, the off-hire clause stipulates events beyond the charterers' control, for example, time lost because of ship's deficiency, breakdown, unnecessary deviation from the course of the voyage, and so on. The intent behind an off-hire clause is to alleviate the charterer's hire payment obligation, while the ship's seaworthiness, functional, operational, or structural integrity may be compromised.

In claims cases such as *"The Athena"* (EWHC 2012c), the charterers had to prove a net loss of time directly associated with the ship. The vessel was fixed on an NYPE 45 time charter party, and in January 2010, the ship was ordered to proceed to the anchorage of Benghazi port in Libya, and anticipate charterers' instructions. In order to resolve certain problems with the Bills of Lading, the ship commenced a "drifting period" as it navigated to a point 50 nautical miles off Libya, while the master sorted out the B/L problems. The charterers filed for an "off-hire claim" for loss of time. The court examined the case and verified that the ship's failure to appear at port was a "default of the Master," despite the fact that the same time would be idle if the ship had waited at the port. Pursuant to the arbitration award, the owners appealed to a commercial court. Focused on Clause 15, both parties had to prove or disprove whether time was "thereby lost." Charterers were unable to prove that the ship would be able to commence her cargo operations even if it had arrived; hence, the owners' appeal was granted (BAILII 2013).

While the basis of the off-hire clause and its interpretation are reasonable, an increasing number of off-hire claims are related to the ship's seizure by pirates. The question that arises is whether ships' piracy and involuntary stoppage of operations should be considered as the owners' fault and whether charterers must exercise their right to cease hire payment under the off-hire clause, while the ship is in danger.

The case of *COSCO Bulk Carrier Co. Ltd. vs. Team-Up Owning Co. Ltd. (The "Saldanha")* (EWHC 2010) is a typical piracy case. *M/V Saldanha* is a Panamax bulk carrier ship, and in 2008, she was fixed under an NYPE time charter party form. In February 2009, as the ship was performing her designated route from Indonesia to Slovenia, she was seized by Somali pirates while navigating through the Gulf of Aden. The ship had to drift off Somalia and remain there for two months prior to the pirates' allowing her release. The master navigated the vessel back to her original preseizure position, ready to assume her voyage. The charterers announced that the ship was off-hire for the 69 days that the ship was captured; hence, the owners would not be reimbursed for these days.

The high court observed that the pertinent clauses were a "patchwork" of amendments, which explained why they could not be easily interpreted by the parties concerned. Despite the fact that the ship's seizure was a recognized peril, the contractual agreement among the charterers and the shipowners did not provide for piracy. In fact, the C/P included a modified clause pertaining to the ship's potential "Seizure, Arrest, Requisition and/or Detention," all of which mainly pertained to third-party claims against the ship; however, the wording did not stipulate for a potential seizure by pirates. Hence, the shipowners' appeal was dismissed.

4.7 MULTIMODAL TRANSPORTATION

Multimodal transportation and the seaport–hinterland interconnectivity have reshaped global trade and transport, while facilitating the safe and time-efficient distribution of goods. Since 90% of global trade is being carried by sea, traditional seaports are bound by geographical and infrastructure barriers, whereas hub ports are the interactive nodal points that combine the benefits of a seaport and coordinate with alternative transportation modes by sea, land, and air to facilitate the vast volumes of sea trade.

Intermodal and multimodal transport provides connectivity in their door-to-door services by combining more than one transportation mode. They utilize efficient logistics networks that allocate time and resources in an optimum manner. But first let us distinguish intermodal from multimodal transportation.

Intermodal transportation is usually associated with containerization or with the understanding of one transportation mode hauling the products of some other mode, for example, container boxes carrying vehicles. To a great extent, containerization has enabled the time processes and cargo handling, yet it is only one of the many components of intermodalism. Intermodalism pertains to the direct use of more than one transportation mode (air, waterways, railroads, highways, pipelines), in order to facilitate the movement of commodities. It also encompasses nodes in the supply chain that are connected through intermediate warehousing (Middendorf 1998). Intermodalism encompasses "a holistic view of transportation in which individual modes work together or within their own niches to provide the user with the best choices of service, and in which the consequences on all modes of policies for a single mode are considered" (Feldman and Gross 1996).

Multimodal or combined transportation pertains to the use of a single contract of affreightment within more than one transport mode, including transshipments and the use of complex haulage networks. The carrier, hereby named multimodal transport operator, is legally liable for the entire carriage of goods, which is typically performed by subcarriers. The United Nations Convention on International Multimodal Transport defines international multimodal transport as "the carriage of goods by at least two different modes of transport on the basis of a multimodal transport contract from a place in one country at which the goods are taken in charge by the multimodal transport operator to a place designated for delivery situated in a different country. The operations of pick-up and delivery of goods carried out in the performance of a unimodal transport contract, as defined in such contract, shall not be considered as international multimodal transport" (United Nations 1980).

Containerization and other technological developments have generated revolutionary concepts about logistics and complex cargo networks, through intermodal and multimodal transport. Prior to containerization and multimodal transport, a ship would require more time in port and less time at sea, as cargo loading and discharging operations were laborious and time-consuming. The so-called break-bulk shipping was handled by stevedores (longshoremen) who entered the ship's holds and enclosed spaces to manually handle cargoes in bulk, barrels, pallets, or sacs. In addition to soaring occupational accident rates, the commercial hazards of manually handling the vast majority of global cargo maximized the accidents' rate because of the human factor, particularly that related to cargo loss, cargo deterioration, and pilferage. Before containerization emerged, commodities were forwarded by land to the loading port and were discharged at the port of their final destination. Figure 4.16 shows the multimodal and intermodal transportation networks.

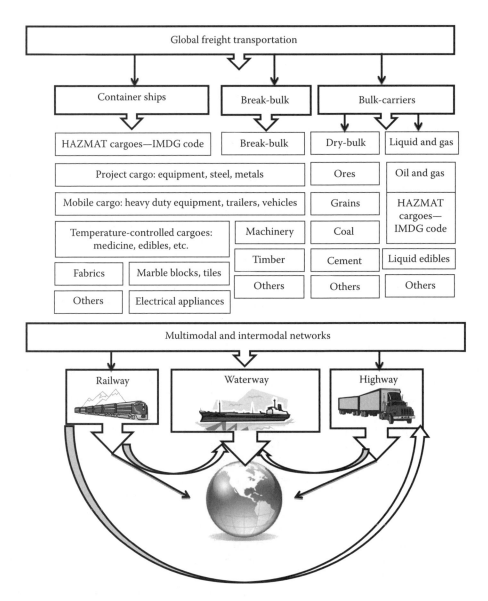

FIGURE 4.16   Multimodal and intermodal transportation networks. (Courtesy of M.G. Burns.)

As stipulated by the INCOTERMS (i.e., the commercial terms published by the International Chamber of Commerce [ICC]), shipowners were responsible for the cargo from port to port, whereas the vessel's rail was the critical point where responsibility for the cargo was shifted from the cargo forwarders to the shipowners (ICC INCOTERMS 2010).

Over the past years, multimodal transportation has rapidly grown and has facilitated globalization and trade growth. Yet, there is a continuous need to improve global logistics services. The relative easiness and agility in establishing substitute gateways is a major area of concern that forces modern ports and multimodal key players to strive for

CASE STUDY: PORT FREEPORT, TEXAS

Port Freeport (Figure 4.17) has been operating for over 100 years, since the first jetty system was designed in Freeport, Texas. Since then, the port has grown to be one of the most rapidly expanding seaports on the Gulf Coast. Port Freeport is a major US seaport, ranking sixteenth in the United States in foreign tonnage, and one of the fastest developing seaports on the US Gulf. It is an autonomous governmental entity sanctioned by an act of the Texas Legislature in 1925. Because of its vast land capacity (7723 acres of undeveloped land), it has tremendous potential for development. It currently has 14 operating berths, both public and private, and a climate-controlled facility (Figure 4.18). With a 45-foot-deep Freeport Harbor Channel and a 70-foot-deep berthing area, the port can accommodate the deeper draft mega-ships.

Its rapid growth in terms of trade flow has necessitated its future expansion, which entails the development of a 1300-acre multimodal facility, two multipurpose 1200-foot berths on 50 feet of water, and two dockside 120,000-square-foot transit sheds. The port provides immediate access to the Gulf Intracoastal Waterway, Union Pacific Railroad, State Highway 36, State Highway 288, and a state-of-the-art Brazos River Diversion Channel. Its efficient multimodal network employs rail, highway, ship, or barge transportation, to feed the heartland of America.

Since 1940, the port has been the landlord of the Dow Chemical Company's largest facility and is one of the greatest built-in chemical manufacturing sites worldwide. The port's distinguished clients also include ConocoPhillips Terminals, Kirby Inland Marine, and wind energy majors such as BP Wind, Suzlon Wind Energy Corp., Clipper Windpower, and Vestas Wind Systems. The port's long-standing partnerships provide a framework of security and mutual prosperity.

Its foreign trade zone (FTZ) allows firms to defer or even eliminate customs duties on import goods. The FTZ Program helps zone users enhance their competitiveness while supporting local business development.

Their three-mile vicinity to deep water, with land availability and purposefully designed transportation infrastructure, enables them to offer their clients easy access, plenty of space, and tailor-made services. The port is surrounded by a Category 4 Hurricane Protection Levee.

FIGURE 4.17 Port Freeport, Texas.

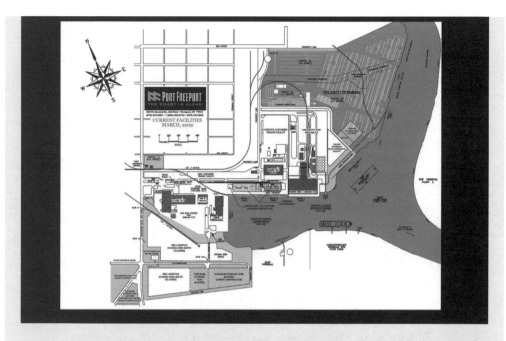

FIGURE 4.18   Port Freeport facility.

Their five-year plan (2011–2016) pursues a capital-intensive expansion through deepwater dredging and the expansion of the Velasco Terminal. An overall expansion of their heavy-lift corridors for point-to-point transportation, Ro-Ro freight, and container barge services is anticipated. The port's culture is focused on satisfying their clients' needs while keeping abreast of the global and regional market changes (Port of Freeport, Texas 2013).

low-cost, time-efficient cargo flows (UNCTAD 2003). The prevailing factors that affect the expenditures and quality of multimodal systems include a nation's geographical situation, its factors of production, and infrastructure. These factors are based upon the rules of supply and demand, and they have the power to determine a nation's trade balance and growth.

Seaports and multimodal key players focus on demand management tactics that increase efficiency. To achieve this, a balance in strategies is needed between overinvestment and optimization of resources. Among the methods that modern decision makers use to control the trade flow, the most prevailing ones are creative traffic control and ships' cargo handling through IT, optimum service time distribution, ashore cargo handling equipment and infrastructure, land optimization, and changes in the terminals' design. A strategy that would focus on purchasing ashore cargo handling equipment, for example, cranes, would require rather large amounts of investment. Alternatively, another strategy would seek to restructure the warehousing and storage facilities and increase the storage density. Yet, any superstructure or infrastructure expansion will be limited by land availability. Ports and logistics companies also seek to install innovative solutions and methods, such as gate and truck scheduling systems, and congestion pricing fees based on peak period pricing and time-of-day pricing. All these paths eventually lead to, or require, the reengineering of their entire logistics processes.

Whichever strategy is finally pursued should always be in line with the frequent trade route realignments and a port's need to reposition itself. As an example, the Panama Canal expansion has generated a rearrangement of trade flows between the Atlantic and Pacific oceans, thus bringing great benefits to the proactive ports and logistic chains of the US Gulf. It is exactly this type of trade flow changes that provides a seaport the golden opportunity to grow in a fast-paced environment.

REFERENCES

ASBA. 2013. Asbatankvoy, issued by ASBA. Available at http://www.asba.org. Accessed July 11, 2013.

BIMCO Charterparty. 1987. TankerVoy 87 issued by BIMCO. Available at http://www.bimco.org. Accessed July 12, 2013.

BIMCO Charterparty. 2013. Gencon, issued by BIMCO. Available at http://www.bimco.org. Accessed July 1, 2013.

Bloomberg Businessweek. 2012. COSCO Says Vale Shuns Its Vessels on China Mega-Ship Ban., 9 May 2012. Available at http://www.businessweek.com/news/2012-05-08/cosco-says-vale-shuns-its-vessels-on-china-mega-ship-ban. Accessed on July 15, 2013.

Bloomberg News. 2013a. China's State Council Urges Credit Support for Shipbuilders. Available at http://www.bloomberg.com/news/2013-08-04/china-issues-plan-to-overhaul-troubled-shipbuilding-industry.html. Accessed August 16, 2013.

Bloomberg News. 2013b. US shipbuilding is highest in almost 20 years on shale energy. Available at http://www.bloomberg.com/news/2013-09-18/u-s-shipbuilding-is-highest-in-almost-20-years-on-shale-energy.html. Accessed August 16, 2013.

British and Irish Legal Information Institute (BAILI). 2013. Available at http://www.bailii.org/. Accessed August 21, 2013.

British Petroleum Shipping Ltd. 2013. BPVoy4, issued by British Petroleum Shipping Ltd. Available at http://www.bp.com. Accessed July 12, 2013.

Burns, M. 2013. Estimating the impact of maritime security: Financial tradeoffs between security and efficiency. *Journal of Transportation Security*, Springer. doi: 10.1007/s12198-013-0119-x). Accessed July 1, 2013.

Burns-Kokkinaki, ABS. 2009. ABS Surveyor Magazine. A fleet asleep: As lay-ups increase, industry gauges its options. American Bureau of Shipping. Available at https://www.eagle.org/eagleExternalPortalWEB/ShowProperty/BEA%20Repository/News%20&%20Events/Publications/Quarterly/Surveyor/2009/Surveyor-Summer2009. Accessed July 1, 2013.

Cargill. 2013. Available at http://www.cargill.com. Accessed July 12, 2013.

CESA. 2011. Community of European Shipyards' Associations. ANNUAL REPORT 2010-2011 Available at http://www.cesa.eu/presentation/publication/CESA_AR_2010_2011/pdf/CESA%20AR%202010-2011.pdf. Accessed July 6, 2013.

CESA. 2013. Community of European Shipyards Associations. Available at http://www.cesa.eu. Accessed July 5, 2013.

Chevron. 2013. Chevron Pascagoula Refinery. Available at http://pascagoula.chevron.com/home/abouttherefinery/whatwedo/crudetrans.aspx. Accessed July 10, 2013.

CIA World Factbook. 2013. Available at http://www.cia.gov. Accessed July 10, 2013.

China Shipowners' Association. 2011. Vale's Construction of Transshipment Hub and Distribution Centre in Philippines and Malaysia to Transport Iron Ore Imported by China. China Shipowners' Association, 6 December 2011. Available at http://eng. csoa.cn/Reports/201112/t20111206_1162695.html. Accessed May 13, 2013.

DP World Southampton. 2013. DPW Media. Available at http://www.dpworldsouthampton. com/media/2013. Accessed May 16, 2013.

EWHC. 2006. Independent Petroleum Group Limited v Seacarriers "Count" Pte Limited (The "Count") English Commercial Court: Toulson J: [2006] EWHC 3222 (Comm). Available at http://www.bailii.org. Accessed July 10, 2013.

EWHC. 2010. COSCO Bulk Carrier Co. Ltd. v Team-Up Owning Co. Ltd. (M/V "Saldanha") [2010] EWHC 1340. Available at http://www.bailii.org. Accessed July 10, 2013.

EWHC. 2012a. Osmium v Cargill (M/V "Stefanos") Neutral Citation Number: [2012] EWHC 571. (Comm). Case No: 2011 FOLIO 1008. Available at http://www.bailii. org/ew/cases/EWHC/Comm/2012/571.html. Accessed July 1, 2013.

EWHC. 2012b. SARL v Five Ocean Corporation (M/V "The Merida") [2009] EWC 3046 (Common Law). Available at http://www.bailii.org/ew/cases/EWHC/Comm/2012/571. html. Accessed July 1, 2013.

EWHC. 2012c. Minerva Navigation Inc. v Oceana Shipping AG (M/V "The Athena") [2012] EWHC 3608 (Common Law). Available at http://www.bailii.org/ew/cases/ EWHC/Comm/2012/571.html. Accessed July 1, 2013.

Federal Maritime Commission. 2013a. FMC, 46United States Code (46 USC). Available at http://www.fmc.gov. Accessed July 10, 2013.

Federal Maritime Commission. 2013b. 46 CFR Parts 515, 520, and 532, Docket No. 11-22, RIN: 3072-AC51, Non-Vessel-Operating Common Carrier Negotiated Rate Arrangements; Tariff Publication Exemption. Available at http://www.fmc.gov. Accessed July 10, 2013.

Feldman, R. and Gross, M. 1996. *Transportation Expressions*. Washington, DC: Bureau of Transportation Statistics, U.S. Department of Transportation. Available at http:// www.dot.gov. Accessed July 3, 2013.

FONASBA. 1993. NYPE 93 (New York Produce Exchange). Available at http://www. fonasba.com. Accessed July 5, 2013.

FONASBA. 2013. Available at http://www.fonasba.com. Accessed July 3, 2013.

FONASBA Annual Report. 2010. Available at http://www.fonasba.com/wp-content/ uploads/2012/02/ANNUAL-REPORT-2010-FINAL.pdf. Accessed July 2, 2013.

FONASBA Governance Handbook. 2012. Available at http://www.agentimar-fvg.it/attach ments/article/145/Allegato%20Circolare%20001-2013%20FONASBA. Accessed July 5, 2013.

Forbes. 2013. Available at http://www.forbes.com. Accessed July 12, 2013.

French International Ship Register (RIF, Registre International Francais). 2012. Available at http://www.rif.mer.developpement-durable.gouv.fr/en/about-the-tonnage-r40.html. Accessed July 12, 2013.

Highstar Capital. 2013. Available at http://www.highstarcapital.com/home.html. Accessed July 1, 2013.

ICC INCOTERMS. 2010. Rules published by the International Chamber of Commerce (ICC). Available at http://www.iccwbo.org. Accessed July 5, 2013.

IMO. 2012. International Shipping Facts and Figures—Information Resources on Trade, Safety, Security, Available at http://www.imo.org/KnowledgeCentre/Ships AndShippingFactsAndFigures/TheRoleandImportanceofInternationalShipping/

Documents/International%20Shipping%20-%20Facts%20and%20Figures.pdf. Accessed July 5, 2013.

KOSHIPA. 2013. Korean Shipbuilding Association. Available at http://www.koshipa.org. Accessed July 5, 2013.

Lloyd's Rep. 1968. Tage v Montoro SS (M/V Dagmar) [1968] 2 Lloyds Rep 563).

Lloyd's Rep. 1979. Unitramp v Garnac Grain Co Inc. (M/V "Hermine") [1979] 1 Lloyd's Rep 212 (Court of Appeal). Accessed July 5, 2013.

Maine DOT. 2007. Container terminal parameters. A White Paper Prepared by: The Cornell Group, Inc. Available at http://www.maine.gov/doc/initiatives/SearsIsland/ContainerTerminal.pdf. Accessed July 10, 2013.

MARAD. 2013. The Economic Importance of the U.S. Shipbuilding and Repairing Industry Maritime Administration (MARAD), May 30, 2013. Available at http://www.marad.dot.gov/documents/MARAD_Econ_Study_Final_Report_2013.pdf. Accessed July 11, 2013.

Maritime Advocate. 2013. The Chartered Institute of Arbitrators. Keeping ADR in the limelight. Available at http://www.maritimeadvocate.com/arbitration/keeping_adr_in_the_limelight.htm. Accessed August 21, 2013.

Middendorf, D. 1998. Intermodal terminals database: Concepts, design, implementation, and maintenance. Center for Transportation Analysis (CTA) in the Oak Ridge National Laboratory (ORNL). Prepared for Bureau of Transportation Statistics, U.S. Department of Transportation. Available at http://cta.ornl.gov/transnet/terminal_doc/#fig2-1. Accessed July 1, 2013.

OECD. 2008. OECD Council Working Party on Shipbuilding (WP6). The shipbuilding industry in China, June 2013. Available at http://www.oecd.org. Accessed July 1, 2013.

OECD. 2011. OECD Council Working Party on Shipbuilding (WP6). The shipbuilding industry in Turkey, September 2013. Available at http://www.oecd.org. Accessed July 1, 2013.

OECD Korea. 2012. Market distorting factors from Korea Shipbuilders' perspective. Organisation for Economic Co-operation and Development. Available at http://www.oecd.org/sti/ind/PART%20B%20-%20KOSHIPA%20-%2021%20June%202012.pdf. Accessed August 16, 2013.

OECD. 2013. OECD Council Working Party on Shipbuilding (WP6). The shipbuilding industry in Japan, April 2013. Available at http://www.oecd.org. Accessed July 1, 2013.

PACLII. 1984. Rabaul Stevedores Ltd v Seeto [1984] PGNC 43; [1984] PNGLR 248; N483, October 5, 1984. Available at http://www.paclii.org/maritime-law/case-summaries-ports-and-harbours/index.html and http://www.oecd.org. Accessed July 8, 2013.

PND Engineers. 2013. Cruise Ship Facilities: Planning and Design. Available at http://www.pndengineers.com/Modules/ShowDocument.aspx?documentid=417. Accessed August 21, 2014.

Port of Freeport, Texas. 2013. Available at http://www.portfreeport.com. Accessed July 15, 2013.

Port of Haifa, Israel. 2013. Available at http://www.haifaport.co.il. Accessed July 15, 2013.

Port of Honolulu. 2012. Available at http://www.kapalamaeis.com/wp-content/uploads/2012/12/App-B-Container-Terminal-Analysis.pdf. Accessed July 1, 2013.

Port Strategy. 2009. Finance and tanker operators less boxed in by downturn. Available at http://www.portstrategy.com/news101/administration/finance-and-investment/bulk_and_tanker_operators_less_boxed_in_by_downturn. Accessed July 18, 2013.

Ports America. 2013. Available at http://www.portsamerica.com/operations.html and http://www.oecd.org. Accessed July 5, 2013.

Reuters. 2011. Exclusive: Vale in talks to sell giant ships to China. 5 September 2011. Available at http://www.reuters.com/article/2011/09/05/us-vale-shipping-idUSTRE78434 U20110905. Accessed June 8, 2013.

SAJ. 2013. The Shipbuilders' Association of Japan. Shipbuilding Statistics, March 2013. Available at http://www.sajn.or.jp. Accessed July 1, 2013.

SMANY. 1988. M/V Adamastos SMA 3416 (Arb. at N.Y. 1998), Society of Maritime Arbitrators (SMA 3416). Available at http://smany.org. Accessed July 10, 2013.

SMANY. 2003. Slebent Shipping v. Associated Transport (M/V "Star B") [2003]. Award of the Society of Maritime Arbitrators of New York. Available at http://www.smany.org. Accessed July 10, 2013.

SMANY. 2013. Partederiet "Primo" vs. "Crispin Co Ltd" (M/T "Primo"). Society of Maritime Arbitrators (SMA) Award Service (Ref 3335). Available at http://smany.org. Accessed July 10, 2013.

Solvere Market Intelligence. 2013. Available at http://www.solveremarketintel.com/uploads/Sample_Ports_America.pdf. Accessed July 12, 2013.

Statutes and Laws, New Jersey Statutes and Codes. 2013. 12:11a-21—Lease agreement with private marine terminal operator; Maintenance and repair. Available at http://www.njleg.state.nj.us/. Accessed July 12, 2013.

UK PANDI. 2013. LP Bulletins. Tanker matters. Available at http://www.ukpandi.com/fileadmin/uploads/uk-pi/LP%20Documents/LP_Bulletins/Tanker%20matters%20-%20LP%20news%20supplement.pdf. Accessed July 16, 2013.

UNCTAD. 1985. (UNCTAD/SHIP/185) Nov. 1. Manual on a uniform system of port statistics and performance indicators. Available at http://r0.unctad.org/ttl/docs-un/unctad-ship-185-rev2/en/UNCTAD_SHIP_185_Rev2e.pdf. Accessed July 12, 2013.

UNCTAD. 2003. United Nations Conference on Trade and Development. Trade and Development Board. Commission on Enterprise, Business Facilitation and Development Expert Meeting on the Development of Multimodal Transport and Logistics Services. Geneva, September 24–26, 2003. TD/B/COM.3/EM.20/2. July 15, 2003. Available at http://unctad.org/en/Docs/c3em20d2_en.pdf. Accessed July 12, 2013.

United Nations. 1980. UN Convention on International Multimodal Transport of Goods. Geneva, May 24, 1980. Available at http://treaties.un.org/doc/Publication/MTDSG/Volume%20I/Chapter%20XI/XI-E-1.en.pdf and http://www.admiraltylawguide.com/conven/multimodal1980.html. Accessed July 1, 2013.

Vale. 2013. Iron Ore and Pellets. Available at http://www.vale.com/australia/EN/business/mining/iron-ore-pellets/Pages/default.aspx. Accessed September 3, 2013.

World Steel Association. 2013. Available at http://www.worldsteel.org. Accessed July 12, 2013.

CHAPTER 5

Port and Terminal Investment

> In investing, what is comfortable is rarely profitable.
>
> **Robert Arnott**

5.1 PUBLIC VERSUS PRIVATE INVESTMENT

Over the past decade, the maritime industry has enjoyed unprecedented growth in terms of trade volume and tonnage of the global commercial fleet. Since the demand for ports is a derived demand, it is only logical to anticipate the equally rapid growth of ports. The new generation of mega-ships requires capacity expansion such as deeply dredged ports, mega-berths, mega-cranes, and ample land availability for mega-storage areas and infrastructure development. Since the 1990s, when Chinese companies joined the New York Stock Exchange, and commenced a large-scale purchase of second-hand ships, the global sea trade and the world economy have never been the same: previously, small regional ports specialized in ship scrapping were radically transformed into mega-ports, capable of handling the vast cargoes produced in the regional trade zones.

Chapter 2 of this book explicitly discusses the reasons and outcome of this unparalleled trade growth, and the conclusions clearly explain the growing numbers of port investment. In fact, 10 years ago, a port investment of $100 million was considered as generous, whereas in the mid-2010s, major ports are typically involved in billion-dollar investments, especially for greenfield investment projects. Figure 5.1 demonstrates the potential port investors.

In view of a port investment, pertinent *appraisal* and *feasibility* studies are conducted in order to examine parameters such as the following:

- Country-specific port planning
- Regulatory and legal analysis
- Environmental, technical, and operational analysis
- Accessibility, networks, and connectivity to hinterland
- Property, control, and administrative analysis (World Bank 2010)

It is worth noting that port investment does not only pertain to strategic expansion of tangible assets such as land purchase, port architecture, engineering, dredging and development, retrofitting, planned maintenance systems, and so on. Financing may also include intangible assets and input such as human resource development, business relationships, and promotion/marketing, to name a few.

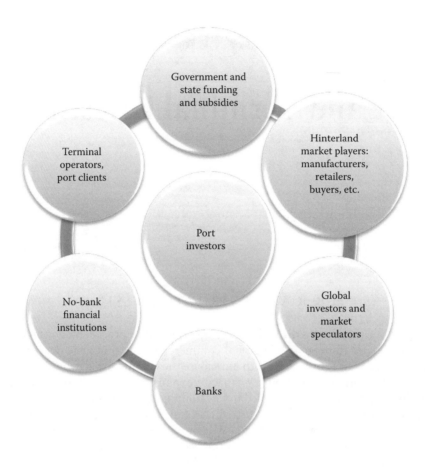

FIGURE 5.1   Port investors and areas of investment.

Port authorities make investment decisions generally based on three key issues: (i) funding opportunities available, (ii) return on investment, and (iii) ability to repay loan.

Over the past few years, several terminals within international public ports have been financed or purchased by private organizations. Furthermore, waterfront property adjacent to ports is frequently purchased by investors, oil majors, and liner conglomerates in an effort to enhance their operational and logistics output. Alternatively, land in the vicinity of ports is frequently purchased by speculators confident about the port's future expansion and market growth.

Since the 1980s, port investment has been substantially intensified in terms of (i) investment amounts; (ii) funding institutions/sources, for example, government and state funding, banks, shareholders, global investors, local industries, and so on; and (iii) funding applicants/recipients, for example, ports, terminal operators, logistics companies, and so on. As modern port investment literally involves billion(s) of dollars, port investment involves the increasing participation of nongovernment financial entities, whose primary concern is high profit and not necessarily national interests. As a result, hedging and the financial instruments that diminish risk derived from financial deficits, commercial losses, or obligations incurred, for example, forward contracts, derivatives, futures, and so on.

Hence, port funding practices become increasingly speculative, with higher risk, higher turnover ratios.

Because of the high returns that port and terminal financing offer to shareholders and speculators, port infrastructure is an appealing financing alternative throughout the market cycles, even at times of market uncertainty. Nevertheless, the highly bureaucratic processes involved with port and infrastructure investment require extended completion processes. Since the element of time is critical for any type of financial commitment, the delays involved in port and terminal funding may prove detrimental to the estimation of the investment time horizon and the level of risk exposure.

Financing methods of public port capital expenditures include the following:

a. *Port Internal Revenues (Earned Income)* encompasses a port's taxable income including tariffs, leasing, sale and purchase funds; excluding interest rates and dividends.

b. *General Obligation (GO) Bonds* in the United States are municipal, state, and local government bonds supported by the municipality's ability to reimburse the debt via legitimately accessible funds such as tax payment or earnings from other financial assignments. Properties and financial assets are not utilized as security. GO Bonds in most cases create funding for ventures related to service in the general population.

c. *Revenue Bonds* are the second best option after municipal bonds, in terms of repayment security. These are employed to create funding for ventures that will assist particular communities or public entities, that is, ports, highway, bridges, and so on. These funding projects have the ability to generate income through tariffs, fares, and tolls, and therefore will pay back the loan through a proportion of their income, for example, through fares, as well as taxation.

d. *Industrial Development Revenue Bonds (IDRBs)* or *Industrial Development Bonds (IDBs)* are municipal or state debt securities released by a federal government entity on the part of a nonpublic agent with the purpose of constructing, purchasing, or developing industrial units, infrastructure or superstructure, machinery, and so on. The purpose is to financially support the development of the nonpublic sector and thus (i) generate taxes and (ii) contribute in the regional growth, trade, transport, employment markets, and so on, which may not have alternative investment opportunities. The private-sector company achieves certain gains from this option, owing to the tax exemption involved, as well as the option of issuing debt obligations (asset-backed securities, i.e., the guarantee of the investment's repayment) at tax-exempt costs. In consequence, this financing option helps ports and supplier of producers to attain a reduced interest rate in relation to the alternative options of standard investment and taxation.

e. *Port Loans* are provided by federal government or state sources, banking institutions, private-sector investors, port clients and partners, and so on.

   The loan agreement terms stipulate among others the amount, repayment terms, currency, interest rates, the balloon repayment option, bank or loan charges, and so on.

   The main loan types are as follows:

   i. Closed-end loan, where the borrowing entity has agreed to specific and irrevocable repayment terms, in particular related to the following:

      A. Credit terms, that is, loan duration, annual installments and fees, balloon payments, collaterals, loan adjustments, bankruptcy, and so on

B. Loan maturity date, that is, the date when the full loan repayment including interest is due to the investment entity. This is the deadline after which a loan should be fully repaid. The loan agreement may have stipulations for covenants, for example, the time duration before the loan payment is demanded.

ii. Open-end loan, which offers flexibility in terms of both the funding amount and the repayment date, installments, and so on. The borrowing entity can obtain the funds required at a specific time, with the option of obtaining the remaining funds at a future time, as required. Consistency in the repayment of installments as well as compliance as to the loan agreement will secure the long-term availability of funds and may increase the credit limit.

f. *Government Grants*: A government grant is a monetary accolade offered by the local, state, or federal authorities to qualified recipients, without a repayment requirement. Typically, grants aim to enhance the national or regional business competitiveness, innovation, technological know-how, employability, and so on. For this reason, the grantees are bound to develop a grant proposal for approval, followed by progress reports submitted at regular intervals throughout a pre-determined period. Ports are typically entitled to apply for any and all of the following grand types:

- Green Infrastructure Grant Programs by the US Environmental Protection Agency (EPA)
- Economic Infrastructure Grants, aiming at regional development and new employment opportunities
- Regional Development Grants, encompassing the elements of transportation, community, tourism, culture, and so on

Port investment in the United States. From 1946 to 2005, investment capital for US public port development exceeded $30.1 billion, which backed up ameliorations to port facilities and connected infrastructure. Over these 60 years, the median annual funding amounted to $501 million, whereas from 2001 to 2005, the median annual funding leaped to $1.5 billion per annum (MARAD 2009). Most of this funding was allocated to spending for the design and building of modern amenities, infrastructure, and super-structure and the alteration and restoration of facilities previously built (MARAD 2013). Figure 5.2 demonstrates the necessity for investment in the US seaports, courtesy of AAPA (2008, 2013).

Since the 1980s, port investment has been substantially intensified in terms of (i) investment amounts; (ii) funding institutions/sources, for example, government and state funding, banks, shareholders, global investors, local industries, and so on; and (iii) funding applicants/recipients, for example, ports, terminal operators, logistics companies, and so on. As modern port investment literally involves billion(s) of dollars, port investment involves the increasing participation of nongovernment financial entities, whose primary concern is high profit and not necessarily national interests. As a result, hedging and the financial instruments that diminish risk derived from financial deficits, commercial losses or obligations incurred, for example, forward contracts, derivatives, futures, and so on.

Hence, port funding practices become increasingly speculative, with higher risk, higher turnover ratios. The positive aspects of the private sector seem to be the highly energetic, competitive, motivated, and commercially focused attitude as opposed to the public sector. On the other hand, the public sector is closely connected with the national

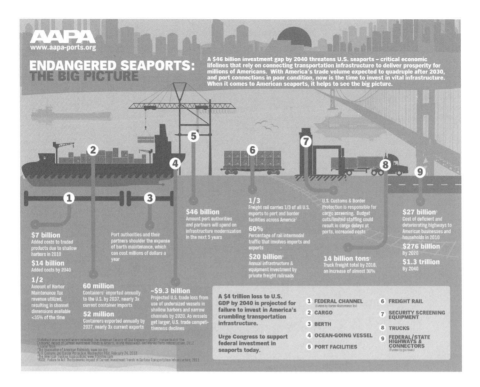

FIGURE 5.2   Endangered seaports: the big picture.

and regional development plans and is more reasonable in its investment policies, and its expansion strategy frequently benefits the entire nation, state, or community—not just the port authorities. All told, an equilibrium among public and private port management would be the optimum solution.

Until the 1980s, national funds were allocated for ports, mainly service and tool port types. A different financing mechanism applied for landlord ports, as funds for their infrastructure were obtained by both federal and port authority sources. Investment for port and terminal superstructure was proportionally divided among the port lessees, that is, terminal operators and other companies that leased port space. Hence, each lessee (terminal or port space user) was responsible for the superstructure of their own leased space.

A port's investment needs typically fall into three categories:

- Management-related investment, which serves as a link between the port authorities' strategic goals and the other two segments, that is, infrastructure and superstructure investment. Managerial investment includes research, port planning, marketing, information technology, and so on.
- Infrastructure investment, which pertains to navigational connectivity, port safety and security protection, inner-port and logistics investment.
- Superstructure investment, which entails administration buildings, cargo handling equipment, cargo storage, and distribution.

Figure 5.3 depicts the three major port investment areas.

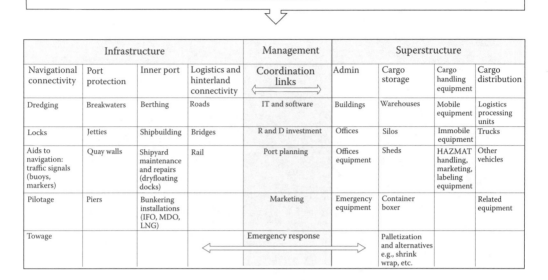

Port investment areas								
Infrastructure				Management	Superstructure			
Navigational connectivity	Port protection	Inner port	Logistics and hinterland connectivity	Coordination links	Admin	Cargo storage	Cargo handling equipment	Cargo distribution
Dredging	Breakwaters	Berthing	Roads	IT and software	Buildings	Warehouses	Mobile equipment	Logistics processing units
Locks	Jetties	Shipbuilding	Bridges	R and D investment	Offices	Silos	Immobile equipment	Trucks
Aids to navigation: traffic signals (buoys, markers)	Quay walls	Shipyard maintenance and repairs (dryfloating docks)	Rail	Port planning	Offices equipment	Sheds	HAZMAT handling, marketing, labeling equipment	Other vehicles
Pilotage	Piers	Bunkering installations (IFO, MDO, LNG)		Marketing	Emergency equipment	Container boxer		Related equipment
Towage				Emergency response		Palletization and alternatives e.g., shrink wrap, etc.		

FIGURE 5.3   Port investment areas.

The overall conditions and opportunities entailed in port investment heavily rely upon each country's federal and economic plans, which are closely associated with the national political regime. For example, port financing in the Western Hemisphere (United States, Canada, European Union) involves higher capital amounts, higher levels of competition at the national and local levels, and decreasing levels of national control and commercial intervention. At the same time, the Eastern Hemisphere, that is, Asia, entails higher government intervention and centralized controls over port specialization and port development plans. In fact, several rapidly developing economies such as China, India, Brazil, and so on, consider ports as a mere component of the national trade and transportation network, and financing covers the entire logistics chain. These highly centralized plans aim for port differentiation; hence, competition is limited and optimum utilization of resources is achieved. Interestingly enough, countries with lower port investment levels are typically the ones with less national port competition owing to higher specialization.

Public ports in the United States typically have more attractive investment terms and conditions as opposed to the private entities. Ports in the public sector benefit from tax-free earnings and general obligation bonds.

Landlord ports can benefit from long-term (i.e., 20 years) lease agreements with terminal operators.

Finally, while joint ventures among port authorities and terminal operators could possibly be a convenient investment option, it might act as a deterrent to the competing terminal operators who may select not to enter in this powerful alliance, but enter another market/port instead.

Research studies by World Bank (2007, 2010) and UNCTAD (2011) as well as more recent industrial developments reveal the condition of modern ports in each continent as follows:

Africa and the Middle East

For several decades, the Middle Eastern countries have been heavily dependent upon domestic oil production and trade agreements that determine their national port development strategies. A number of major tanker ports have adopted the landlord model and have undergone concession agreements with major terminal operators.

Before the 2010s, the most developed African ports were located in North Africa, in particular Libya and Egypt, as well as in South Africa. The vast oil and gas findings initially in Nigeria and later all across the Gulf of Guinea have attracted the attention of oil majors; governments, that is, the United States and China (see the port of Dar Es Salaam, Tanzania); and terminal operators (e.g., APM investments in Nigeria). These are the regions where the first mega-transshipment hubs have been established.

The prior condition of underdeveloped ports is radically changing as more African ports are adopting the landlord port arrangement; hence, in the decades to come, African ports are increasingly dredging their ports to accommodate larger ships, and multinational investors will modernize their ports' infrastructure and superstructure.

Asia

The contribution of the Asian economies to the global production can hardly be exaggerated: in less than a decade, Asian ports have surpassed Western global ports in size and trade growth. China, India, Japan, Singapore, South Korea, the Philippines, Indonesia, Taiwan, and Bangladesh are countries with a strong trade and maritime potential, with port investments in the years to come. Most of these countries' political regimes exercise control on the ports' investment. Some of these governments, for example, China, ensure differentiation by assigning the trade of specific trade zones to specific ports. On the other hand, a characteristic of many Asian countries is that their economies strongly depend on exports to non-Asian nations (as opposed to the United States, whose domestic markets absorb almost half of their production). Without strong domestic consumption, these economies are forced to invest in port infrastructure.

Europe

Ports in Western Europe have long been ranked among the largest ports in the world. They offer the intermodal advantage that inland water transport is possible deep into the European hinterland. Despite the fact that water allows for the cheapest way of transport, Europe has seen a revival of cargo transport by rail, implementing the so-called dedicated block train, and even of multimodal transport corridors and terminals. The landlord port management model is the predominant model in Western European ports and is increasingly implemented in East European and Central Asian ports.

Central and South America

Over the past few years, Central and Southern American ports are increasingly adopting the landlord option and the long-term leasing arrangements to terminal operators and logistics companies. Strategic locations such as the Panama Canal, and major exporting countries such as Brazil, Venezuela, and Mexico, enjoy port development, sustainable growth, and state-of-the-art technologies and designs. Nevertheless, the majority of Central and South American countries handle limited cargo volumes, depending on their national markets, trade agreements, and their level of development.

The principal financial and commercial advantage of port investment is based on the port's capacity to diminish vessels' turnaround time frame.

Initially, an instantaneous advantage from the investment will derive from the port's customers that will see their cargoes being handled and distributed in time efficiency, safety, and reliability.

Second, the port's success will be measured in terms of return on investment acquired from the increased customers' satisfaction and increased market share.

Third, the national and regional benefits that arise from port investment will have an impact on economic and trade growth, employment, and so on.

Prior to the approval of a port's funding venture, port development plans and evaluation studies should be conducted, to assess the commercial and financial feasibility of the project, which should not be limited to the port's benefits, but should include the regional, state, and national benefits resulting from such a venture.

Port investment evaluations, such as the practical utility and estimation methodology of investment-related formulas, will be duly discussed in Section 5.2.

5.2 RETURN ON INVESTMENT AND THE IMPACT ON TRADE GROWTH

> When investing, or buying shares, ask yourself,
> would you buy the whole company?
>
> **Rene Rivkin**

Based upon the principle that open economies are inclined to expand faster than closed economies, the development of international trade and transport markedly elevates the global productivity growth. Arguably, port investment has an effect on regional growth, as it enables nations and states to cash in on domestic production and consumption. At the same time, port growth frequently signifies a country's exposure to larger markets and larger trade volumes.

Port investment and global trade growth seem to go hand in hand. As services signify 70% of world output but only 20% of world trade, it seems that ports are the answer to the trade-related services. The geographic routes of global trade and investment seem to spread technological advancement around the globe. Hence, modern ports need to overcome any barriers that will inhibit their position within a rapidly transforming market.

The level of port development is clearly demonstrated in James Wolfensohn's notion of a "four speed" world, which divides the globe into Affluent, Converging, Struggling, and Poor nations, on the basis of their income and growth rate per capita, compared to the developed countries (OECD 2010). Port development clearly reflects the new world map of economic development, which ranges from one extreme of high growth, abundance of resources, and innovation, to the other extreme of low income and limited resources.

Port investment as perceived by investors and the private sector is guided by the anticipation of profitability and gains. On behalf of the port authorities and the public sector, investment is related to (i) trade and financial growth, (ii) national/local market benefits, and (iii) employability and benefit to the community (EU 2010).

Since the element of risk exists in every commercial and financial transaction, port development plans and evaluation studies should be conducted prior to the approval of a port's funding venture, in order to assess the commercial and financial feasibility of the project. These studies should not be limited to the port's benefits, but should include the regional, state, and national benefits resulting from such venture(s).

Investment appraisals involve the assessment of the financial risk, costs, and benefits/ profitability of a funding proposition and encompass both tangible investment assets and intangible sectors such as research and development, marketing, and so on. The investment evaluation process should incorporate techniques such as the following:

  i. Average rate of return
 ii. Payback period
iii. Net present value
 iv. Internal rate of return
  v. Benefit–cost ratio

  i. *Average rate of return* is used to measure an investment's earnings and therefore viability. The average rate of return estimates the return on investment while taking into account the element of time, yet it will not reveal the annual fluctuations, with possible losses over a period of time.

  There are two primary methods of estimating the mean of consecutive rates of return:

  a. The Arithmetic Average Rate of Return (AAROR):

$$\text{AAROR} = \frac{1}{n}\sum_{t=1}^{n} r_t \tag{5.1}$$

  where t is time, n is factorial notation, measurement of time cycles, and r_t is the return on investment during each time cycle.

  b. The Geometric Average Rate of Return (GAROR or Gn):

$$\text{GAROR} = (1 + r_c)\ 1/n - 1 \tag{5.2}$$

  where r_c is cumulative rate of return on the investment over the whole period and n is the measurement of time cycles.

 ii. *Real rate of return* is used to determine the return on investment following inflation adjustments. The "Fischer" or rate of return equation is measured as follows:

$$r = \frac{1 + nr}{1 + ir} - 1 \tag{5.3}$$

  where r is real rate of return, nr is nominal rate of return, and ir is inflation rate.

iii. *Payback period* involves the period needed to fully retrieve the investment amount. Its estimation will ascertain the attractiveness of a specific financial venture; generally, shorter payback cycles are more attractive investment options. The equation is estimated as follows:

$$\text{Payback Period} = \text{Cost of Initial Investment/Annual Cash Inflows} \qquad (5.4)$$

An alternative formula may read as follows:

$$\text{Payback Period} = (p - n) \div p + n_y = 1 + n_y - n \div p \qquad (5.5)$$

where p is the cash flow amount where the initial positive value of collective earnings takes place, n is factorial notation, the cash flow amount where the final negative value of collective earnings takes place, and n_y is the period (number of years) following the initial funding where the final negative value of collective earnings takes place.

This method is only useful in estimating the years of payback of investment. However, it does not take into consideration the currency fluctuations, inflation, deflation, interest rates, and so on; neither does it reveal the profitability of the project after the payback year.

iv. *Net present value* (NPV) is an estimation employed to establish an investment's present value by the discounted amount of all cash flows (inflows less outflows) obtained from the financed venture. The investment's financial value is estimated in the long run, by calculating present versus future value, while also estimating inflation and earnings.

$$\text{NPV} = -C0 + \sum_{i=1}^{n} \frac{Ci}{(1+r)^i} \qquad (5.6)$$

where $-C0$ is the initial investment amount, C is the cash flow amount, n is factorial notation, measurement of time cycles, i is interval of time in years, from year 0, and r is rate of return.

An alternative formula reads as follows:

$$\text{NPV} = \sum_{i=1}^{n} \frac{NB}{(1+r)^i} \qquad (5.7)$$

where n is factorial notation, measurement of time cycles, i is interval of time in years, from year 0, r is rate of return, and NB is net annual income.

Based on the NPV formulas (Equations 5.6 and 5.7), the value of NPV reads as follows:

Should NPV > 0 = financially rewarding investment
Should NPV = 0 = breakeven investment
Should NPV < 0 = financial loss

Despite the straightforward and simple method of estimating the NPV formula, the overall concept lacks flexibility, as the outcome is based on the supposition that the financing project is either revocable, that is, allowing the investor to cancel the investment during unfavorable economic patterns, or irrevocable, based on the premise that the investor should act promptly in order to achieve favorable rates of return.

v. *Internal rate of return (IRR)* or *economic rate of return (ERR)* is the growth rate estimated in capital budgeting while comparing several investment options, in order to verify which option is more profitable. This estimation considers the NPV of all earnings within a specific investment equal to zero. The greater a venture's IRR, the more profitable it is. The disadvantage of IRR is that it does not take into account the initial investment amount.

$$\text{NPV} = \sum_{t=0}^{n} \frac{C^t}{(1+r)^t} = 0 \tag{5.8}$$

where n is factorial notation, t is time, C is the cash flow amount, and r is rate of return.

Based on the assumption that NPV = 0, IRR is estimated as follows:

$$\left[\frac{\text{CIF}_1}{(1+r)^1} + \frac{\text{CIF}_2}{(1+r)^2} + \ldots + \frac{\text{CIF}_x}{(1+r)^x} \right] - \text{C0} = 0 \tag{5.9}$$

where $-\text{C0}$ is the initial investment amount, r is internal rate of return, CIF_1 is the first cash inflow time, CIF_2 is the second cash inflow time, and CIF_x is the x cash inflow time.

vi. *Benefit–cost ratio (BCR)* signifies the ratio of the overall benefits versus the overall costs of the investment, with their present values suitably discounted; it is estimated as follows:

$$\text{BCR} = \frac{\sum_{i=1}^{n} B/(1+r)^i}{\sum_{i=1}^{n} \frac{C}{(1+r)^i}} \tag{5.10}$$

where BCR is the benefit–cost ratio, n is factorial notation, measurement of time cycles, i is interval of time in years, from year 0, B is annual gross income, C is annual cost, and r is rate of return.

Based on the BCR formula (Equation 5.10), the value of BCR reads as follows:

Should BCR > 1 = financially rewarding investment
Should BCR = 1 = breakeven investment
Should BCR < 1 = financial loss

5.3 PORT SUBSIDIES AND INVESTMENT: CHALLENGES AND OPPORTUNITIES

5.3.1 Port Subsidies

Seaport subsidies may be defined as the funding offered by the federal government to a seaport in the form of direct public subsidies (e.g., grants), as well as indirect public subsidies (such as tax exemptions, loans with minimal or no interest rates, etc.).

Subsidies may be distinguished according to their purpose and usefulness, that is:

a. *Port development subsidies*, which are frequently offered by the state and federal government, in order to increase the port's competitive edge, as well as its strategic significance at a global level.

b. *Energy and green energy subsidies* pertain to oil, clean-diesel, gas, nuclear, fossil fuel, and renewable energy subsidies, which are offered with the purpose of achieving "close to zero" effect on emissions. Alternative or energy subsidies such as subsidies for solar and wind power are used for this purpose. Energy tax funds are allocated to optimize fuel emissions and invest in "green" technology. As an example of energy subsidies, the European Commission has granted energy subsidies to the Port of Antwerp, Belgium, to be utilized for the design, development, and construction of a liquefied natural gas (LNG) bunkering station for barges. The LNG bunkering project for seagoing ships will be in place in 2015 and will comply with the most rigorous environmental regulations of the IMO for sulfur emissions of fuels (Port of Antwerp 2013).

As an example of "clean air subsidies," the seaports of Seattle, Tacoma, and Vancouver have a partnership agreement in order to noticeably diminish energy emissions, while retaining their current volume of cargoes.

c. *Incentive subsidies for shipowners* are offered by the port authorities in collaboration with the state and central government in order to attract new concession agreements for leasing the ports' terminals. As an example, the Port of Portland authority (Oregon, USA) has sanctioned the "Container Carrier Incentive Program" designed to encourage liner shipowners in the container industry, in order to boost port utilization.

d. *Incentive subsidies for cargo owners* under national or state strategies are directed toward boosting cargo traffic. For example, the arctic Port of Churchill in Manitoba via the Port of Churchill Utilization Program provides greater monetary inducements to grain companies.

e. *Maintenance, repair, and shipbuilding subsidies* are offered to shipyards and shipowners.

Port-financing arrangements are generated by state or central government funding. This option suggests that the public sector will bear the financial and commercial risk of the port's expansion to a large extent. Key sources of revenue used for subsidies, may include the following:

a. *Cross-subsidies*, which entail the utilization of revenue from another venture, service, or corporate entity for port funding, aiming business expansion and earnings maximization. Certain port authorities use earnings from the port's infrastructure or superstructure, for example, bridges, tunnels, buildings, other assets, and so on.

b. *Local property taxes*, where a dedicated percentage of taxation is allocated to the local port(s) for a certain period.

c. Local oil, gas, and vehicle taxes, where a predetermined proportion of taxation is apportioned to the local port(s) for a designated period.

d. State or federal government's general funds. Typically, this public sector arrangement is based on the presumption that the port's growth will benefit the regional economy, the industries, the job market, and so on.

Both overinvestment and underinvestment will lead to financial and commercial losses. Hence, risk analysis and thorough investment selection are required. The modern trends of logistics and intermodal systems integration facilitate the transfer of risk and the financial responsibility from ports and the public sector to the private sector, such as the terminal operators, logistics companies, charterers, and so on. This is a win/win arrangement, as the private sector may be closer to drilling down to the markets.

5.3.2 Foreign Investment

A port's ability to attract foreign direct investment (FDI) will maximize its competing strategy.

FDI represents foreign investment opportunities where the "direct investors" aim to penetrate overseas markets and thus exercise control over the venture, that is, the direct investment corporation, and consequently the foreign market (UNCTAD 2013). These investors have the potential to set up intricate systems to gain maximum advantages from their investment strategies and for effective handling of the capital and associated ventures (OECD 2008). According to the International Monetary Fund (2007), a ceiling of 10% of equity ownership is required for a company to meet the criteria of a foreign direct investor.

The most common methods of FDI may normally include the following:

a. The ownership and operation of a local business entity

b. The merger or acquisition of a local company

c. The investor's active participation in the local stock market, that is, the purchase of stocks, bonds, and shares

d. The strategic alliance, partnership, or joint venture of the investor with a local business entity.

REFERENCES

American Association of Port Authorities (AAPA). 2008. Public Seaport Agencies in the United States and Canada, By Rexford B. Sherman, AAPA Director of Research and Information Services. 1995, 3rd Revision 2008.

American Association of Port Authorities (AAPA). 2013. Interview with Mr Aaron Ellis (AAPA Public Affairs Director) and Dr. Rexford B. Sherman, AAPA Director of Research and Information Services.

EU. 2010. Trade, growth, and world affairs. Trade policy as a core component of the EU's 2020 strategy. Available at http://trade.ec.europa.eu/doclib/docs/2010/november/tradoc_146955.pdf. Accessed August 5, 2013.

IMF. 2007. *Balance of Payments Manual: 6th Edition (BPM6)*. Washington, DC: International Monetary Fund. Available at http://www.imf.org/external/pubs/ft/bop/2007/pdf/bpm6.pdf. Accessed August 5, 2013.

MARAD. 2009. Port expenditure, 2006–2011. US Department of Transportation, Maritime Administration. Available at http://www.marad.dot.gov/documents/2006_port_expenditure_rpt_—_final.pdf. Accessed August 1, 2013.

MARAD. 2013. Port development and finance. US Department of Transportation, Maritime Administration. Available at http://www.marad.dot.gov/ports_landing_page/infra_dev_congestion_mitigation/port_finance/Port_Fin_Home.htm. Accessed August 8, 2013.

OECD. 2008. *Detailed Benchmark Definition of Foreign Direct Investment: 4th Edition (BD4)*. Paris: Organisation for Economic Co-operation and Development. Available at http://www.oecd.org/daf/inv/investmentstatisticsandanalysis/40193734.pdf. Accessed August 5, 2013.

OECD. 2010. Perspectives on global development 2010: Shifting wealth. ISBN 92-64-084650 © OECD 2010. Available at http://www.oecd.org/dev/pgd/45451514.pdf. Accessed August 2, 2013.

Port of Antwerp. 2013. Port Antwerp gets EU subsidies for LNG project. Available at http://www.portofantwerp.com/en/news/port-antwerp-gets-eu-subsidies-lng-project. Accessed August 5, 2013.

UNCTAD. 2011. World development report. United Nations Conference on Trade and Development. Accessed August 5, 2013.

UNCTAD. 2013. Foreign direct investment. Available at http://unctad.org/en/Pages/DIAE/Foreign-Direct-Investment-%28FDI%29.aspx. Accessed August 5, 2013.

World Bank. 2007. *Port Reform Toolkit*, 2nd edition. Available at http://www.ppiaf.org/sites/ppiaf.org/files/documents/toolkits/Portoolkit/Toolkit/pdf/modules/00_TOOLKIT_FM_Vol1.pdf. Accessed August 12, 2013.

World Bank. 2010. Ports and waterborne transport. Available at http://web.worldbank.org. Accessed August 4, 2013.

CHAPTER 6

Ships' Size, Ports' Size:
A New Era Ahead

If you're thinking already,
you might as well think Big!

Donald Trump

In the aftermath of the 2008 financial and trade implosion, the global maritime industry observed the well-designed commercial and economic structure subside. The maritime community was not immune to the financial crisis but proved to be resilient. With persistence and determination, port and maritime executives saw the industry rebound and as a result became wiser in the process. Among the many lessons learned, three of the top focal points that currently capture the industry's undivided attention are duly discussed in this chapter: (i) *Economy*, (ii) *the Environment*, and (iii) *Energy*.

Economy is associated with both trade agreements (Section 6.1) and larger size of ports and ships (Section 6.2). The growth of global trade has necessitated the increase in ships' size and subsequently the increase in ports' size. After all, the demand for sea transport is a derived demand.

Furthermore, the latest industry's interest in *Energy and Environmental* protection have spurred a new generation of innovation technologies that make this era truly unique (Section 6.3).

6.1 INTERNATIONAL TRADE AGREEMENTS AS TOOLS TO GROWTH

The twenty-first century's global trade system is characterized by the development of mega-ports and mega-ships, technological advances, mass industrialization, outsourcing, historic mergers, and the development of multinational conglomerates as well as complex and versatile supply chains and the elimination of the restrictions of geographical borders. International trade aims to create an efficiency matrix where the world's factors of production are utilized in an optimum, cost-effective, and value-added manner.

Three interesting side effects of globalization have been established in pursuit of economic growth:

a. The amalgamation of both protectionism and free trade principles in most countries' fiscal and trade policies

    b. The increasing economic integration and the formation of trade market blocs, such as economic and monetary unions (European Union, Caribbean Single Market and Economy), common markets (e.g., EFTA, CES, EEA), customs unions (e.g., Mercosur, EUCU, KUBKR, SACU, EAC, etc.), and multilateral free trade areas (NAFTA, GAFTA, TPP, CEFTA, etc.)

    c. The active collaboration of governments, port authorities, national trade zones, and supply chains, in order to fulfill the nation's contractual obligations with these trading blocs

Leverage and the ability of a nation and all parties concerned to enhance their negotiating position will define the final outcome, which may range from economic growth to conflict of interests and serious financial, political, and diplomatic repercussions.

Once a trade agreement has been signed, it is the job of national ports and supply chains to materialize the terms and conditions stipulated, whereas the outcome entirely depends on the success of the national and supply chain collaboration.

A trade agreement's outcome is measured in terms of sustainable growth, rectification of a country's balance of trade, improving the jobs market, opening new markets, and ameliorating a nation's economic status.

Other key factors that typically affect the results in trade agreements include the following:

    a. *Economic factors*, that is, the country's *bargaining power*, global economic cycles, achieving optimum economies of scale.

    b. *Supply and demand*, that is, the equilibrium between commodity *scarcity and market needs*.

    c. *Time efficiency* achieved as a result of satisfactory supply chains and efficient transport infrastructure.

    d. Avoidance of complications pertaining to *time, money, safety, and quality* within the supply chain.

    e. Compatibility between *national legislations and compliance to regulations*, that is, safety, environmental protection, customs' clearance rules, charges, and clearance times.

    f. All the above factors are based on the *negotiations* between countries and the effective leveraging techniques that are needed for a country to achieve its competitive edge.

Trade agreements formed by global trading blocs are indispensable for a nation and its ports since they (a) exert political and economic influence, (b) allow the growth and expansion of markets, (c) benefit from economies of scale and regional comparative advantages, (d) eliminate inflation, (e) stabilize balance of trade, (f) stabilize currency power, (g) generate think tanks, and (h) promote innovation in strategies, designs, and technologies.

The role of trade agreements is exactly to reduce trade barriers and therefore form long-term relationships of mutual benefits. The trade agreements formed may be bilateral, that is, among two countries, or multilateral, that is, between more than two countries.

The global and regional agreements that follow entail useful facts and figures that will help the reader comprehend the impact of these trading blocs to seaports and national economies.

Since capital, land, and labor are significant factors of production, the regional agreements of this section (i.e., excluding the global agreements of the General Agreement on

Tariffs and Trade [GATT] and the World Trade Organization [WTO]) contain combined socioeconomic facts and figures such as (a) GDP, (b) land, (c) population, and (d) unemployment rates. These combined data highlight the comparative advantages and disadvantages of each trade bloc and thus will enable decision and policy makers to search for areas of growth and development. Figure 6.1 shows a concise matrix of selected free trade partnerships.

The terms and conditions of trade agreements are initiated by governments, yet the transportation aspect is subsequently carried out by supply chains, including the global ports. Numerous trade agreements have been conducted, on grounds that are regional, bilateral, multilateral, and so on. For the sake of practicality and efficiency, a selection among the most influential or the most distinguished agreements was presented in this section.

6.1.1 The General Agreement on Tariffs and Trade

GATT was established in 1947. It is a fundamental multilateral trade agreement that controlled commerce among its 153 nation-members. Since its inception, GATT managed to diminish the global tariff levels and significantly expanded global trade. The aim of GATT is the "substantial reduction of tariffs and other trade barriers and the elimination of preferences, on a reciprocal and mutually advantageous basis." It has carried out eight discussion rounds handling numerous vital commercial issues to be resolved and remedying global trade conflicts. In 1993, the well-known Uruguay Round led to memorandums of understanding among the United States and 116 other nations to mutually establish more viable global trade rules and decrease the existing commercial barriers. GATT created the WTO, which in 1995 succeeded GATT (WTO 2013).

6.1.2 The World Trade Organization

WTO was formally launched in 1995 under the Marrakech Agreement, as a successor of the GATT. Its mission is to regulate, monitor, and remove the restrictions of global trade. It offers a global platform for settling and discussing global trade agreements. The WTO is targeted on items addressed in former trade agreements, in particular from the Uruguay Round (WTO 2013).

6.1.3 Multiregional Partnerships

Trans-Pacific Strategic Economic Partnership (TPSEP)

The 2005 TPSEP or P4 is a state-of-the-art multiregional agreement geared toward effectively managing trade issues in the twenty-first century. It was established in 2005 with the aim of opening and developing the Asia-Pacific trade and economic agreements among its first members, that is, Brunei, Chile, New Zealand, and Singapore. The latest proposal entails a considerably broadened version of TPSEP comprising new members, that is, Australia, Canada, Malaysia, Mexico, Peru, the United States, and Vietnam. There are current discussions for Japan and South Korea to join TPP (Foreign Trade Information System 2013).

Trade agreements in the Americas and Europe

NAFTA	TAFTA or TTIP	CAFTA	SICA	CA-4	MERCOSUR	European Union (EU)	European Union (EU)	EU Candidates	CEFTA
USA	USA	USA	Panama		Argentina	Austria	Luxembourg	*EU Candidates:	
Canada	EU	Costa Rica	Costa Rica		Brazil	Belgium	Poland	Iceland	Poland
Mexico		El Salvador	El Salvador	El Salvador	Bolivia	Cyprus	Malta	Montenegro	Montenegro
		Guatemala	Guatemala	Guatemala	Paraguay	Greece	Netherlands	Serbia	Serbia
		Honduras	Honduras	Honduras	Uruguay	Portugal	Denmark	FYROM	FYROM
		Nicaragua	Nicaragua	Nicaragua	Venezuela	Romania	Estonia	Turkey	Romania
		Dominican Republic	Dominican Republic		Equador	Slovakia	Finland		Slovakia
"G2" FTA			Belize		Peru	France	Slovenia		Slovenia
Mexico			*Mexico (obs)		*Mexico (obs)	Germany	Lithuania	*EU Potential Candidates:	Albania
Colombia			*Chile (obs)		Chile	Ireland	Spain	Albania	Moldova
			*Brazil (obs)		Colombia	Italy	Sweden	Bosnia and Herzegovina	Bosnia and Herzegovina
			**China (obs)			Latvia	United Kingdom	Kosovo	Kosovo
			**Spain (obs)			Hungary			Hungary
			**Germany (obs)			Czech Republic			Czech Republic
			**Japan (obs)			Bulgaria			Bulgaria
						Croatia			Croatia

Trade Agreements in Australasia and Africa

APTA	SAFTA	ASEAN	SADC	COMESA	African Union	African Union	GAFTA	GCC
China	Afghanistan	Brunei	Angola		Angola	Algeria	Bahrain	Bahrain
India	India	Indonesia	Botswana		Botsuana	Benin	Egypt	
S. Korea	Bhutan	Malaysia	Congo	Burundi	Burundi	Burkina Faso	Iraq	
Mongolia	Maldives	Philippines		Comoros	Comoros	Cameroon	Jordan	Kuwait
Bangladesh	Bangladesh	Singapore		Congo	Congo	Cape Verde	Kuwait	
Sri Lanka	Sri Lanka	Thailand		Djibouti	Djibouti	Central Africa Republic	Lebanon	

Laos	Nepal	Laos			Egypt	Egypt	Egypt	Chad	Libya	
	Pakistan	Vietnam	Myanmar		Eritrea	Eritrea	Eritrea	Cote d'Ivoire	Morocco	Oman
			Cambodia		Ethiopia	Ethiopia	Ethiopia	Equatorial Guinea	Oman	Qatar
					Kenya	Kenya	Kenya	Gabon Rep.	Palestine	Saudi Arabia
TPSEP				Lesotho	Libya	Libya		Gambia	Qatar	United Arab Emirates
Brunei					Madagascar	Madagascar	Lesotho	Ghana	Saudi Arabia	Yemen (*2015)
Chile		Malawi	Malawi	Malawi	Malawi	Malawi	Madagascar	Guinea Rep.	Sudan	
New Zealand		Mauritius	Mauritius	Mauritius	Mauritius	Mauritius	Malawi	Guinea-Bissau	Syria	
		Mozambique			Mozambique		Mauritius	Liberia	Tunis	
Singapore		Namibia		Namibia	Namibia	Namibia		Mali	United Arab Emirates	
Proposed new members:				Rwanda	Rwanda	Rwanda	Mozambique	Mauritania	Yemen	
USA		Seychelles	Seychelles	Seychelles	Seychelles	Seychelles	Niger			
Canada		South Africa	South Africa		South Africa		Nigeria			
Australia					Sudan Rep.	Sudan Rep.	Saharawi Rep. Arab Democracy			
Malaysia			Sudan Rep.				Sao Tome and Principe			
Mexico		Swaziland	Swaziland	Swaziland	Swaziland	Swaziland	Senegal			
Peru		Tanzania		Tanzania			Seychelles			
Vietnam		Uganda		Uganda	Uganda	Uganda	Sierra Leone			
Japan		Zambia	Zambia	Zambia	Zambia	Zambia	Somalia			
South Korea		Zimbabwe	Zimbabwe	Zimbabwe	Zimbabwe	Zimbabwe	South Sudan Republic			
								Tongolese Republic		

Notes:
(*) Regional Observer; (**) Extra-Regional Observer.

☐ Country belonging to 2 Trade Agreements

☐ Country belonging to 3+ Trade Agreements

FIGURE 6.1 The Global Trade Matrix. M.G. Burns, based on data from WTO, 2013. World Trade Report, 2013: Factors shaping the future of world trade. Available at http://www.wto.org/english/res_e/publications_e/wtr13_e.htm (accessed on August 30, 2013); USTR, 2013. Office of the US Trade Representative. Available at http://www.ustr.gov (accessed on July 12, 2013); MERCOSUR Trade Center, 2013. Available at http://www.mercosurtc.com (accessed on July 12, 2013); UNESCAP, 2013. Asia-Pacific Trade Agreement, APTA. Available at http://www.unescap.org (accessed on July 8, 2013); SAARC, 2013. South Asia Association for Regional Cooperation. Available at http://www.saarc-sec.org/ (accessed on July 8, 2013); African Union (AU), 2013. Available at http://www.au.int (accessed on July 8, 2013); COMESA, 2013. Common Market for Eastern and Southern Africa. Available at http://www.comesa.int (accessed on July 8, 2013); European Union (EU), 2013. Available at europa.eu (accessed on July 12, 2013); US Department of State, 2013. Available at http://www.state.gov (accessed on July 10, 2013); and Arab Monetary Fund, 2013. Available at http://www.amf.org.ae (accessed on July 8, 2013).

Transatlantic Free Trade Area (TAFTA) or Transatlantic Trade and Investment Partnership (TTIP)

TAFTA or TTIP is a proposed free trade area between the United States and the European Union. Its negotiations commenced in the 1990s and in 2007. Finally, in 2013, US President Obama announced the commencement of formal negotiations on TAFTA, which was confirmed by the European Council and European Commission. This historical agreement appreciates the vast potential to establish a United States–European Union economic relationship, establishing a partnership for the world's largest trade markets, which covers one-third of global trade and 50% of global economic output. The TTIP is an aspiring, superior trade and investment agreement that will significantly boost economic growth, trade, investment, and employment opportunities in both sides of the Atlantic. The agreement will also seek to promote entrepreneurship, remove nontariff barriers, and streamline trade regulations and standards (USTR 2013).

6.1.4 The Americas

North American Free Trade Agreement (NAFTA)

NAFTA is a trilateral trade bloc within the American continent, that is, between the United States, Canada, and Mexico. Its implementation began in 1994. Its aims are to eliminate the majority of the existing trade and investment barriers (NAFTA 2013). Meanwhile, it has grown to become the world's largest free trade area, which now connects 450 million people producing $17 trillion worth of goods and services. The United States concluded a $918 billion multilateral trade with its partners Canada and Mexico in 2010. Commodities' exports totaled $412 billion, while imports amounted to $506 billion (USTR 2013).

Dominican Republic–Central America–United States Free Trade Agreement (CAFTA-DR)

In 2004, the United States signed the CAFTA-DR agreement with five Central American countries—Costa Rica, El Salvador, Guatemala, Honduras, and Nicaragua—and the Dominican Republic, which was considered as a single, common market (USTR 2013). This is the first agreement of the United States joining developing economies and aims at developing regional trade growth by removing tariffs, opening trade markets, decreasing service barriers, and supporting regional integration through investment, trade, and visibility. The combined trade between the CAFTA-DR partners in 2009 was $37.9 billion (US FAS 2013). Trade growth between the CAFTA-DR members is anticipated to further increase and create new economic opportunities among the members.

Central American Integration System (SICA)

SICA is the economic and political organization of Central American states and was established in 1993 in search of economic development, democracy, and regional peace. SICA's eight full members include Guatemala, El Salvador, Honduras, Nicaragua, Costa Rica, Panama, Belize, and, recently, the Dominican Republic. Its structure includes three regional observers—Mexico, Chile, and Brazil—as well as four extra-regional observers—the Republic of China, Spain, Germany, and Japan (SICA 2013).

The Central America Four (CA-4)

CA-4 was inaugurated by the four founding members of SICA, that is, Guatemala, El Salvador, Honduras, and Nicaragua, which has declared a common border, a border control policy, and the Central America-4 passport, which is of a similar type yet its cover depicts the country of issuance. Canada is under negotiations with the CA-4 for a free trade agreement initially entailing the multilateral trade of agricultural products. The remaining four SICA members, that is, Panama, Belize, Costa Rica, and the Dominican Republic, have registered as members to resolve economic issues and promote regional solidarity.

The Southern Common Market (Mercosur)

Mercosur is a powerful economic and political catalyst in South America, uniting the region's two wealthiest countries and presenting a prospective springboard for Latin America's integration. It was founded in 1991 pursuant to the signing of the Treaty of Asuncion and a long history of economic cooperation agreements between Argentina and Brazil since the 1980s. It consists of six full member states: Argentina, Brazil, Bolivia, Paraguay, Uruguay, and Venezuela, which have entered a customs union that will eventually develop into a common market similar to the European Union. Its associate members are Colombia, Chile, Ecuador, and Peru; Mexico retains an observer's status. The organization involves 240 million people and possesses a combined GDP of approximately $3 trillion (MERCOSUR Trade Center 2013).

G-2 Free Trade Agreement

This free trade agreement was established in 1995 and was initially named "G3" as it was formed by the three most populous nations of the Greater Caribbean area, that is, Venezuela, Colombia, and Mexico. When Venezuela decided to exit in order to join Mercosur, Colombia and Mexico proceeded with a bilateral agreement. The terms covered several concerns such as public-sector investments, eliminating trade barriers and intellectual property rights. Commodities traded between the two countries are considered as duty-free, with certain exclusions.

6.1.5 Asia

The Association of Southeast Asian Nations (ASEAN)

ASEAN was established in 1967 and consists of 10 members, that is, Brunei, Indonesia, Malaysia, the Philippines, Singapore, Thailand, Vietnam, Laos, Myanmar, and Cambodia (ASEAN 2013). Its multidimensional aims and purpose include (a) stimulating economic, social, and technical collaboration; (b) promoting Southeast Asian science, studies, and culture and enhancing training, research, and educational collaboration; and (c) ensuring optimum utilization of the region's agriculture and industries, trade, and transport expansion, and amelioration of the standard of living.

Asia-Pacific Trade Agreement (APTA)

APTA is the former Bangkok Agreement and was founded in 1975 as a program launched by ESCAP (Economic and Social Commission for Asia and the Pacific). The aim of this tariff agreement is to significantly increase and support intraregional trade through mutual concession agreements on the promotion, protection, and liberalization of investment and the framework agreement on trade facilitation. Its seven participating countries are China, India, Republic of Korea, Mongolia, Lao People's Democratic Republic, Bangladesh, and Sri Lanka. The agreement also aims at establishing common negotiating positions in WTO (UNESCAP 2013).

South Asia Free Trade Agreement (SAFTA)

The governments of the SAARC (South Asian Association for Regional Cooperation) signed for SAFTA's inception in 2004 and enforcement in 2006. Its eight members as of 2013 are Afghanistan, Bangladesh, Bhutan, India, Maldives, Nepal, Pakistan, and Sri Lanka. Their vision was to create a free trade area of South Asia's largest economies and to eliminate customs duties of all traded commodities by 2016 (SAARC 2013; SAFTA 2013).

6.1.6 Africa

African Union (AU)

The AU was established in 1999 and it is considered as an event of "great magnitude" in Africa's historical development. Its 54 members are Algeria, Angola, Benin, Botswana, Burkina Faso, Burundi, Cameroon, Cape Verde, Central African Republic, Chad, Union of the Comoros, Congo, Cote d'Ivoire, Congo, Djibouti, Egypt, Equatorial Guinea, State of Eritrea, Ethiopia, Gabonese Republic, Gambia, Ghana, Republic of Guinea, Republic of Guinea-Bissau, Kenya, Liberia, Libya, Kingdom of Lesotho, Madagascar, Malawi, Mali, Mauritania, Mauritius, Mozambique, Namibia, Niger, Nigeria, Rwanda, Republic Arab Saharawi Democratic, Sao Tome and Principe, Senegal, Seychelles, Sierra Leone, Somalia, South Africa, Republic of South Sudan, Republic of Sudan, Kingdom of Swaziland, Tanzania, Togolese Republic, Tunisian Republic, Uganda, Zambia, and Zimbabwe. It aims at advancing the political and socioeconomic integration of Africa; sustaining economic, social, and cultural development and trade integration; resolving issues that hinder the region's growth, including political, economic, customs, and immigration challenges, natural resources, desertification, and environmental and production issues; and promoting investment and resource mobilization, economic integration, financial growth, and private sector advancement (AU 2013).

Common Market for Eastern and Southern Africa (COMESA)

COMESA is a free trade region and a pillar of the African economic network. It was formed in 1994 and currently has 19 member states: Burundi, Comoros, Democratic Republic of the Congo, Djibouti, Egypt, Eritrea, Ethiopia, Kenya, Libya, Madagascar, Malawi, Mauritius, Rwanda, Seychelles, Sudan, Swaziland, Uganda, Zambia, and Zimbabwe. In

2008, COMESA initiated the expansion of the free trade zone including members of two other African trade blocs, the East African Community (EAC) and the Southern African Development Community (SADC). A standard visa arrangement is also under consideration, within an effort to promote Pan-African tourism (COMESA 2013).

Southern African Development Community (SADC)

SADC was formed in 1980, as a result of joint efforts aimed at terminating racial segregation and colonial rule in southern Africa, through diplomatic, political, and military means. It is an intergovernmental community aiming to expand socioeconomic cooperation and integration as well as political and security cooperation among its 14 southern African states, namely: Angola, Botswana, Democratic Republic of the Congo, Lesotho, Malawi, Mauritius, Mozambique, Namibia, Swaziland, Tanzania, Zambia, Zimbabwe, South Africa, and Seychelles. Its actions complement the role of the African Union (SADC 2013).

6.1.7 Europe

The European Union (EU)

The EU is an economic and political union of 28 member states. It was founded after WWII in 1948 as the European Economic Community, aiming to stimulate economic synergy and common financial goals among member states. The EU is a rather distinguished entity, as it has achieved international economic integration in which the EU states share a single market, a single currency, and a common trade and budget policy. At the same time, its member states still retain their national autonomy, political regimes, and so on. Inside the Schengen area of 22 EU and four non-EU member states, a free transfer of people (i.e., abolished passport control), commodities, services, and capital (i.e., subsidies, investment, research funding, etc.) is secured. The EU is represented at all the global conventions such as the United Nations (UN), the WTO, the G8/G20, and so on. Populated by over 500 million inhabitants, or 7.3% of the global population, the EU produced a nominal GDP of $16.584 trillion in 2012, accounting for almost one-fifth of the global GDP, when measured in terms of purchasing power parity (European Union 2013; US Department of State 2013).

Central European Free Trade Agreement (CEFTA)

CEFTA was signed in 1991 and was in effect as of 1993, as per the Visegrad Agreement. CEFTA is an organization of seven member states: Poland, Hungary, the Czech Republic, Slovakia, Slovenia, Romania, and Bulgaria. In 2006, Albania, Bosnia and Herzegovina, Croatia, Macedonia, Moldova, Serbia, Montenegro, and the UN Interim Administration Mission in Kosovo joined CEFTA. This treaty provides for the abolition of customs for interstate trade and aims at practicing growth and trade competition, regulatory harmonization, the simplification of the procedures pertaining to the documentation on the origin of merchandise, the elimination of monopoly arrangements, trade collaboration, and exchange of information pertaining to subsidies (CEFTA 2013).

6.1.8 Middle East

Greater Arab Free Trade Area (GAFTA)

GAFTA was established in 1957 and was declared within the Social and Economic Council of the Arab League as an executive program to activate the Trade Facilitation and Development Agreement that has been in force since January 1, 1998. It consists of 17 Arab countries, namely: Jordan, United Arab Emirates, Bahrain, Saudi Arabia, Oman, Qatar, Morocco, Syria, Lebanon, Iraq, Egypt, Palestine, Kuwait, Tunis, Libya, Sudan, and Yemen.

GAFTA is one of the most significant economic accomplishments in the Arab world, with noteworthy endeavors toward building the Arab Common Market. Since 2005, GAFTA has achieved total trade liberalization of commodities through the abolition of customs duties and charges having comparable effect among the vast majority of the Arab member states, apart from Sudan and Yemen (Arab Monetary Fund 2013).

Gulf Cooperation Council (GCC)

The GCC was founded in 1981 and consists of six Arab states located in the Persian Gulf region, that is, Bahrain, Kuwait, Oman, Qatar, Saudi Arabia, and United Arab Emirates, with Yemen to join by 2015 (Arab Business 2007). The council was designed to function as a complex network of collaborations, principles, and common action, based on the common religion and geographic vicinity of the members. It serves as a common market with a defense planning council and an economic and political platform. Furthermore, it is founded on the Islamic principles that encompass religion, investment, funding, and legal doctrines. Influenced by their common grounds pertaining to security and their principles that hostility against any member is considered as hostility against all members of the council, military alliance lays within the nature of the GCC states. The security concerns in the volatile gulf area have triggered the GCC states' common policies and mobilization of forces (Gulf Cooperation Council 2013).

6.1.9 Ports' Growth and the Global Trade Agreements Matrix

Trade agreements represent the *Strength of Unity*, as they have the power to influence a nation's well-being, and a port's strategy. In fact, the world's most influential trade agreements mold the global sea routes and build a port's reputation.

A port manager's strategy aims to penetrate global markets by agreements from representatives of both the private and the public sector. Port executives' basic principles pertain to effective negotiations, keeping abreast with the industry's trade developments, and critically evaluating the data available. A marketing executive within a port, a supply chain, a terminal, or a liner company would need to follow up with the latest developments of the trade agreements that either are regional or are directly related to trade routes that involve the port. For example, the Panama Canal is a passage not only for the American trade agreements but also for many trans-Atlantic and trans-Pacific agreements.

Table 6.1 illustrates a concise list of the world's major trade agreements. Since port traffic is directly influenced by these markets, a simple and efficient methodology should drill down crucial information about our potential government-clients.

TABLE 6.1 The Global Trade Agreements Matrix

1. NAFTA

Country	Geography as a Factor of Production				Economy (2012)				Trade Balance (2012)		Socioeconomic Data and Labor as Factors of Production		
	Total Area (km²)	Land (km²)	Coastline (km)	Water (km²)	GDP Purchasing Power Parity ($)	Per Capita GDP ($)	Inflation Rate (%)	External Debt ($)	Import ($)	Export ($)	Population	Labor Force	Unemployment Rate (%)
USA	9,826,675	9,161,966	19,924	664,709	15.66 trillion	49,800	2	14.71 trillion	2.357 trillion	1.612 trillion	316,668,567	154.9 million	8.2
Canada	9,984,670	9,093,507	202,080	891,163	1.446 trillion	41,500	1.8	1.181 trillion	480.9 billion	481.7 billion	34,568,211	18.85 million	7.3
Mexico	1,964,375	1,943,945	9330	20,430	1.761 trillion	15,300	3.6	125.7 billion	379.4 billion	370.9 billion	116,220,947	50.7 million	5

2. Asia-Pacific Trade Agreement (APTA)

Country	Geography: Land and Coastlines as Factors of Production				Economy (2012)				Trade Balance (2012)		Socioeconomic Data and Labor as Factors of Production		
	Total Area (km²)	Land (km²)	Coastline (km)	Water (km²)	GDP Purchasing Power Parity ($)	Per Capita GDP ($)	Inflation Rate (%)	GDP Real Growth Rate (%)	Import ($)	Export ($)	Population	Labor Force	Unemployment Rate (%)
China	9,596,961	9,569,901	14,500	27,060	12.61 trillion	9300	2.6	7.8	1.735 trillion	2.057 trillion	1,349,585,838	798.5 million	6.5
India	3,287,263	2,973,193	7000	314,070	4.784 trillion	3900	9.2	6.5	500.3 billion	309.1 billion	1,220,800,359	498.4 million	9.9
South Korea	99,720	96,920	2413	2800	1.611 trillion	32,400	2.2	2	514.2 billion	552.6 billion	48,955,203	25.5 million	3.2
Mongolia	1,564,116	1,553,556	0 (landlocked)	10,560	15.44 billion	5500	14.2	12.3	6.739 billion	4.385 billion	3,226,516	1.037 million	9
Laos	236,800	230,800	0 (landlocked)	6000	19.16 billion	3000	4.9	8.3	2.645 billion	2.28 billion	6,695,166	3.69 million	2.5
Bangladesh	143,998	130,168	580	13,830	311 billion	2100	8.7	6.1	34.56 billion	26.25 billion	163,654,860	77 million	5
Sri Lanka	65,610	64,630	1340	980	128.4 billion	6200	7.5	6.4	17.32 billion	9.785 billion	21,675,648	8.194 million	5.2

TABLE 6.1　(Continued) The Global Trade Agreements Matrix

3. ASEAN

Country	Geography: Land and Coastlines as Factors of Production				Economy (2012)				Trade Balance (2012)		Socioeconomic Data and Labor as Factors of Production		
	Total Area (km²)	Land (km²)	Coastline (km)	Water (km²)	GDP Purchasing Power Parity ($)	Per Capita GDP ($)	Inflation Rate (%)	GDP Real Growth Rate (%)	Import ($)	Export ($)	Population	Labor Force	Unemployment Rate (%)
Brunei	5765	5265	161	500	21.94 billion	50,500	1.2	2.7	3.02 billion	12.75 billion	415,717	205,800	2.6
Indonesia	1,904,569	1,811,569	54,716	93,000	1.212 trillion	5000	4.5	6	179 billion	188.7 billion	251,160,124	118 million	6.1
Malaysia	329,847	328,657	4675	1190	492.4 billion	16,900	1.9	4.5	202.4 billion	247 billion	29,628,392	12.92 million	3.2
Philippines	300,000	298,170	36,289	1830	423.7 billion	4300	3.2	6.6	65 billion	50.96 billion	105,720,644	40.36 million	7
Singapore	697	687	193	10	325.1 billion	60,900	4.4	1.3	379.7 billion	408.4 billion	5,460,302	3.618 million	2
Thailand	513,120	510,890	3219	2230	645.7 billion	10,000	3	5.5	213.7 billion	226.2 billion	67,448,120	39.77 million	0.6
Vietnam	331,210	310,070	3444	21,140	320.1 billion	3500	6.8	5	114.3 billion	114.6 billion	92,477,857	49.18 million	4.5
Laos	236,800	230,800	0 (landlocked)	6000	19.16 billion	3000	4.9	8.3	2.645 billion	2.28 billion	6,695,166	3.69 million	2.5
Myanmar (Burma)	676,578	5876	1930	23,070	90.93 billion	1400	1.5	6.3	7.477 billion	8.23 billion	55,167,330	33.41 million	5.4
Cambodia	181,035	176,515	443	4520	36.64 billion	2400	2.9	6.6	8.84 billion	6.148 billion	15,205,539	7.9 million	0

4. Transatlantic Free Trade Area (TAFTA) or Transatlantic Trade and Investment Partnership (TTIP)

Country	Geography as a Factor of Production				Economy (2012)			Trade Balance (2012)		Socioeconomic Data and Labor as a Factor of Production		
	Total Area (km²)	Land (km²)	Coastline (km)	Water (km²)	GDP Purchasing Power Parity ($)	Per Capita GDP ($)	Inflation Rate (%)	Import ($)	Export ($)	Population	Labor Force	Unemployment Rate (%)
USA	9,826,675	9,161,966	19,924	664 709	15.66 trillion	49,800	2	2.357 trillion	1.612 trillion	316,668,567	154.9 million	8.2
EU	4,324,782	12,440.8	65,992.9		15.97 trillion	35,100	2.6	2.397 trillion	2.17 trillion	503,890,016	230 million	10.3

5. Central American Integration System (CICA)

Country	Geography: Land and Coastlines as Factors of Production				Economy (2012)			GDP Real Growth Rate (%)	Trade Balance (2012)		Socioeconomic Data and Labor as Factors of Production		
	Total Area (km²)	Land (km²)	Coastline (km)	Water (km²)	GDP Purchasing Power Parity ($)	Per Capita GDP ($)	Inflation Rate (%)		Import ($)	Export ($)	Population	Labor Force	Unemployment Rate (%)
Panama	75,420	74,340	2490	1080	58.02 billion	15,900	5.7	10.7	24.69 billion	18.91 billion	3,559,408	1.517 million	4.4
Guatemala	108,889	107,159	400	1730	79.97 billion	5300	3.8	3	15.84 billion	10.09 billion	14,373,472	4.359 million	4.1
El Salvador	21,041	20,721	320	307	47.09 billion	7600	1.8	1.6	9.912 billion	5.447 billion	6,108,590	2.593 million	6.9
Honduras	112,090	111,890	200	832	38.42 billion	4700	5.2	3.3	11.18 billion	7.931 billion	8,448,465	3.437 million	4.5
Nicaragua	130,370	119,990	910	10,380	27.1 billion	4500	7.2	5.2	6.45 billion	3.655 billion	5,788,531	2.961 million	7.4
Costa Rica	51,100	51,060	1290	40	59.79 billion	12,800	4.5	5	16.75 billion	11.44 billion	4,695,942	2.182 million	7.8
Belize	22,966	22,806	386	160	3.048 billion	8900	1.2	5.3	808.3 million	548.5 million	334,297	120,500	11.3
Dominican Republic	48,670	48,320	1288	350	100.4 billion	9800	3.7	3.9	17.76 billion	9.08 billion	10,219,630	4.806 million	14.7

Source: M.G. Burns, based on data from WTO, 2013 World Trade Report, 2013: Factors shaping the future of world trade. Available at http://www.wto.org/english/res_e/publications_e/wtr13_e.htm (accessed August 30, 2013); CIA, The World Factbook, 2013. Available at http://www.cia.gov/library/publications/the-world-factbook/geos/ch.html (accessed on July 9, 2013); USTR, 2013. Office of the US Trade Representative. Available at http://www.ustr.gov (accessed on July 12, 2013); MERCOSUR Trade Center, 2013. Available at http://www.mercosurtc.com (accessed on July 12, 2013); UNESCAP 2013. Asia-Pacific Trade Agreement, APTA. Available at http://www.unescap.org (accessed on July 8, 2013); SAARC, 2013. South Asia Association for Regional Cooperation. Available at http://www.saarc-sec.org/ (accessed on July 8, 2013); African Union (AU), 2013. Available at http://www.au.int (accessed on July 8, 2013); COMESA, 2013. Common Market for Eastern and Southern Africa. Available at http://www.comesa.int (accessed on July 8, 2013); European Union (EU), 2013. Available at http://europa.eu (accessed on July 12, 2013); US Department of State, 2013. Available at http://www.state.gov (accessed on July 10, 2013); and Arab Monetary Fund, 2013. Available at http://www.amf.org.ae (accessed on July 8, 2013).

The facts and figures are self-explanatory as to the comparative advantages of each country. Numerous comparisons and combinations can be made as to the principal factors of production, geographic particularities; for example, an extended coastline may suggest the potential for sea transport, warehousing, and the cargo's accessibility, as opposed to a hinterland that would require a more complex multimodal logistics network. Indeed, a nation's coastline is its platform to growth and prosperity. Coastlines are not simply about ports and sea transport: they typically serve as a nation's logistics control tower. Even powerful nations with limited coastlines tend to grow disproportionately, with a part of their population thriving, and the landlocked regions gradually migrating toward the coast.

An example of the inconvenience of landlocked regions is the *Caspian Sea* in Eastern Europe, whose proven *reserves* surpass 2.9 billion tons of *oil* and 3 trillion cubic meters of *gas*. Despite the vast reserves discovered in the region, the logistics complexities and the relevant cost of extracting and pipelines' installation make this endeavor a challenging one.

National population figures can be contrasted to the actual labor force, to initially identify the availability of labor. When drilling down the facts and figures, for example, the population's composition, GDP per capita, inflation, and so on, one can reach valuable conclusions about the local markets, balance of trade, opportunities for exports, and so on. Some of the factors of production demonstrated in Table 3.1 include land, labor population, GDP, and unemployment, and other factors should be cross-checked to define the nation's factors of production.

Most important, it should be strongly suggested that globalization has disassociated a nation's economic growth with the national entrepreneurial activities in the private sector. The reason for this is that the private sector's activities are global and are not always reflected in their countries' economic status. For example, global entrepreneurs have the privilege of selecting offshore corporate tax havens or shipowners can select flags of convenience for their ships, and although these are perfectly legal business decisions, they do not count as national growth indicators.

Greece, for example, is one of the countries that has suffered the most severe effects of the 2008 global financial crisis; an event that brought a dichotomy between the public sector's collapse and the maritime industry's enormous growth. As of 2013, the country's unemployment rate has *scaled* a new record of 27.2%, a real growth rate of −6.4%, and a public debt that exceeds 156.9% of its GDP. At the same time, Greek shipping tycoons have become the second global power in fleet ownership, right after Japan, and followed by economic mega-powers such as the United States, Germany, China, and so on. This is an example of how a nation's private sector and the market niches may also merit consideration when considering national trade opportunities.

Another observation pertains to the real growth rate, which demonstrates most accurately the country's dynamic and future trends. However, this should always be compared to the country's past and current economic level, as a developed and highly sustainable economy with a 2% growth rate enjoys higher standards of prosperity compared to a less developed economy with a growth rate of 8%.

The economic analysis of the trade agreements is a rather extensive subject to be discussed. To conclude, some other factors to consider would be the national currency rates, fixed pegged arrangements, monetary policies, products per country, and so on.

6.2 PORTS AND SHIPS AS DERIVED DEMAND: TRAFFIC FORECASTING

6.2.1 Ports and the Principles of Derived Demand

By nature, the shipping industry is characterized by derived demand: the demand for a specific commodity or service generates the demand for ships and consequently ignites the demand for ports. Due to the fact that seaports' activities and the maritime industry as a whole are considered as factors of production, the demand for ports and ships is considered as derived demand, that is, the demand is not direct for the service itself, but instead it is triggered by the consumers' demand for the final product transported by ships and distributed to the final consumers through port operations and logistics' activities. Figure 6.2 demonstrates how a shift in demand for cargoes affects the demand for ships and ports.

The demand for ports and ships derives from the demand for the commodities they handle. Figure 6.3 illustrates that during the past three decades, that is, from 1980 to 2012, the weight of the global fleet has grown threefold.

Ninety percent of global commodities are being carried by water, and as world trade increases in a sustainable manner, the demand for ports and maritime operations also increases. Figure 6.4 demonstrates the steady growth of the world fleet from 1980 to 2011, classified as per ship type (UNCTAD 2011).

Ports belong to the same group of service providers within the transportation industry, that is, logistics companies, carriers, freight forwarders, and so on, that need to adjust their services, infrastructure, and superstructure arrangements, on the basis of what the market dictates. This implies that ports' fluctuation of demand is associated with their geographic location and their vicinity to other parts of the logistics chain, such as warehousing centers, multimodal transport centers/hub ports (sea–land–air), and commercial centers triggered by increased customers' demand for specific commodities.

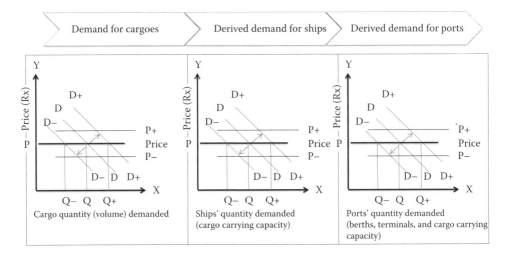

FIGURE 6.2   Derived demand for ships and ports. (Courtesy of M.G. Burns.)

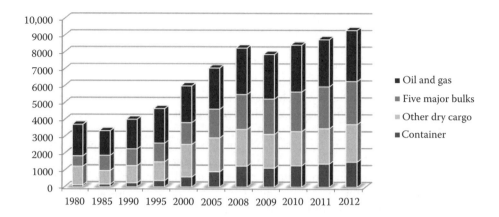

Ship types	1980	1985	1990	1995	2000	2005	2008	2009	2010	2011	2012
Oil and gas	1871	1459	1755	2050	2163	2422	2742	2642	2772	2796	3033
Five major bulks	608	900	988	1105	1295	1709	2065	2085	2335	2477	2547
Other dry cargo	1123	819	1031	1125	1928	2009	2173	2004	2027	2090	2219
Container	102	152	234	371	598	909	1249	1127	1275	1385	1498

FIGURE 6.3 World commodities carried by major ship types (global fleet, millions of DWT). (From UNCTAD, 2012. Trade and Development Report 2012. Available at http://unctad.org/en/PublicationsLibrary/tdr2012_en.pdf. Accessed on July 22, 2013.)

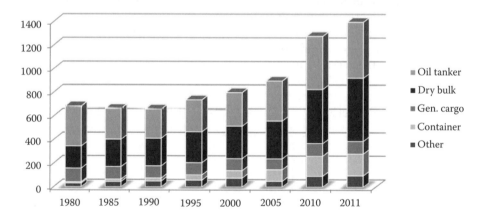

	1980	1985	1990	1995	2000	2005	2010	2011
Oil tanker	339	261	246	268	282	336	450	475
Dry bulk	186	232	235	262	276	321	457	532
Gen. cargo	116	106	103	104	101	92	108	109
Container	11	20	26	44	64	98	169	184
Other	31	45	49	58	75	49	92	96

FIGURE 6.4   World fleet by key ship types (millions of DWT).

Over the past few years, shipboard and logistics automation as well as technological innovations have significantly reduced the number of seafarers and transport employees. In practice, this means that shipboard crewmembers can handle daily operations and navigation. However, the minimum shipboard manning arrangements do not allow contemporary crewmembers to undertake full and regularly planned maintenance. Therefore, the role of ports increasingly requires that they offer full ships' maintenance and repair services in their repair yards. This signifies the beginning of an increasing demand for new port services.

6.2.2 Shipping, Ports, and the Ripple Effect

The ripple effect in transportation suggests that any changes within a key player in the industry will affect all the supply chain businesses in one or more sectors such as operations, economics, technology used, regulatory compliance, and so on.

In terms of modernization, the ripple effect suggests that when new technology (e.g., ships burning natural gas as opposed to fuel oil) or increased ships' sizes are designed, ports need to keep abreast of these developments in order to retain and increase their market share. If they fail to keep pace of the new market trends, their market share will always be threatened by alternative ports and transportation modes, that is, by rail, trucks, pipelines, and so on.

Ports should also be able to adapt to global and regional political or social changes. In examining maritime history, it becomes apparent that unexpected sociopolitical developments have demanded ports to rapidly change their role, services, and magnitude. As an example, wars and sociopolitical turmoil in oil-producing regions have resulted into canal closures, that is, Suez, pipeline closures, and bombing of port terminals. When one trade route or navigation path is restricted, alternatives need to be found. Typically, this redirected business goes to the most prepared ports.

Interestingly enough, shipping market forecasters are usually capable of predicting the major industry's trends at a macro-level, yet a regional unforeseen disaster or even a short-term disturbance may ripple through the entire global market. Isolated events such as a maritime accident, regional social turmoil, strikes, and so on, often trigger large-scale reactions, including regulatory reforms, amending long-established charter party clauses, insurance and arbitration arrangements, and so on.

An effective port development plan should be designed to examine the market trends and proactively address likely future developments, using its resources in a manner that will satisfy developing market demands. These include the demands of the government, shareholders and investors, the ports' multiple logistics' chains, cargo forwarders and receivers, shipowners, and so on. Ports also need to meet the industry requirements driven by changes in regulatory compliance.

As an example, the 1975-built tanker vessel, *M/V Erika, sank off the coast of France* in 1999. Thousands of tons of oil were released into the sea causing an environmental disaster. The incident ignited a wave of regulations pertaining to the phasing out of single-hull oil tankers and the expeditious construction of double-hull ships. In April 2001, the IMO adopted a revision of its Regulation 13G of Annex I to MARPOL 73/78, and in 2002, the EU Commission adopted and applied this measure at a global level.

Marine environmental regulations also require ports to adopt their services, for example:

- In 2012, IMO's STCW Convention (Seafarers' Training, Certification, and Watchkeeping) underwent major amendments, including increased environmental awareness and protection (IMO 2013a).
- In 2015, the US Environmental Protection Agency will enforce new environmental regulations evaluating several air pollution sources from oil and gas operations as "one combined source" (EPA Green Building 2013; EPA Green Power Partnership 2013).
- Between 2013 and 2016, the global and national ballast water requirements and shipboard technology also changes. Port pollution and invasion of invasive species will be closely monitored.

These are just a few characteristic examples that substantiate the position that regulatory measures are not triggered merely by the industry's need to change its operational, technological, economical, or other systems. Instead, the ripple effect confirms the industry's *interconnectivity*.

6.2.3 Optimum Size and Economies of Scale

Increased demand for cargoes eventually results in the increase of ships' size, that is, growth in terms of draft, overall length, beam, deadweight, and so on. This effect is due to the financial and efficiency benefits of economy of scale. To accommodate this ever-changing market need, that is, increased volumes of cargo, ship sizes grow. Furthermore, ports' designs, expansion, and development plans change in terms of ports' size, design, draft, and berth types and focus toward specific ship types. Intermodal and cargo handling arrangements should also be included in port expansion plans. Their tariffs and prices fluctuate according to demand and market fluctuations.

Hence, ports change in terms of the following:

  i. Repairs and maintenance services are shifted from the ships toward the ports.
 ii. Ports increase their size to accommodate the larger vessels built (e.g., the *Panama Canal* expansion).
iii. Ports change their design, services, and operations, to accommodate the ship types and cargoes they handle.
 iv. Ports offer the latest technological innovations and IT systems.

Optimum profitability for ports, ships, and canals arises from utilizing their full capacity, with low cost or high profit margin. Namely, optimum capacity may be distinguished into the following:

  a. *Handling capacity* (for ships, ports, and canals) measures the cargo volume to be loaded or discharged in a given period.
  b. *Hauling capacity* (for ships and other transportation modes) measures the ship's size multiplied by its speed.

A port or ship manager's aim is to achieve economies of scale, that is, minimize cost per metric ton carried. It is important to remember that an increase in size does not result in profitability unless the asset's full capacity is duly utilized, at an attractive hire or freight rate (for ships) or tariff (for ports and canals). In fact, economies of scale and the increase of global fleet size can clearly be seen in the world's major canals, that is, the Panama Canal and the Suez Canal.

The Suez Canal

The Suez Canal (قناة السويس) is the first canal globally that was built in order to boost trade and transport. The Suez Canal Authority (SCA) of Egypt owns and operates the canal. Egypt was the first nation to dredge an artificial canal thousands of years ago, in order to join the Red Sea and the Mediterranean Sea. In 1874 BC, Pharaoh *Senausert III* commenced the dredging operations, which were continued in 1310 BC by Pharaoh *Sity I*. Numerous Egyptian leaders empowered the canal's expansion, as well as the Greek *Ptolemy II* and the Romans. In 1869, the Suez Canal was opened pursuant to a decade of designing, engineering, and dredging operations. Because of the extreme geopolitical significance of the canal, any period of closure affects the shipping industry to a great extent. The canal was constructed in 1869 under the administration of a French nobleman and developer of the canal, Ferdinand de Lesseps. Seven percent of world sea transport passes through the Suez Canal.

The Suez Canal is situated in the vicinity of the oil-producing Middle East nations, and therefore any disruptions, delays, or closures have a severe impact especially in the global oil trade. During the five times that the canal has closed so far, ships have had to navigate around Africa via the Cape of Good Hope. Global fleet size has grown during the canal's prolonged closures, in order to eliminate cargo traffic disruptions and accommodate the demand for global transport.

In 1967, the Yom Kippur War (Six-Day War) caused the Suez Canal to close for eight years, that is, until 1975. The voyage to the oil markets was then prolonged as ships had to navigate around Africa and the Cape of Good Hope.

In case of a canal closure, ships will need to increase their speed at around 20–22 knots, and still, the detour around Africa will take them approximately an extra week. A typical example would be a voyage from China, Asia, to Belgium, Western Europe: the distance from Shanghai port to Antwerp terminals is 10,476.8 miles.

At a speed of 14 knots, the voyage through the Suez Canal would have a duration of 49 days and 4 hours. Hence, should the ship need to navigate through the Cape of Good Hope, the distance would be 26,630.8 miles. At a speed of 14 knots, the voyage would last 119 days and 17 hours, whereas if the vessel increases its speed to 20 knots, the voyage would last 107 days and 7 hours. It is worth noting that any increase in speed, or any detour, will significantly increase the fuel oil consumption and the voyage's operational costs. This example demonstrates the significance of the Suez Canal in global shipping and verifies the interconnection between ship's size and the canal.

Figure 6.5 illustrates the Suez Canal expansion and the growth of the global fleet since 1869.

For shipowners to undergo investments of multiple hundreds millions of dollars, they need to have the certainty and some sort of security of an increased demand for their services. For example, the securing of long-term chartering contracts (contracts of affreightment, charter parties) would justify this capital-intensive investment. A typical example of high demand for larger ships was the development of VLCCs (very large crude carriers) and ULCCs (ultra large crude carriers) with a carrying capacity ranging from 250,000 DWT to 560,000 DWT. During the canal's closure, there was an "immense demand" for these ships. By the time the Suez Canal reopened in 1975, the demand for these ships was significantly reduced, as the shorter voyages created an oversupply of fleet that dropped

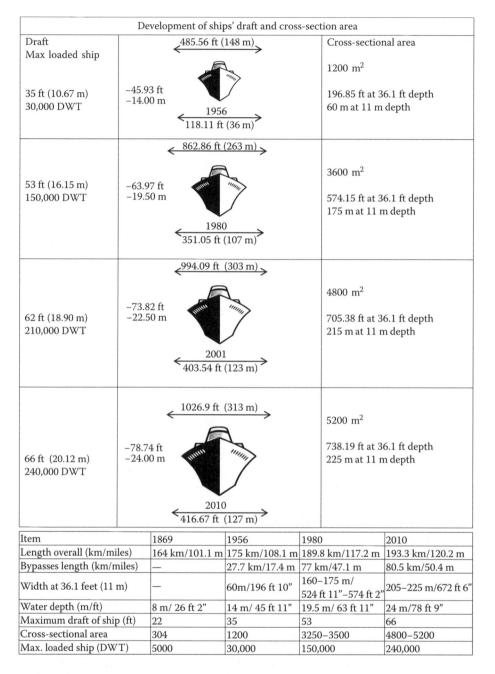

	Development of ships' draft and cross-section area	
Draft Max loaded ship 35 ft (10.67 m) 30,000 DWT	485.56 ft (148 m) −45.93 ft −14.00 m 1956 118.11 ft (36 m)	Cross-sectional area 1200 m² 196.85 ft at 36.1 ft depth 60 m at 11 m depth
53 ft (16.15 m) 150,000 DWT	862.86 ft (263 m) −63.97 ft −19.50 m 1980 351.05 ft (107 m)	3600 m² 574.15 ft at 36.1 ft depth 175 m at 11 m depth
62 ft (18.90 m) 210,000 DWT	994.09 ft (303 m) −73.82 ft −22.50 m 2001 403.54 ft (123 m)	4800 m² 705.38 ft at 36.1 ft depth 215 m at 11 m depth
66 ft (20.12 m) 240,000 DWT	1026.9 ft (313 m) −78.74 ft −24.00 m 2010 416.67 ft (127 m)	5200 m² 738.19 ft at 36.1 ft depth 225 m at 11 m depth

Item	1869	1956	1980	2010
Length overall (km/miles)	164 km/101.1 m	175 km/108.1 m	189.8 km/117.2 m	193.3 km/120.2 m
Bypasses length (km/miles)	—	27.7 km/17.4 m	77 km/47.1 m	80.5 km/50.4 m
Width at 36.1 feet (11 m)	—	60m/196 ft 10"	160–175 m/ 524 ft 11"–574 ft 2"	205–225 m/672 ft 6"
Water depth (m/ft)	8 m/ 26 ft 2"	14 m/ 45 ft 11"	19.5 m/ 63 ft 11"	24 m/78 ft 9"
Maximum draft of ship (ft)	22	35	53	66
Cross-sectional area	304	1200	3250–3500	4800–5200
Max. loaded ship (DWT)	5000	30,000	150,000	240,000

FIGURE 6.5 Suez Canal expansion and fleet growth (1869–2010). (Based on data from Suez Canal Authority.)

the freight rates and created a market imbalance. Figure 6.6 demonstrates how the global trade is facilitated by the Suez Canal, the Panama Canal, and the proposed Nicaragua Canal.

This is a timely case study, as the Egypt crisis may affect the accessibility to the Suez Canal in the near and medium-term future.

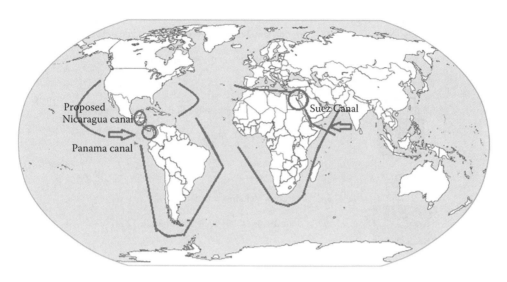

FIGURE 6.6   Suez Canal, Panama Canal, and the proposed Nicaragua Canal.

The lesson learned here is that while profitability is the ultimate objective of building larger ships and dredging or expanding ports and canals, sometimes expansion may result into diseconomies of scale, leading to oversupply, capital losses, and reduced market power.

The Post-Panamax Era

Vast amounts of investment are required for global ports and the global fleet to increase in size.

The Panama Canal expansion has exceeded $6.2 billion. The project's completion in 2015 will enable the global post-Panamax ships to transit the canal and initiate a new era of mega-ships and mega-ports:

Global Fleet Size

- The top shipowning companies aim to further increase their market share, as they are on top of the global order book: (i) APM-Maersk, (ii) Mediterranean Shipping, (iii) CMA-CGM Group, (iv) Evergreen Line, (v) COSCO Containers, (vi) Hapag Lloyd, (vii) Hanjin Shipping, (viii) APL, (ix) CSCL, (x) NYK Line, and so on.
- Maersk has ordered 20 Triple-E Class mega-containers. As their name suggests, their advantages include economy of scale, being environmentally friendly, and being energy efficient.
- The ships have a carrying capacity of 18,000 TEUs, an LOA of 400 meters (1312 feet), a width of 59 meters (193 feet), and a height of 73 meters (240 feet). The ships are being built in Daewoo Shipbuilding, South Korea, and their delivery is spread between 2013 and 2015 (Maersk Press 2013).
- Figure 6.7 illustrates the container generations and the Panama Canal (1960s–2020).
- Valemax is a new generation of 35 ships designed by the Brazilian ore conglomerate Vale. They are VLOCs or very large ore carriers, with a deadweight of 400,000 tons and a length of 360–362 meters (1181–1188 feet). They are the world's largest bulk carriers and among the world's largest ships of all categories. The ships are being built in Jiangsu Rongsheng Heavy Industries, China, and Daewoo Shipbuilding and Marine Engineering, South Korea (Vale 2013).

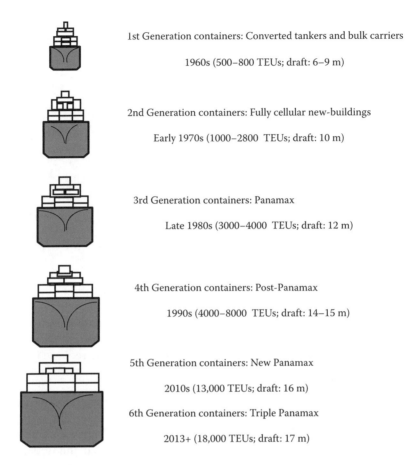

FIGURE 6.7   Container generations and the Panama Canal (1960s–2020). (Courtesy of M.G. Burns.)

Global Ports' Size

The Panama Canal expansion combined with the new-generation mega-ships has ignited a "neck-and-neck" competition among global seaports and logistics chains.

- The US ports' morphology and water draft varies: ports in the West Coast United States enjoy a naturally deep draft, whereas East Coast and Southern ports require dredging operations. The US Senate Subcommittee has approved a record amount of $1 billion for the fiscal year 2014, allocated to maintain America's federal navigation passages and connected infrastructure. The United States is a major global trader that is located in the vicinity of the Panama Canal; hence, the development of selected mega-ports will further boost the American economy. Some of the ports that were ready to accommodate the new-generation mega-ships include Baltimore, Norfork, Virginia, New York/New Jersey, and so on.
- An interesting alternative has also been adopted by the European Union: in addition to a vertical development, that is, the enlargement of existing ports such as Antwerp, Rotterdam, Felixstowe, Southampton, Hamburg, and so on,

a horizontal development has been established: the financing of medium-sized ports aims to relieve the congestion and traffic from Europe's major ports, that is, Rotterdam, Antwerp, and Hamburg, which amount to 20% of European sea trade (BBC 2011; Port of Rotterdam 2013; DP World Southampton 2013).

Rapidly growing economies are full steam ahead with port dredging:

- In the Far East and the Middle East, the key mega-ports that are ready for the post-Panamax fleet include Hong Kong, Singapore, Colombo, Dubai, and Salalah. In order to retain their competitive advantage, further investments and ongoing dredging plans are in place (Maritime and Port Authority of Singapore 2013). Sri Lanka's major port, Colombo, underwent a $400 million expansion plan in order to facilitate contracts with the new-generation mega-ships. Colombo's major share of earnings derives from handling Indian containers. Low-priced port tariffs, a strategic geographical position, and larger and deeper berths contribute to enticing larger vessels that provide economies of scale to charterers and profits to the port.
- China's port expansion plans include the development of Tianjin, Dalian, Wenzhou, Yichang, Guangzhou, and so on. Dalian in particular is a priority for China and its elite business partner, Vale of Brazil (Vale 2013).
- Brazil aims to invest $1.3 billion to expand the ports of Santos and Para.
- India's objective is to reclaim most of its cargo moved through motherships that call nearby ports of deeper drafts. Hence, four major ports will be dredged, two in the East Coast and two in the West Coast.

Mega-ports: A New Era and the Threat of Oversupply and Overinvestment

With expectations of increased business, global ports actively pursue the dredging operations that will enable their berths to have capacity for greater ships and increased cargo volumes. Naturally deeper ports enjoy a competitive advantage over more shallow ports: the ports that are first ready for the new era are most likely to conclude lucrative contracts with the larger ships. While competition is both healthy and rewarding, it becomes obvious that the first ports to conclude the contracts will be the great winners. At the same time, the "last come, last served" mega-ports will create a dangerous oversupply, which over time is likely to reduce port tariffs. The game is not over for the mega-ports that will be dredged at a later stage: they can easily become the "alternative ports," an attractive solution for terminal operators. Nevertheless, in order for them to increase their trade growth, they will have to drive the port tariffs down, as well as promise—and deliver—exceptional services.

The preparations for the post-Panamax era are based on positive expectations and scenarios for a global economic growth. This is a feasible scenario mainly for the oil and gas industry, as the newly discovered reserves in multiple global regions such as the United States, China, Russia, Algeria, Scandinavia, Australia, Brazil, the Eastern Mediterranean, and so on, are estimated to increase sea trade and offshore activities.

The dry bulk and container ship sectors are heavily reliant upon long-term contracts with charterers, that is, manufacturers, major commodity traders, and so on. There are also strong expectations that rapidly developing economies such as China, India, and Brazil, to name a few, will boost global trade at a sustainable level.

In essence, there are two major points that port authorities should consider prior to investing:

a. In reality, the larger container vessels were built by shipowners in order to benefit from economies of scale. This may result in fewer, larger ships and not a dramatic increase in cargo volumes.

b. Furthermore, terminal operators are typically conglomerates that control large segments of the global trade. Hence, instead of bargaining with established ports in existing trade routes, they are likely to create new mega-terminals in countries with low-cost, high-value factors of production (raw materials, oil and gas, land, labor). Ports in advanced and rapidly developing economies may not be affected, as they are crucial components of the production and consumption chains. Terminal operators will not bypass global key players. However, previously thriving and currently struggling economies with a shrinking market may be surprised to see poor returns-on-investment, as their long-established clients increasingly shift the market toward less developed economies. In observing the trade agreements between China and multiple African countries, as well as the investment of global conglomerates in African ports and land, two things are now apparent:

   i. Africa will become the new China.

  ii. Whoever controls Africa will become the next global leader (Burns 2012).

African Investment by APM Terminals: The Badagry Port, Nigeria

APM Terminals is a major terminal operator and a key port and terminal investor in the Gulf of Guinea, West Africa. The Badagry Port venture in Lagos State, Nigeria, is an exemplary public–private collaboration between APM and the State of Lagos: the free trade zone and Greenfield hub-port venture will facilitate trade in the rapidly growing Gulf of Guinea and will ascertain the role of Nigeria as a leading mega-port in West and Central Africa. Infrastructure investment for inland transportation has been secured through the Benin–Lagos Expressway, a 10-lane highway. The vast reserves of oil in Nigeria and the nations within the Gulf of Guinea will ascertain regional growth and demand for containerized cargoes.

African Investment by China: Dar es Salaam, the Largest African Port

China's trade agreements and heavy investment in African ports, terminals, and land commenced in the 1990s. China is the second largest partner of Africa, after the United States. Approximately 1000 Chinese companies have been established in Africa, mostly related to trade, transport, energy, and financial services. In terms of port management, China is investing $10 billion in order to establish the port of Dar es Salaam, Tanzania, as the largest African container port. Because of the port's strategic geopolitical location, the port will serve as a major cargo distributor in the landlocked nations of Western and Central Africa. Owing to the port's vicinity to the Suez Canal and Middle Eastern ports, *Dar es Salaam* was designed to compete with well-established ports and trade routes in the region.

Based on these examples, it becomes apparent that heavy investment in developing regions is likely to create a shift of well-established trade routes. These radical, often unpredictable changes in sea trade routes, combined with the overinvestment of ports in weakening trade routes, are factors that global ports should consider prior to being committed to irrevocable and intensive investment plans.

6.3 CAPACITY UTILIZATION, CAPACITY MANAGEMENT, AND CAPACITY PLANNING: PORTS' TECHNOLOGY AND INNOVATION

6.3.1 Capacity Utilization, Capacity Management, and Capacity Planning

Capacity utilization is a vital constituent of *capacity management* and *capacity planning* that focuses on the market, the port's services, and resources. *Capacity utilization* measures the degree to which a port uses its constructive capacity and its factors of production. In particular, it evaluates and compares the ratio among the actual materialized production generated by the port, and the potential production that could have been materialized, subject to increased efficiency, or alternative allocation of resources. Certain focal points of capacity utilization include the number and length of berths, total terminal area, handling capacity, accessibility, warehousing, technology, and innovation.

Capacity utilization rate or *operative rate* measures to which extent the port's actual output has reached its full potential. It is measured by the formula:

$$CUR = \frac{Real\ Output - Potential\ Output}{Potential\ Output} \times 100 \qquad (6.1)$$

In order to estimate the average capacity utilization rate for a port's *overall output*, the following econometric formula may be used:

$$CURx = \sum_{y=1}^{n} CURd \times W(y) \qquad (6.2)$$

where CURx is the average capacity utilization rate for a port's *overall output*; CURd is the capacity utilization rate for a specific port's department, berth, and terminal of facilities; W(y) is the weighting function for the specific port's department, berth, and terminal of facilities; and n is the number of departments, berths, and terminals of facilities.

Both capacity management and capacity planning aim to ensure that the port's infrastructure, services, and superstructure meet the market's demand. In other words, they aim to monitor and control the ships' time costs and the port's costs. While capacity management works with the existing resources, capacity planning entails the elements of strategic forecasting and proactiveness; it plans ahead based on future market trends, as well as clients' needs. Both capacity management and capacity planning are strategic tools that seek for an attractive market equilibrium between supply and demand.

Capacity management involves the following:

- Analyzing the operational efficiency and production metrics, based on the port's input and output
- Performance engineering, monitoring, and controlling
- Evaluating internal processes and external factors, that is, within the supply chain, the market, nationwide, or regionally
- Continuously examining methods that would best utilize the port's capacity potential

Aggregate capacity management is a technique of capacity management and planning of all the port's assets, funds, services, and facilities. While prioritizing cost efficiency, its application seeks to achieve equilibrium among demand and supply (port capacity) through incorporating all resources and factors of production. Its implementation entails three simple, yet effective steps: (i) weighing port capacity versus aggregate demand for a given period, (ii) determining substitute plans suitable for any market shifts, and (iii) deciding upon the most suitable option.

Capacity planning entails the following:

- Conducting market forecasts that would identify shifts in demand, future traffic, and market trends.
- Comprehension of the market's dynamics will help port managers formulate the strategies for the port's future.
- Considering reallocation of resources, funds, methods, and processes, in order to best promote the port's services, facilities, and marketability.

Port capacity encompasses all the factors of production, that is, entrepreneurship, labor, land, and capital, which also include innovation, technology, infrastructure, and superstructure. Optimum port capacity requires the minimum cost per unit (input) and the maximum quality of services provided (output), that is, value for money offered to the clients. It may focus on the port's activities or expand throughout the entire supply chain. Resources and funds' allocation, monitoring, and controlling help port managers to offer optimum capacity and services, while taking into consideration the port's competitive advantages, market opportunities, and threats.

The results of capacity planning eventually influence the entire port as capacity affects all functions, operations, and departments. Decision makers should understand that any choices made are binding, and sometimes irrevocable. Based on the presumption that funds and other resources are limited, it is significant for investors to distinguish successful from unsuccessful venture capital investments, and maximize return on investment (ROI). Unsuccessful capacity planning practices imply that the port services are not appealing to the market; hence, the port's supply cannot reach demand. This may lead to loss of resources during the input process, reduced income during the output process, marketing challenges, and reduced market share.

A checklist that includes the key elements for monitoring port capacity performance include the following:

- Optimum occupancy of berths, terminals, infrastructure, superstructure
- Time:
  - Swift turnaround time
  - Eliminated or reasonable delays or bottlenecks
- Economies of scale should be achieved through the following:
  - Input efficiency
  - Minimum cost per unit of output
  - Banking and finance, that is, through attractive investment arrangements, repayment methods, rate of exchange or balloon repayment clauses, and so on
  - Purchasing, that is, through large-scale orders and discounts
- High quality of services will ensure optimum capacity and eliminate delays, corrective action, claims, and so on

- Minimizing errors such as damage or loss of property, of the ships, container boxes, cargoes, and so on
- Engineering, operational, technical, and maintenance efficiency
- Human productivity, skills, training, talent
- Legal, regulatory, and policy compliance
- External dynamics such as the market, the economy, supply chain issues
- Port facilities, availability, connectivity, resource allocation

The main strategic variations of capacity planning are as follows:

a. *Lead strategy*; this is a highly competitive method of increasing capacity in antic-ipation of increased market share and is prevailing among regional and global competitors. Here, the port is confident of the market's growing trends and acts upon the principles "first come, first served," or "high risk, high gains." On the other hand, there is a higher risk of losses or improper investment. Benefits from economies of scale and rapid growth are the port's priorities.
b. *Match or adjust strategy*; this is an intermediate option of pursuing growth with cautious steps and moderate levels of progress. This strategy is preferred during times of market uncertainty and when the port managers wish to take action without risking valuable investment funds and resources.
c. *Lag strategy*; this is the most conservative port action plan, where port managers boost capacity only after the market demand is strongly indicated. This method is usually implemented when prior pessimistic market scenarios prevailed, or when limited resources are available. The total lack of risk brings the port into a stagnant position. Although port managers may select this method in order to minimize losses from misallocation of resources and oversupply, this decision is associated with delay in action, which results to loss of clients and loss of resources.

6.3.2 Port Capacity and Competition

While equilibrium between capacity demand and supply ensures smooth business deal-ings within regional ports, any disproportion in terms of capacity and supply seriously increases competition. Conversely, reduced capacity in regional ports generates the need for funding; hence, ports may again compete while in search of investors or subsidies.

Oversupply is also detrimental to tariffs, as ports may be compelled to reduce their charges in order to keep their existing clients.

Other factors that need to be incorporated in capacity planning forecasts are the ele-ments of accessibility, regional markets and supply chains, trade routes, and so on. It is preferable for ports to differentiate and thus establish their comparative advantages and niche markets, rather than trying to increase their regional influence by competing in saturated markets.

Port capacity planning should focus on port efficiency and optimum utilization regarding berth occupancy, vessel traffic control, cargo handling, and so on. In addition, they should focus on running costs, as well as look into incurring costs of not only the construction but also the opportunity costs, that is, the comparison between the selected investment or resource allocation, versus the alternative choices and their subsequent

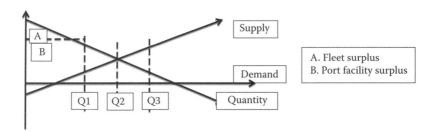

FIGURE 6.8   Matching supply and demand for berth occupancy and port services.

profits versus losses. Figure 6.8 shows how supply and demand for berth occupancy and port services are matching.

6.3.3 Port Technology and Innovation

Large-scale transportation has necessitated larger vessels and ports, as discussed in Section 6.2. At the same time, the increasing growth of global maritime trade would not have been possible without innovative technology at ports, onboard ships, and throughout the entire logistics system.

Innovative technology and innovation boost port capacity and ensure optimum efficiency. New technologies monitor, control, and accelerate cargo flows at berth, at warehouses, and throughout the supply chain. Port traffic, berth scheduling, and cargo handling are processed with the use of state-of-the-art technologies, software, and equipment.

Port gates' technologies, OCRs, and IT enable the fast processing of cargoes and container boxes as well as the clearance of documentation, with minimum or no human intervention. Software and the Internet enable cargo forwarders, charterers, receivers, agents, commodity brokers, and so on, to track specific cargoes and follow up with online documentation such as bills of lading, contracts of affreightment, bank guarantees, and so on.

IT enables information exchange, interactive systems, and smart software that not only can track cargoes throughout the entire supply chain but also can identify any discrepancies, errors, and inconsistencies.

As discussed in the introduction of this chapter, the industry's focal points are Energy, the Environment, and Economics or Economies of scale.

   i. Economy + Larger ports + Larger ships = Economies of scale:
      Port and maritime managers seek to benefit from economies of scale, that is, the cost benefits achieved owing to larger size or cheaper unit cost. Hence, this is an era for larger ships and larger ports. This means that a modest global trade growth will now be reshuffled among fewer and bigger ships and ports. At the same time, medium- and smaller-sized ships and ports can still play a dynamic part based on the ever-changing trade routes, elements of demand and supply, and global political and economic changes that are likely to occur.
  ii. Environmental protection and innovative technologies:
      Traditionally, the maritime industry has been proactive and sensitive toward the environment. Some of the most crucial regulatory requirements for the industry include ballast water management and treatment systems, IMO's requirements for

low sulfur limits, EPA's vessel general permits regulations, and STCW's Manila Amendments for environmental awareness, and so on. The most significant regulatory requirements are briefly stated herewith (also see Chapter 10):

EPA, 2015: "Air Aggregation" or "Single Source of pollution":

New environmental regulations by the US Environmental Protection Agency are effective as of January 2015. Several air pollution sources from oil and gas operations will now be considered as "one combined source."

BWM, 2016:

The IMO Convention has established timelines for the installing of ballast water treatment technologies in order to fulfill the regulatory ballast water treatment standards. The US Coast Guard and many national Coast Guard Authorities have adopted these regulations. These regulations require ships to be installed with innovative technology whose installation alone, excluding annual maintenance, costs from $1 million to $6 million per ship.

IMO (2010–2015) Emission Standards, low sulfur limits:

Sulfur oxides and particulate matter emissions will be further regulated and monitored in all shipboard fuel oil combustion machinery, including main engines and auxiliary engines, along with equipment such as boilers and inert gas generators (IMO 2013).

Although the maritime industry is responsible for less than 12% of global maritime pollution, numerous maritime professionals have been combating the environmental pollution and have voluntarily invested billions of dollars in environmentally friendly ships, ports, and innovative technologies.

An Environmental Legacy:

George P. Livanos (1926–1997) of "Ceres Hellenic" was a Greek tycoon passionately dedicated in environmental protection. In addition to his entrepreneurial spirit and the innovative designs of his 180 ships, his forward thinking led him back in 1982 to sign a declaration of *voluntary* commitment "To Save the Seas," cosigned by five global maritime organizations. This is an example of how a handful of maritime pioneers can indeed change the world.

iii. Energy:

On top of the environmental and regulatory parameters, the element of energy has a severe economic significance: fuel oil consumption is the highest operational cost that shipowners or charterers have to bear. Over the past 20 years, average freight rates have increased by less than 5%, whereas fuel oil price has increased from $20 a barrel in 1993 to over $107 in 2013.

Modern ships are gradually converting from heavy fuel oil to liquefied natural gas (LNG) operations. This is the beginning of an energy shift era, where shipowners are willing to invest in innovative engine designs, in search for a more sustainable and viable energy solution.

Some of the most modern vessels that tackle the elements of size, the environment, and energy efficiency are demonstrated in the following case study of Wärtsilä, a global innovator in the field of ship technology and alternative energy efficiency.

CASE STUDY: WÄRTSILÄ—INNOVATING FOR SUCCESS

Port technology encompasses three major sectors: port infrastructure, port super-structure, and ships' technology. As the demand for ports is a derived demand, ships' technology and innovation drive to a great extent ports' innovation. Wärtsilä is a characteristic example of groundbreaking ship technologies that can transform the maritime industry by achieving alternative energy efficiency and independence, cost-effectiveness, environmental protection through emissions reduction, optimum engine performance, innovative designs, and much more.

THE COMPANY

Wärtsilä is a major global innovator specializing in full life-cycle power alternatives for the maritime and energy industries. While focusing on pioneering technological solutions, Wärtsilä offers to its global customers optimum efficiency achieved through economy, environmental and energy advancements for ships, offshore platform supply vessels, power plants, and other global industrial sectors. The company develops groundbreaking concepts on LNG fuel, engine lubrication management, and ultra large ship propulsion. Wärtsilä's new technologies are especially designed for both newbuilding vessels and conversions to secondhand ships. The company is headquartered in Helsinki, Finland, and recruits over 19,000 employees in over 70 countries (Wärtsilä 2013).

STRATEGIC ALLIANCES

The company strengthens its global leadership position by establishing strategic partnerships with influential companies from the maritime, offshore, and other industries. The partnership with Hamworthy combines the exhaust gas scrubbing systems and marine engineering know-how with industry strategists to develop a powerful foundation for future innovations (Letnes 2012).

OFFSHORE SOLUTIONS

Wärtsilä's technological innovations encompass the offshore oil and gas industry, with focus on (i) offshore service vessels, (ii) offshore drilling, and (iii) offshore production. Wärtsilä delivers an extensive series of offshore solutions such as power generation management and vessel dynamic positioning, as well as complete electrical and automation systems, emission control systems, and customized vessel design.

LEADERSHIP

The driving force behind Wärtsilä's success is its highly skilled management team and employees with diverse skills, talents, and a corporate vision.

Captain Paul Glandt is Wärtsilä's Director, Ship Power Business Development for North America, responsible for both commercial and government sector business development. With a lifelong career with the maritime industry, Capt. Glandt possesses the expertise in identifying market opportunities and operating solutions to drive forward revenue and profit growth through complex start-up, turnaround, and high-growth cycles. An esteemed US Navy Veteran, he served for 30 years with the Naval Reserve as an Engineering Duty Officer, and retired at the rank of captain supporting the Naval Sea System Command.

Capt. Glandt's professional experience enables him to grasp the current and future market trends in the most accurate manner. He considers Wärtsilä well positioned in a market with increasing requirements for technological innovation and optimum performance. In particular, Wärtsilä's initiatives offer a wide spectrum of solutions to meet the market's needs:

- Supporting shipping efficiencies—Larger ships with larger and more fuel-efficient engines improving the ships' Energy Efficiency Design Index (EEDI) and the Ship Energy Efficiency Management Plan (SEEMP) as per MARPOL Annex VI. Modern technology allows for larger vessels with reduced costs, such as Maersk's triple E.
- Reducing emissions—Helping ports meet increasing demands to reduce emissions. In the United States with lower priced natural gas, the shipowners will also realize reduced fuel operational costs.
- More environmental efforts to help the owners comply with regulations. As an example, scrubbers reduce the sulfur emissions, allowing shipowners to comply with regulations while using lower-cost fuels. Furthermore, cruise ships/ferries use LNG as a marine fuel, reducing emissions, changing the requirements for port fueling capability, and again in the United States reducing operating expenses.

WÄRTSILÄ'S INNOVATIONS

LNG Fuel Consumption to New and Converted Ships

It has been almost 10 years since Wärtsilä introduced its first innovative design for a large Ro-Ro passenger vessel (Ro-Pax) using LNG as the main fuel, and the company's partnership with STX Europe, which launched a post-Panamax cruise ship design using LNG as the main fuel (Figures 6.9 and 6.10).

FIGURE 6.9   The concept features a short superstructure with a wide and long hull. (Courtesy of Wärtsilä.)

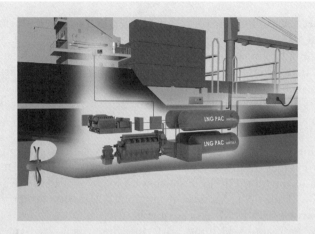

FIGURE 6.10   Design using LNG as the main fuel. (Courtesy of Wärtsilä.)

M/V Viking Energy of Eidesvik Offshore Company, Built in 2004

M/V Viking Energy (Figure 6.11) is a "new generation support vessel" designed by Wärtsilä ship design and built by Kleven Verft shipyard, Norway. It is the first LNG-driven ship globally: the ship has a diesel–electric propulsion plant consisting of four Wärtsilä 6L32DF dual-fuel engines, that is, burning both LNG and diesel oil. The alternative LNG option offers the ship an overall 30% fuel cost reduction. From an environmental perspective, this leads to a 90% reduction in the NOx emissions—that is, almost 200 MT annually—and a 30% decrease in CO_2 emissions. The ship's particulars are as follows: LOA: 94.9 m; LBP: 81.6 m; breadth molded: 20.4 m; draft (max): 7.9 m; cargo deck area: 1030 m³ (Eidesvik 2013).

FIGURE 6.11   The *M/V Viking Energy.* (Courtesy of Wärtsilä.)

M/T Bit Viking of Tarbitt Shipping, 2011 Conversion

M/T Bit Viking (Figure 6.12), the 25,000 DWT product tanker ship, was the very first ship globally that was converted by Wärtsilä from heavy fuel oil to LNG consumption (Figure 6.13) (Tarbit 2013).

M/V Viking Line of STX Europe, 2011 Conversion

Wärtsilä has now produced a unique solution for medium-sized cruise ships that emphasizes the benefits of effective machinery functioning on ecofriendly LNG. Wärtsilä received significant global recognition in 2010 after the launching of LNG as a marine fuel for the 60,000 GT Cruise Ferry *Viking Line* by STX Europe (Figure 6.14). The ship will be furnished with four Wärtsilä dual-fuel engines and

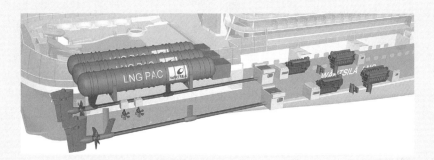

FIGURE 6.12   Dual-fuel electric machinery concept. *M/V Viking Energy*. (Courtesy of Wärtsilä.)

FIGURE 6.13   The *Bit Viking* owned by Tarbit Shipping after becoming the world's first merchant ship to undergo an LNG conversion. (Courtesy of Wärtsilä.)

FIGURE 6.14   *M/V Viking Line* for STX Europe. (Courtesy of Wärtsilä.)

an LNGPac gas storage and handling design. The distinguished advantages pro-vided include the following:

- Significant emission reduction, environmentally friendly.
- Low-cost consumption.
- Super-efficient LNG storage capacity onboard: the ship can make a 12-day voyage with no need for refueling.
- Designed to accommodate the large volumes of LNG cylindrical tanks in compliance with the IMO interim guidelines.
- Larger hull beam increases stability and the option for additional cabin decks.
- High percentage of balcony cabins (Levander 2011).

LNG Fuel Consumption to New and Converted Ships

Wärtsilä's SCHIFFKO CV 7300

The *SCHIFFKO CV 7300* is a product of Wärtsilä ship design. This post-Panamax multipurpose/container ship of over 7300 TEUs has over 1300 reefer plugs and is designed with all cellular cargo holds. Length: 325.40 m; GT: 77,500; service speed: 25.50 K.

The following are the vessel's special features:

- Low-speed main engine
- Low fuel consumption, waste heat recovery
- Optional design with dual-fuel auxiliary engine(s)
- Optionally available LNG-powered auxiliary gen-sets
- Modular deckhouse
- Optimum safety and ecological specifications (Antonopoulos 2012)

FIGURE 6.15   The *Wärtsilä X92* offers superior propulsion performance for large and ultra large container vessels. (Courtesy of Wärtsilä.)

The Wärtsilä X92

The *Wärtsilä X92* (Figure 6.15) is a two-stroke diesel marine engine that is designed to supply effective propulsion power to the new-era post-Panamax container ships. The blend of a large bore, long stroke, and low shaft speeds, along with the sophisticated and well-established common-rail technology, leads to an engine with particularly high efficiency and outstanding environmental performance.

   This innovative engine offers distinct tuning features for delivering optimum fuel consumption with diverse operating load profiles, for example, as part and low load. Extremely low, steady running speeds at 12% of nominal speed can be performed.

BLENDING ON BOARD (BOB): A MAERSK–WÄRTSILÄ PARTNERSHIP

BOB is a breakthrough design by Maersk Fluid Technology Inc. while in partnership agreement with Wärtsilä in order to jointly undertake the system's marketing and sales. BOB maximizes the general lubrication performance of large-bore diesel engines while at the same time it boosts its operational versatility and autonomy (Jumaine et al. 2012).

INTERVIEW

Captain Paul Glandt, Wärtsilä's Director, Ship Power Business Development for North America.

REFERENCES

African Union (AU). 2013. Available at http://www.au.int. Accessed July 8, 2013.

Antonopoulos, D. 2012. The new Wartsila X92 for propulsion of modern large and ultra large container vessels. *Wartsila Technical Journal*, February 2012.

Arab Business, Roberts, L. 2007. Yemen to join GCC by 2015. Available at http://www.arabianbusiness.com/yemen-join-gcc-by-2015-57086.html. Accessed January 1, 2013.

Arab Monetary Fund. 2013. Available at http://www.amf.org.ae. Accessed July 8, 2013.

ASEAN. 2013. The Association of Southeast Asian Nations. Available at http://www.asean.org. Accessed July 21, 2013.

BBC, Scott, R. 2011. Felixstowe opens new berths for giant container ships. Available at http://www.bbc.co.uk/news/business-15082311. Accessed July 12, 2013.

Burns, M. 2012. The global crude oil market: Shifting trade flows with the current financial crises. Asian Logistics Round Table, ALRT 2012, June 2012, Canada.

CEFTA. 2013. Central European Free Trade Agreement CEFTA. Available at http://www.cefta.int. Accessed July 14, 2013.

CIA The World Factbook. 2013. Available at http://www.cia.gov/library/publications/the-world-factbook/geos/ch.html. Accessed July 9, 2013.

COMESA. 2013. Common market for Eastern and Southern Africa. Available at http://www.comesa.int. Accessed July 8, 2013.

DP World Southampton. 2013. Available at http://www.dpworldsouthampton.com/why-southampton/new-deep-water-berth. Accessed July 15, 2013.

Eidesvik. 2013. m/V "Viking energy" Available at http://www.eidesvik.no/getfile.php/norske%20dokumenter/Datablad/ Viking%20energy%281%29. pdf. Accessed August 15, 2013.

EPA Green Building. 2013. Environmental Protection Agency Green Building. Available at http://www.epa.gov/greenbuilding. Accessed July 10, 2013.

EPA Green Power Partnership. 2013. Environmental Protection Agency Green Power Partnership. Available at http://www.epa.gov/greenpower/communities/index.htm. Accessed July 15, 2013.

European Union (EU). 2013. Available at europa.eu. Accessed July 12, 2013.

Foreign Trade Information System. 2013. Available at http://www.census.gov. Accessed July 12, 2013.

Gulf Cooperation Council. 2013. Available at http://www.globalsecurity.org. Accessed July 8, 2013.

IMO. International Maritime Organization. 2013a. Ballast water management. Available at http://globallast.imo.org. Accessed August 3, 2013.

IMO. International Maritime Organization. 2013b. Available at http://www.imo.org. Accessed July 8, 2013.

Jumaine, S., Zehnder, M. and Wiesmann, A. 2012. Blending on board–innovative engine lubrication management. *Wartsila Technical Journal*, January 2012.

Letnes, M. 2012. Combining complementary strengths to meet global environmental requirements. *Wartsila Technical Journal*, February 2012.

Levander, O. 2011. Dual Fuel Engines, latest developments. Available at http://www.ship-efficiency.org/onTEAM/pdf/PPTLevander.pdf. Accessed August 5, 2013.

Maersk Press. 2013. First Triple-E named Mærsk Mc-Kinney Møller. Available at http://www.maerskpress.com/NEWS-ROOM/first-triple-e-named-mrsk-mc-kinney-mller/s/f2b9b688-00f3-4ff6-86b0-5a834f48912a. Accessed July 15, 2013.

Maritime and Port Authority of Singapore. 2013. Available at http://www.mpa.gov.sg. Accessed July 8, 2013.

MERCOSUR Trade Center. 2013. Available at http://www.mercosurtc.com. Accessed July 12, 2013.

North American Free Trade Agreement (NAFTA). 2013. US Dept. of Agriculture: Available at http://www.fas.usda.gov/itp/policy/nafta/nafta.asp. Accessed July 2, 2013.

Port of Rotterdam. 2013. Your chemical port of choice. Available at http://www.portofrotterdam.com/en/News/pressreleases-news/Documents/Your_chemical_port_of_choice-PDF_tcm26-20160.pdf. Accessed July 8, 2013.

SAARC. 2013. South Asia Association for Regional Cooperation. Available at http://www.
    saarc-sec.org/. Accessed July 8, 2013.
SADC. 2013. Southern African Development Community. Available at http://www.sadc.
    int. Accessed July 15, 2013.
SAFTA. 2013. South Asia Free Trade Agreement. Available at http://saarc-sec.org. Accessed
    July 11, 2013.
SICA. 2013. Central American Integration System. Available at http://www.sica.int.
    Accessed July 8, 2013.
Suez Canal Authority. 2013. Available at http://www.suezcanal.gov.eg/. Accessed July 22,
    2013.
Tarbit. 2013. M/T "Bit Viking". Available at http://www.tarbit.se. Accessed August 15,
    2013.
UNCTAD. 2011. Trade and Development Report 2011. Available at http://unctad.org/en/
    Docs/tdr2011_en.pdf. Accessed July 22, 2013.
UNCTAD. 2012. Trade and Development Report 2012. Available at http://unctad.org/en/
    PublicationsLibrary/tdr2012_en.pdf. Accessed July 22, 2013.
UNESCAP. 2013. Asia-Pacific Trade Agreement, APTA. Available at http://www.unescap.
    org. Accessed July 8, 2013.
US Department of State. 2013. Available at http://www.state.gov. Accessed July 10, 2013.
US FAS. 2013. Federation of American Scientists. Available at http://www.fas.org/.
    Accessed July 8, 2013.
USTR. 2013. Office of the US Trade Representative. Available at http://www.ustr.gov.
    Accessed July 12, 2013.
Vale Brazil. 2013. Available at http://www.vale.com. Accessed July 22, 2013.
Wärtsilä. 2013. Available at http://www.wartsila.com. Accessed August 15, 2013.
World Trade Organization (WTO). 2013. World Trade Report, 2013: Factors shaping the
    future of world trade. Available at http://www.wto.org/english/res_e/publications_e/
    wtr13_e.htm. Accessed August 30, 2013.

CHAPTER 7

Strategic Alliances, Market Positioning, and Differentiation

Business has only two functions:
Marketing and Innovation.

Milan Kundera

7.1 DEVELOPING HOLISTIC MARKET POSITIONING AND DIFFERENTIATION

Within the marketing industry, a "brand promise" is the commitment to deliver made between that brand and its audience. The "brand" in this instance is a port and the "audience" is the terminal operators, visiting ships, and others who trade and do business with a port. Additionally, "holistic marketing" is a marketing strategy based on the collective development, design, and implementation of marketing programs, processes, and activities that recognize their breadth and interdependencies.

Market competition among global and regional ports encourages port managers to alter their clients' acquisition and retention strategies, by augmenting the port's brand promise. This section examines how holistic marketing can be utilized as an effective tool for maximizing port clients' retention by delivering or exceeding the brand promise.

7.1.1 Port Strategies and the Components of Holistic Marketing

A port manager's marketing strategies entail multiple tiers of marketing commencing with internal marketing and expanding to external and interactive marketing, through holistic marketing. The intense competition among global ports necessitates the application of holistic marketing in all its industrial sectors: manufacturing, commercial, and service:

a. *Internal Marketing*, which pertains to a port's commitment to motivate, train, and support its employees in order to meet its clients' needs.
b. *External Marketing*, where emphasis is placed in communication and systematic gathering of clients' input as a means of improving performance. The port's market expansion and customers' retaining strategy are based on the information gathered on regional and global markets.

FIGURE 7.1   Holistic marketing applied in port management. (Courtesy of M.G. Burns.)

   c. *Interactive Marketing* involves social responsibilities and relationship market-
      ing. It is a port's powerful shift toward enhanced communication with clients,
      by using advanced interaction with the entire supply chain through technologies
      and systematic data gathering and assessment.
   d. *Holistic Marketing* is a strategy that considers the port as an entity and encom-
      passes the different components of marketing. Information gathered and internal
      brainstorming techniques lead to action, such as a revised port strategy, changes
      in port planning, and development and amelioration of services, infrastructure,
      and superstructure. Figure 7.1 shows how holistic marketing encompasses the
      different levels of marketing.

7.1.2 Aligning Workforce with the Port's Brand Promise

Pursuant to the technological evolution of the past few decades, the role of technology
seemed to overshadow the significance of the workforce. A distinction was made between
the professionals that would give instructions to high-tech machinery and labor that

would receive instructions from the machinery. The recent economic crisis and the persistent market fluctuations proved the anthropocentric nature of the maritime industry in general, and ports in particular. Humans are able to attract clients, and humans are able to retain them by delivering the port's brand promise.

In December 2012, 14,500 American dockworkers from Baltimore to Texas announced their intention to strike, a decision that could disable the nation's leading ports and interrupt the movement of goods throughout the country. This strike could be compared to the 2002 strike in the West Coast that lasted 11 days and caused losses of $1 billion per day, or about 4% of the US output (Plumer 2012).

During the 1980s, strategic marketing in the maritime industry introduced a more persuasive advertisement form, occasionally aggressive or highly competitive. The brand promise was often exaggerated in comparison with the actual service delivered. Over the years, the global market brought about more intense competition and global key players had a wider choice of alternative ports, transportation networks, and service providers. Consequently, ports decided to amplify the brand promise, in an effort to gain market leverage and increase revenue.

Nevertheless, clients feel betrayed when the maritime services rendered do not meet the brand promise. Clients' satisfaction seems to be proportional to the degree it meets or exceeds their expectations on utility and the "value for money" it offers.

Although billions of dollars are invested annually into ports' marketing, for the brand message to be effective, it is necessary to consider its alignment with employees. Port executives and workforce that operate among corporate strategy and brand development are more likely to act consistently and align their corporate goals to their clients' aspirations. It is important to remember that most global ports deal with public, not private commodities; therefore, their corporate goals and aspirations may be of a "public service" and "social responsibilities" character, and not strictly related to profitability and aggressive competition. This belief is in agreement with the principles of a service-profit chain, where a port's efforts are focused on customers' satisfaction; this in turn brings loyalty, trust, and long-term partnerships, which result in sustainable productivity on behalf of the port.

Motivated port employees have the power to provide the desired product or service. Factors like "internal service quality," "employee satisfaction," and "employee productivity" determine the degree of clients' satisfaction when representing the port authority and its services.

By establishing efficient and effective systems for the human resources management, a port should involve its personnel in its strategic, tactical, and operational mission, vision, and goals. The degree of employees' involvement in the product leads to their empowerment and development of brand qualities to external components. It is frequently referred to as internal marketing, employee branding, or internal branding.

In order to identify the possibilities of internal branding and port or maritime employees' empowerment, it is necessary to examine the branding procedures and how employees are involved therein.

Figure 7.2 shows a port's branding process from its inception throughout its commercial life span.

After the idea generation on a specific port service or process improvement, the brand's image is developed and tested. Research and development (R&D) procedures are necessary in order to examine how the port's services would gain market share and which services it will replace. The feedback obtained by marketing R&D is perused in order to develop a solid maritime brand strategy.

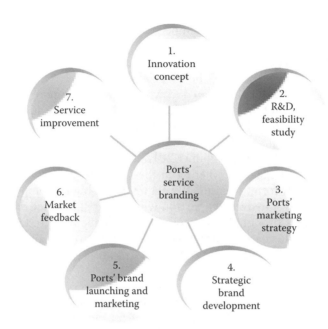

FIGURE 7.2   Ports' service branding. (Courtesy of M.G. Burns.)

Holistic marketing is employed in order to establish a satisfactory level of communication between the port and its clients. The new or upgraded service is then ready to be launched into the market: effective and efficient distribution is required. Marketing is an ongoing procedure, and thus continuous efforts are required, in order to ensure high client retention. This means that *interactive marketing* serves as a link between a port's clients and its employees. At this stage, *internal marketing* undertakes to implement corrective action, either by improving or by replacing the product. New services, methods, or processes may need to be generated in order to attain client satisfaction.

Figure 7.3 reveals the interrelation between clients' satisfaction and port employees' commitment, whereas holistic marketing serves as a facilitator to both sides.

Developing and maintaining a highly productive corporate culture is the port manager's main priority and goal. From the port's interior, the scope of holistic marketing is to establish a culture that reinforces desired work patterns and enhances the port's competitive edge in terms of stakeholders' relations, customers' retention, loyalty, trust, transparency, and improvement.

Previous marketing strategies in the maritime industry tended to be of a narrow perspective: they communicated the port's message to their clients, yet there was limited provision as to the gathering of information and improving the port's efficiency based on this feedback. Modern, holistic approaches in marketing offer some great benefits to the industry, the greatest benefit being a "reality check."

Modern port marketers have the opportunity of receiving feedback from their clients that will help them assess and redefine their role in the market.

In addition, when ports implement holistic marketing, it is possible to monitor any changes in the industry or with the port's clients and their corporate needs. This newly acquired knowledge can provide ports with new business opportunities.

Port clients' satisfaction

Delivery of brand promise, that is, achieving

service quality, utility, or "value for money"

Interactive communication between clients and the port

Holistic marketing

Inside-out management approach

Alignment of port's corporate goals with employees' behavior

Sustainable productivity and improved port performance

Optimum employees' motivation and port's commitment

FIGURE 7.3 Alignment between clients' satisfaction and port's commitment. (Courtesy of M.G. Burns.)

7.2 PORT MARKETING STRATEGY AND COMPETITIVE POSITIONING

Know thyself

Ancient Delphic Maxim

7.2.1 Market Strategy

Marketing strategy is an ongoing process that aims to best utilize the port's factors of production, existing clientele, and geographic significance as a means of expanding and increasing revenue.

The quote "Know Thyself" should be a marketer's constant reminder that in order to successfully promote the port's services, infrastructure, and superstructure, a realistic and thorough evaluation of the port's position, status, and possibilities should be undertaken. Once the port's profile is evaluated, the entire marketing strategy can be established and marketing tools and leveraging techniques can be designed.

Marketing strategies are classified into four principal divisions:

1. *Mass marketing*: A major market division encompassing large components of the market. It is based on undifferentiated services and low tariffs.
2. *Segment marketing*: A narrow market division where differentiated services are offered.
3. *Niche marketing*: Concentrated market where highly differentiated services are offered.
4. *Micromarketing*: An even smaller market segment of local, tailor-made services.

As a first step, the port's current market positioning must be established, revealing its ranking, capacities, and past marketing strategies, errors, and achievements. While the objectives of a port's marketing strategy are to boost berth occupancy and increase sales and annual contracts, its preliminary study should be based on the port's capabilities, assets, services, and geographic position.

A *port's strategic planning* process serves the purpose of designing a solid marketing strategy, through the following tools: (i) marketing analysis, (ii) segmentation, (iii) targeting, and (iv) positioning.

i. Market Analysis

The first step of a port's marketing strategy is to conduct a detailed research covering a multitude of areas and market indexes. The key principle here is to remember where the port currently is, and where it wants to expand, which current or future projects are feasible and desirable, and which partnerships no longer serve their purpose.

a. *Geography*: The port's geographic location, size, and function within its existing local and global networks. Any changes in terms of accessibility, trade routes, supply chain integration, infrastructure, market players, and so on, should be highlighted.

b. *Supply*: Development of a matrix with the port's factors of production, assets, resources, services, talent and innovations and a listing of the port's specialization areas, for example, tanker ships, break bulk, and so on.

c. *Demand*: Since the demand for ports is a derived demand, the commodity markets and their business cycles, elements of production/consumption, imports/exports, and demand/supply should be taken into consideration.

d. *Cost–benefit analysis (CBA)*: CBA is an efficient, systematic framework of assessing and contrasting costs and benefits during the decision-making process of one or more business development projects or ventures. CBA is an essential tool for port managers, as it unleashes the business possibilities available in the current market: its applications encompass all port aspects, that is, strategic and tactical, administrative, HR, engineering, operational, technical, network and accessibility related, and so on.

e. The component that CBAs need to be accurate is consistency in evaluation, that is, in terms of time period, money, services, and other common denominators. The money measurement should be precise as to the currency, time of value, exchange rate (fixed vs. flexible), interest rate, inflation versus deflation, purchasing power at a specific time period, real versus nominal value, and so on.

ii. Market Targeting

The ultimate goal of any global seaport is profitability, growth, and the facilitation of local and national trade. Sustainable growth can better be achieved by establishing long-term partnerships, as opposed to conducting short-term contracts. For ports to secure a solid and loyal clientele, it is necessary for them to pinpoint their target market groups, on the basis of which they will design their promotion strategy.

Systematic data gathering and analysis will provide the port managers crucial information about the trade patterns. A thorough examination of the market conditions will help the port define its target markets. The next step will be to define its marketing strategy.

Marketing Mix: "The Six P's"

The 6 P's in port marketing consist of selected marketing tools that can be used to differentiate the port from its competitors and measure or pinpoint the port's competitive advantages. The marketing mix (i.e., "the six P's" includes the following:

a. Place

Place refers to the port's accessibility in terms of (i) geography, (ii) the port's navigation channels, and (iii) the port's hinterland connectivity and logistics networks, that is, regional markets, distribution channels, inter-modal transportation, rail, highways, and so on. The port is the location where supply and demand have to meet. Port design, infrastructure accessibility, land availability, and strategic location are highly desired features.

b. Product or service

Product or service includes all the port's facilities, characteristics, features, competitive advantages, incentives, and so on, that will satisfy the target markets' needs. Therefore, it is necessary for port marketers to offer tailor-made services to their segment markets, that is, maintenance, repairs, bunkering, shipbuilding, berths' layout, Hazmat (dangerous goods) handling services, warehouses, cargo handling equipment, and so on.

c. Package

Transportation segments include ship types and cargo types. Package refers to cargo handling services required, that is, containers, dry bulk, liquid bulk, palletized goods, break bulk, and so on.

d. People

This factor of production will greatly determine the quality of services, human error, berth scheduling, delays, and so on. Ports can establish their reputation on the basis of their special entrepreneurial skills and reliable workforce.

e. Promotion

Promotion notifies the port's clients about the other five components, that is, price, place, people, product, and package. Chronologically, promotion is the last element that is presented to the market in order to make known information about the port's marketing mix. A sound knowledge of the market, the port's clients, and competition is essential for effective promotion, media relations, advertising, and so on. The aims and objectives of promotion are to announce to the market the port's developments, competitive advantages, and market differentiation; to distinguish its services from other ports; and to boost sales and increase its market share. The principal tools of promotion and public relations include the following:

i. *Advertising* can be defined as the action of promoting the port's or company's profile and commercial activities through connecting with a large target market. The ultimate goal of advertising is to (re)claim the port's market share by assertively ensuring the port's achievements and current and potential growth.

ii. *Financial and commercial promotion and public relations* is the mass communication of the port's commercial and financial data, for example, through the port's web site, as a means of announcing its achievements and potential.

iii. *Partnership or affinity marketing* is a tool of target marketing, where two or more partners that belong to similar or different markets can jointly brand their services and market share.

iv. *Direct response marketing*, where the clients' feedback and business concluded are monitored, evaluated, and linked to specific marketing movements.

v. *Sponsorship and event marketing*, in addition to being an effective marketing tool, is also a highly efficient builder of public image, corporate value, and social responsibility.

vi. *Trade show marketing* for port managers that wish to communicate their new corporate messages to a large number of prospective clients.

A port's promotion strategy has a direct impact on the "price" component, that is, the port's revenue. A port's selection of promotion tools and their subsequent expenses seem to be a source of conflict within ports and with shareholders. Promotion costs and the port's total expenditure on marketing activities should be estimated as a percentage of revenue.

f. Price

In a competitive market such as the maritime industry, pricing arrangements such as tariffs, leasing, and so on, are vital determinants of the negotiations' outcome. Ideally, the pricing derives from a supply/demand interaction. While the other five "P's" are cost based, pricing is the only component that is profit related. While the price mechanism reflects the supply/demand connection, the element of pricing is widely used in corporate finance and quantitative finance applications. The key elements of pricing are incorporated in a port's financial analysis and cash flow forecasting, while its applications encompass input/output ratios, service and product standardization, competitive analysis, market analysis, and so on.

Modern port managers must have a clear idea of the local, regional, and global tariffs of similar ports, that is, ports of a similar geographical and economic significance, ports of similar size, ship and berth type, and similar services offered. The price setting process needs to take into consideration the current and future market with the respective demand/supply factors and competitors' tariffs and services. Finally, a clear outline of the port's competitive advantages is required, as well as the offering of special incentives and quality services.

As a conclusion, a port's marketing mix strategy can be modified according to the market's supply and demand fluctuations. The 6 P's are tools that port managers can reshape as required, in order to fulfill a market demand.

iii. Market Segmentation

Target marketing encompasses the process of aiming a port's promotional ventures in the direction of specific local, regional, or global markets. Market segmentation entails the process of a port's clustering its existing and potential clients into special classifications, based on demand, corporate profile, trade patterns, geography, economic and commercial characteristics, and so on.

The benefits of market segmentation include the following:

a. The port can promote its business directly to the most appropriate markets, by meeting the demand of specific clients.

b. Optimum utilization of resources and budget monitoring and controlling are better achieved in focused investments and market segments.

   c. Segmentation implies controlled competition; hence, the port can monitor the most attractive markets with less competition and eventually find its niche market(s).

   d. Port clients are most likely to conclude agreements with ports that offer specific services and belong to a specific market, rather than ports that may be broadly available, with no serious investment or commitment to specialization areas.

Segmentation practices enable port managers to fully utilize their ports facilities and expand their activities by satisfying different markets. Ports, unlike ships and transportation companies, are bound by geographic restrictions. In addition, their clients need to be part of the regional supply/demand or imports/exports. The positive side to the geographic limitations is that ports are surrounded by local markets, belong to specific trade agreements, and can incorporate in their marketing strategy their existing clientele and regional networks. This can be clearly seen in the case of privatized ports or terminal operators, where the new owners seem to "inherit" the port's previous clients and automatically become part of the local networks.

Market segments can be distinguished in terms of certain common characteristics:

   a. *Corporate structure similarities*, that is, terminal operators, liner shipping companies, tramp shipowners, logistics companies, rail companies, and so on.

   b. *Contractual needs* in terms of contract type, duration, financial terms, and so on.

   c. *Services required* and potential for future business growth. For example, a terminal operator that is also the owner of a liner company has different needs to be covered by a port, compared to a rail company. Also, tanker operator services frequently include and typically require the use of refineries at the discharging port(s).

A "zoom in" of target marketing is market segmentation, which further drills down the selected markets into subcategories. Ports can offer superior services and better meet their clients' needs, when classifying their clientele according to their status, services required, trade segment(s), cargo volume, commercial presence in specific geographical locations, and so on.

4. Marketing Positioning

Marketing positioning is a port's strategic tool that creates a distinctive corporate profile that will appeal to the port's existing and potential clients. It is the blueprint that port managers and marketers design in order to differentiate themselves from competition and therefore retain, or increase, their market share. In fact, the maritime community frequently exchanges information pertaining to port positioning, including tariffs and quality of services, among others.

5. Competitive Positioning

Modern ports seek to design and sustain their cutting-edge position that will help them retain and expand their market share. However, on the basis of the principle that the demand for ports is a derived demand, it seems that ports have to face both internal and external challenges, many of which are beyond their control.

This becomes apparent while examining the leading global ports over the past decades: the ever-changing trade routes, the global market, trade agreements, and supply/demand equilibrium are given factors that port managers cannot avert. Hence, Eastern and Central European ports cannot control the shift of trade; neither can they control the annual change in GDP at a national or continental level.

In the second decade of the twenty-first century, global ports seem to be divided in surplus regions and deficit regions, where port managers have little or no control. What can global ports do about the unmanageable losses occurred since the 2008 debt crisis at a national and global level?

Can ports attract regional and global clients through promoting their comparative advantages? They surely can, and competitive positioning is the tool that will help them achieve their full commercial potential.

Competitive positioning seems to gain significance in a rapidly changing maritime industry, where commercial antagonism is the driver of growth and economic development. It enables port managers to achieve a sweet spot in the market, that is, find a space where ports can achieve the best possible advantages, most lucrative markets, and minimum costs.

Competitive positioning endeavors to "sell" a port's competitive advantages and key features, while prominently differentiating from its market rivals. Port managers need to conduct a smart and meticulous market research and overall trade conditions. Based on these findings, they will discover the port's "sweet spot":

a. *Demand: Market demand and clients' need for specific functions, utilities, services, and characteristics*; market segmentation and target marketing tools will serve this purpose.

b. *Supply: Port's accomplishments*, output, and track record verifying that the port is capable of meeting and exceeding the market's demand. Marketing mix tools led by promotion will enable the port to deliver this special message to its markets.

c. *Competitive Intelligence: A thorough analysis of the competitor's positioning* should be evaluated by port managers. At a first stage, an honest and realistic appraisal should reveal the competitors' challenges and weaknesses. At a second level, port managers should identify and build their own port's strengths and opportunities on the basis of their competitor's weaknesses. For this analysis, objective evaluation is needed, coupled with inside information. Moreover, port managers should consider in which ways and to which extent their rivals' current positioning, assets, and resources could shape their future growth and commercial possibilities. This is where competitive intelligence fits in.

Competitive positioning is about differentiation, competitive advantage, and a clear-cut brand name that ideally is in great need by the port's target markets. Each port enjoys its own reputation, areas of specialty, and advantages. The port's areas of distinction and excellence should be both plausible and long lasting.

7.3 STRATEGIC ALLIANCES, JOINT VENTURES, MERGERS, AND ACQUISITIONS: THE ECONOMIC PATTERNS OF MARKET RESPONSE

> The forces of a powerful ally can be useful and good
> to those who have recourse to them...
> but perilous to those who become too dependent on them.
>
> **Niccolo Machiavelli**
> *"The Prince", 1882*

The international marketplace with its complex financial and commercial networks can impose a great amount of concern for port managers, as to the ways in which they can mitigate the market's unpredictability. Modern ports are as strong as they are united, and as weak as they are commercially isolated.

Competition among ports, just like in any business sector, is a natural consequence of growth. Ports can seize opportunities for expansion through (i) adaptation to the new market realities, (ii) port reengineering, and (iii) pursuing alternative configurations such as strategic alliances, joint ventures, and so on.

Port managers increasingly find the structure of port networks insufficient to accommodate the ports' strategic commercial goals. Hence, they seek to accomplish business partnerships with national or regional ports. An example of the latest projects has been undertaken in Belgium, that is, the Flemish port network consisting of the northern ports in Flanders.

7.3.1   Ports' Strategic Alliances

Strategic alliances are the most common corporate arrangements that port authorities prefer in order to strengthen their competitive edge. It is common for port-related companies to form strategic alliances such as among inland terminals, stevedoring companies, rail terminals, barge terminals, and so on.

In the case of port authorities, frequently these schemes are formed among (i) rival ports, (ii) mega-ports and smaller feeder ports, or (iii) ports and other business entities, that is, terminal operators, liner shipping companies, logistics companies, shipyards, and so on. A strategic alliance is a corporate configuration among two ports or companies with the purpose of utilizing their assets in common and achieving an exclusive venture to their common advantage. It is a formalized partnership among two or more parties to engage in a set of agreed upon targets or to satisfy a significant business need while retaining their corporate independence.
Strategic Alliance Examples

The Americas
  a. The Panama Canal Authority and the major US ports
      The Panama Canal has established a strategic alliance with the major US ports. All memoranda of understanding signed among these entities were initiated in 2003, with the purpose of boosting trade and funding opportunities, while enhancing the trade route from the Far East to the United States through the Panama Canal. The so-called "exclusive fraternity" is established among the Panama Canal and the following US ports:
      – The Port of Houston Authority, Texas (Port of Houston 2011)
      – The New Orleans Port Authority, Louisiana (Port of New Orleans 2011)
      – The North Carolina State Ports Authority (NCSPA 2013)
      – The South Carolina State Ports Authority (SCSPA 2013)
      – The Port Authority of New York and New Jersey (Port Authority of New York and New Jersey 2013)
      – The Georgia Ports Authority (Georgia Ports 2011)
      – The Virginia Port Authority (VPA 2013)

- The Massachusetts Port Authority (Massport, Massachusetts Port Authority 2003)
- The Miami Port Authority, Florida (Port of Miami 2013)
- The Tampa Port Authority, Florida (Tampa Port Authority 2013)
- The Manatee Port Authority, Florida (Port Manatee 2009)

Oceania

b. Australia and New Zealand ports

In 2000, New Zealand's Port of Tauranga and Australia's Port of Brisbane signed a broad agreement described as the "first step in establishing a worldwide strategic alliance of leading ports." Initially, the agreement will include sharing of technologies, marketing, trade information, and specialized expertise.

c. New Zealand is ready to welcome the next generation of mega-ships, and for this purpose, a strategic alliance was formed between the Port of Tauranga and PrimePort Timaru.

Over the past years, shipping lines have decided to benefit from economies of scale by sending larger vessels to New Zealand, yet visiting fewer ports. A synergy within the country's most strategic ports will not only reduce competition but also ensure the even and timely distribution of cargoes nationwide. For the past two decades, New Zealand's Port of Tauranga has become the country's most significant and largest sized container terminal. PrimePort enjoys a strategic location, in the vicinity of the nation's industrial regions and most populated areas of the South Island. The Port of Tauranga will offer over $21 million to PrimePort, in order to acquire 50% of its shares, lease its container terminal for a maximum of 35 years, and purchase the terminal's operating assets.

Europe

d. The Netherlands: Rotterdam and the Zeeland ports

In the Netherlands, the Dutch Government controls and regulates all of the country's ports and is in charge of port mergers and strategic alliances, including the established partnership among the major ports of Rotterdam, Amsterdam with Groningen, and the Zeeland Seaports, that is, the united Terneuzen and Vlissingen (Flushing) ports company (also see Section 7.3.3). The partnered ports form a geographic arrow covering the entire coastal line of Holland. In the center, Rotterdam and Amsterdam are the country's major global ports, which are competing in annual traffic and cargo volume with the top Asian mega-ports. On the other hand, Groningen in the north and the historic Zeeland ports in the south are rapidly expanding their container terminals and enjoy excellent logistics networks and hinterland connections. The port of Flushing (Zeeland) hosts the historic De Schelde shipyard. An interesting observation on ports' strategic alliances is that such a contractual arrangement among a mega-port and two rapidly expanding ports may result in conflicts of interest. In the case of the Dutch ports, the strategic partnership among Rotterdam and Zeeland ports resulted in the halting of the Westerschelde Container Terminal development in Vlissingen.

e. In England, the Port of Liverpool has formed a £300 million ($468 million) joint venture with land owners "Peel," an infrastructure, transportation, and real estate investment company, in order to expand the port and dredge the navigational channel, to accommodate the new-generation post-Panamax vessels. The project, cofinanced by the *European Union* Trans-European

Transport, will enable the port's container handling capacity to be doubled from 750,000 TEU to 1.5 million TEU per annum (Peel Holdings 2013).

Transcontinental Partnerships

f. Belgium and India

   The leading Belgian seaport "Port of Antwerp International" has formed a strategic partnership with Essar Ports Limited, a major private sector port corporation of India. This partnership aims to boost the growth of trade volume, investment, and development within the ports. Their partnership encompasses the sectors of port planning, traffic flow, productivity improvement and capacity increase, business consulting, skill set, training, and quality.

7.3.2 Ports' Joint Ventures

A port's joint venture pertains to a business agreement where all parties concerned consent to form a new corporate entity with commonly owned assets. Joint ventures are most common in the shipping industry and all across the supply chain network. In particular, joint ventures are common among ports and terminal operators, charterers, carriers (sea, rail, road), and hinterland key players.

As an example, China's Ningbo Port Co. Ltd., a major port of mainland China, has established a joint venture with the conglomerate A.P. Moller–Maersk (APM) in order to invest and operate berths 3, 4, and 5 in a kilometer-long quay. The Ningbo port will participate with a 67% share, while APM will take up a 33% share. Ningbo is the third largest port in China and the sixth biggest container port globally.

There are different types of joint ventures, which determine the depth, corporate arrangement, level of interdependence, level of authority, and so on: (i) contractual joint venture, (ii) corporate joint venture, (iii) unincorporated joint venture, (iv) equity joint venture, and (v) cooperative joint venture, among others.

The type selection together with the contractual terms of the agreement will stipulate the degree of ownership and control each party will exert on the company, as well as profit sharing, input, output, stock and bonds retribution (if applicable), and so on.

A port can reap the benefits of a joint venture in order to penetrate new markets, obtain investment, launch innovative technologies, and expand its services and competitive edge. This continuous alliance offers to all parties concerned the advantages of innovation, increased market exposure to new industries, enlarged strategic potential, logistics networks, and new clientele.

Additional benefits and opportunities offered in this type of partnership are based on the contrast between the complementary markets that these ports serve and the radical differences, for example, geography, technology, layout, areas of specialization, logistics networks, and so on. The outcome is novelty, risk sharing (or risk distribution), increased market share, and increased possibility of profits because of the new competitive advantages generated therefrom.

7.3.3 Ports' Mergers and Acquisitions (M&A)

Port merger is the legal permanent consolidation or absorption of two or more seaports into a larger entity, typically through providing the stockholders of one port stock options in the purchasing port, as a swap for the forfeit of their asset, inventory, or investment.

Port acquisition takes place when one port (Port A) assumes overall control of another port's assets and services (Port B). While the acquired port (Port B) is still considered as an independent legal entity, it is now fully controlled by its buyers, that is, Port A, and has become an element of its expansion strategy. Acquisitions are frequently settled in funds, stock options, or both.

It is worth noting that within a port's expansion strategy, it is typically more profitable and less risky for the buyers to purchase another port's market share, services, business, clients, and competitive edge, rather than investing the same amount of funds to develop their existing port facilities. The following examples reflect port M&As within the same country and initiated by the respective governments:

a. *The ports of Ningbo and Zhoushan Island, China*

The mega-port of Ningbo, a leading container handler whose cargo throughput volume surpasses 100 million tons on a yearly basis, has been merged with the island Port of Zhoushan, in order to formulate a merged container distribution center. The island of Zhoushan is located 35 nautical miles or 40 kilometers away from Ningbo. This M&A will eliminate bottlenecks in Ningbo and utilize and expand Zhoushan's port dynamics; it will also be useful in terms of logistics and alternative services in case of weather restrictions, traffic, and so on. The combined Ningbo–Zhoushan Port handles approximately 700,000,000 tons of freight cargo each year, rendering it as the second biggest port globally, after Shanghai, China.

b. *The ports of Tauranga and Timaru PrimePort, New Zealand*

The Port of Tauranga Ltd., New Zealand's largest export port, has purchased a 50% shareholding of the PrimePort, Timaru Port Company in the nation's South Island, at a cost of $17.2 million. Under the agreement package, Tauranga Ltd. leases the container terminal of PrimePort for up to 35 years and has purchased its property.

c. *Batumi Port, Georgia, and KazTransOil JSC, Kazakhstan*

Kazakhstan is a landlocked country in Central Asia, rich in oil, natural gas, diamonds, gold, and minerals (uranium, iron, etc.). It is located right between China and Russia; hence, Kazakhstan produces and controls oil and gas movements from Central Asia to China and Eastern Asia (including Iran), and the West to Russia, Turkey, and Europe. The combination of the country's rich energy and mineral reserves with its landlocked status makes it an interesting case study as regards its logistics arrangements. The country's land infrastructure requires vast investments, as the oil and gas reserves are situated sparsely, that is, are widely spaced. Sea transport is the most efficient and cost-effective transportation mode, and yet Kazakhstan's only water access is to the landlocked Caspian Sea. It desperately needs access to the Black Sea, and through the Bosphorus Straits, it will gain access to Turkey, Greece, and the Mediterranean Sea.

KazMunayGas is Kazakhstan's state-owned energy company, established in 2002 through a merger between Kazakhoil and Kazakhstan's oil and gas transportation company. KazTransOil JSC is Kazakhstan's state-owned pipeline and oil and gas transportation company, providing services on oil and petroleum products pertinent to their extraction, storage, and transportation, that is, pumping, transshipment, unloading, loading, storage, blending, and so on.

A strategic movement on behalf of KazTransOil JSC was to acquire the Batumi Port and an oil and gas loading terminal in the country of Georgia. Since 2008, Batumi Oil Terminal functions as a subsidiary company of KazMunaiGas Exploration and Production JSC (Batumi Port 2013; KazTransOil JSC 2013).

7.4 COMPETITION AND CONFLICT PREVENTION

Competition has been shown to be useful up to a certain point and no further, but Cooperation, which is the thing we must strive for today, begins where competition leaves off.

Franklin Roosevelt

7.4.1 Port and Terminal Competition

In port management, competition is perceived as the struggle for prevalence within the market, with the ultimate objectives being higher revenue, optimum occupancy, and long-term partnerships.

Competition is about setting the service price, setting the quality standards, and defining the market share. Competition affects all the key players in the maritime industry, including the port authorities, terminal operators, service providers, shipowners, cargo owners, charterers, and all the components of the supply chain.

Any efforts on behalf of ports, transportation managers, logisticians, and so on, aim to reduce the possibility of their services being replaced by those of their rivals. It is worth noting that a port's services are more likely to be substituted when no single port or transport company has a cost advantage over another.

The core elements of competition that will determine a port's continued competitive success are as follows:

i. Geography
ii. Population density
iii. Supply/demand equilibrium, that is, buyers versus sellers' power (typically unity enhances a party's power, i.e., port associations, government support, consortia, etc.)
iv. Entry barriers to market (UNCTAD 2011)
v. Port accessibility
vi. Land availability
vii. Infrastructure and connectivity
viii. Vicinity to growing markets as well as financial, production, and trade centers

Since this type of rivalry suggests increased supply, competition averts monopolistic markets and reduces tariffs. Interestingly enough, competition among rival service providers remains steady throughout the market's cycles: competition is an ongoing process that seems to intensify through the market fluctuations, for different reasons. In particular, business antagonism increases during market booms as companies struggle for expansion and increased market share. Meanwhile, competition also increases during financial crises, as companies struggle for survival and retention of their existing market share.

A typical example of intense competition is the well-established Panama Canal, as well as the imminent threat arising from the building of the Nicaragua Canal. The possibility of two enormous canals located in the vicinity of each other may not be an attractive return-on-investment option for the canal authorities; however, the element of choice will be profitable for the shipowners and the terminal operators that will be able to bargain for lower canal tolls, improved services, and attractive incentive packages. Furthermore, the opening of the Arctic Circle in 2016 will likely generate new trade routes from North America and Europe to the rapidly developing Asian economies (China, India, Japan, Taiwan, Singapore, Indonesia, South Korea, etc.).

Transportation Competition is classified into the following:

a. *Intermodal competition*, which entails the competition among different transportation modes, that is, sea, land, and air
b. *Intramodal competition*, which breaks down competition as each transportation mode wishes to prolong their trade routes and increase their market power over a certain trade:
   - *Sea*: competition among networks, as well as oceangoing shipping versus short sea shipping
   - *Land*: competition among networks, as well as rail versus trucks versus pipelines
   - *Air*: competition among networks

Geographical constraints usually determine which transportation modes are most appropriate for a specific trade route. Also, the type of cargo, the urgency of delivery time, and the charterer's ideas on freight payment will further define which route(s) and which transportation mode(s) will be selected.

Waterborne transportation enjoys the competitive advantage of low cost owing to the economies of scale, as well as reliability in service and delivery times as ships typically encounter traffic for a small part of their voyage, that is, while at port, until berthing time.

Port competition is distinguished into the following:

a. *Internal port competition*: (i) among neighboring ports, (ii) among ports serving in the same or competing logistics networks, and (iii) among ports serving in the same or competing trade routes
b. *External port competition*: among port authorities and their clients, that is, (i) terminal operators, (ii) commodity traders, (iii) liner and tramp shipowning companies, (iv) logistics companies, etc.

From a terminal operator's perspective, the World Bank distinguishes two levels of competition:

a. *Intraport competition*, which involves the antagonism of two or more terminal operators that wish to dominate the very same marketplace. Typically, the terminal operators control neighboring terminal locations of the same port from dock to port gate. Each terminal operator may pursue better port and logistics services, more efficient scheduling for quicker turnaround time, and greater market impact.
b. *Intraterminal competition* refers to the rivalry among logistics, trade, and transport companies located within the port's premises or in its vicinity that compete for the supply of services to the local terminals (World Bank 2000).

Competition may generate power abuse, which, in the shipping industry and ports in particular, may be expressed as refusal to provide services, time delays, and increased tariffs that do not correspond to the value of services.

Competition determines port tariffs and the overall costing parameters of the industry. Although competition is a resource-depleting and labor-intensive process for the rival companies, from an industry perspective, it is a revival process, offering the opportunity for (i) innovation, (ii) differentiation, (iii) optimum resource utilization, (iv) cuts of superficial costs, and (v) improvement of services.

While competition in the maritime industry generates differentiation, innovation, and new industry players, it forces the repositioning and reengineering of the entire supply chain, that is, ports, terminal operators, shipowners, and so on.

7.4.2 Conflict Prevention

Generally speaking, intense competition reveals that the market is not differentiated enough and most likely contains clusters of identical ports that offer identical services. As the new generation of mega-ships has been introduced in the shipping industry, the winning ports will be the ones that are capable of handling these larger ships and large volumes of cargoes in a most efficient and effective manner.

Typically, the greater the prospective market benefits, the more entrants offering similar services, and this increased level of supply leads to conflict. A solution to avert conflict would be through diversification of services and exploration of alternative, complementary, or supplementary services.

If competition arises in large commercial entities, that is, within a supply chain, or among departments or terminals of a single port, then division of power and segmentation of services provided will help avert any conflict.

7.5 ESTABLISHING LONG-TERM VALUE THROUGH CUSTOMER LOYALTY

> A brand that captures your mind gains behavior.
> A brand that captures your heart gains commitment.
>
> **Scott Talgo**

In a highly competitive, globalized shipping market, with unpredictable, often violent fluctuations, strategic management or corporate wisdom seeks customer loyalty as a means of securing and expanding corporate market share.

The modern shipping market is characterized by intensely competitive, global branding and trade fluctuations in terms of flow and routes. Within this volatile frame, ports seek *customer loyalty* as a means of establishing long-term value. For this reason, most ports and shipping companies tend to meticulously assess the value of customers and their loyalty to the firm, in order to retain their market share and gain a competitive edge.

A brand is the personification of a product, service, or a whole firm. It resembles a person, its physical body being the products or services it provides (Blanchard 1999). Just like people, brands have a name, a personality, a character, and a reputation.

Hence, for a port brand to inspire loyalty, it should have certain competitive advantages, be likeable, and be valuable to its consumers. It is significant for the port to have a well-defined strategy with an explicit message conveyed to its clients. Port marketing executives define marketing strategies with the perception that clients are loyal to ports for their own reasons (e.g., network and market accessibility), which sometimes may be irrelevant to the port strategy conveyed in its marketing messages.

The ultimate objectives in a port marketing strategy should include enhancing port loyalty and increasing market share, that is, by reducing or eliminating competition. The following case study demonstrates the marketing strategy between Central Port, Illinois, and the Port of New Orleans, by means of a marketing agreement.

CASE STUDY: MARKETING AGREEMENT BETWEEN AMERICA'S CENTRAL PORT, ILLINOIS, AND THE PORT OF NEW ORLEANS, LOUISIANA

America's Central Port is located at the center of America's intermodal freight transportation system, with immediate access to rail, river, or road networks. Every year, the port moves over 3 million tons of cargo among river barges, rail wagons, and trucks.

New Orleans, Louisiana, has been playing a leading role as a logistics center because of the hydraulic fracturing activities in the region.

NOLA's new $30 million Gulf Gateway Terminal will transfer as much as 10,000 barrels an hour of crude oil straight into barges or into large capacity storage tanks. Meanwhile, a greater crude oil terminal, Wolverine Terminals, will commence operations by 2014.

PORTS' JOINT MARKETING STRATEGY

The Port of New Orleans, Louisiana, and Central Port, Illinois, have established a marketing contract to investigate and promote joint business opportunities. The agreement was initiated owing to the two ports' strategic location, by the Mississippi River, and the potential of increasing domestic maritime trade through inland waterways.

To execute this joint marketing initiative, common strategies between the two ports are necessary, including the following:

i. Allocation of common funding for cooperative marketing
ii. Linkage between the two port's web sites and sustainable information exchange
iii. Promotion of pertinent news
iv. Designated port employees will jointly work to promote the ports' best interests.

They will also assign designated personnel to initiate a common marketing strategy with synchronized sales prospects and mutual visits with current and prospective clients.

Mutual marketing strategy will encompass business generated from the expansion of the Panama Canal and the construction of the inland port *South Harbor*, which is positioned near Locks No. 27 on the Mississippi River. The growth opportunities for the ports are tremendous.

Over the past few years, the maritime industry has encountered unprecedented changes in terms of contractual arrangements and changes in the supply chain partnerships. Hence, it is deemed necessary to analyze in depth the essence of loyalty, the motivator's attributes that attract loyalty, and its utility in the modern shipping industry toward a port or a trade route.

7.5.1 Redefining Business Loyalty in the Shipping Industry

Business loyalty pertains to "a strong feeling of support or allegiance" (Oxford University Press Dictionary 2013), but it also refers to the clients' consistency in conducting business with a specific port, to seek for specific services.

Business loyalty contains the elements of ethics and justice (Philosophy Dictionary 2013). Hence, once customers have justified their reason for being loyal toward a port, they will pursue long-term partnerships with the said port even if equal or better ports are available in the market.

The element of mutuality is implied within the essence of loyalty, in order to establish a reciprocal business partnership with mutual benefits.

Over the past 25 years, corporate trends select loyalty marketing as a main tactic of customer strategy. Delighted customers become the firm's most effective advertisement.

Every CEO and every marketer would dream of establishing a loyal, steady clientele. Modern marketing and business strategies are designed to serve or reach this principal aim. A brand is considered as successful when it creates financial value for its owner (i.e., either higher sales or premium prices) and value for its consumers.

At this stage, the following questions arise:

1. What are the corporate benefits for practicing loyal marketing?
2. Is customer loyalty measurable? In what ways?
3. Why do brands increasingly fail to inspire customer loyalty? In what ways are these achieved?

These issues shall be duly analyzed in the following sections.

7.5.2 The Benefits of Practicing Loyal Marketing

Vilfredo Pareto (1848–1923) was an Italian economist who, in 1906, observed that 20% of the Italian people owned 80% of their country's accumulated wealth. Over time and through usage in a variety of applications, Pareto's theory was consolidated into "Pareto's principle" or the "80–20 rule." Pareto's rule is a useful theory that applies when there is a question of effectiveness versus diminishing returns on any (or any combination) of the following: effort, expense, or time. In effect, Pareto's rule states that a small number of causes are responsible for a large percentage of an effect, in a ratio of about 80:20. Pareto's principle applied to our consideration of ports suggests that 80% of ports' revenue is achieved by retaining approximately 20% of their current customers. This verifies that loyal marketing enhances brand firm profitability.

The following are some of the reasons why modern ports prefer establishing solid, long-term relationships with loyal customers:

1. Loyal customers actually *confirm the ports' services utility, quality, and value*: they prove that a specific port fully covers the needs of a specific target group.

2. In a market characterized by cutthroat competition, *a steady, periodical flow of customers enables the port to sustain its market position and even expand its market share.* This invaluable clientele will enable the port to withstand even the most volatile market fluctuations.

3. *Loyal customers result in high fixed costs and low variable corporate costs.* They will facilitate ports' financial forecasts, which will be rather accurate, by containing mostly fixed periodical expenditures and mostly fixed periodical turnover.

4. *The ports' logistics/supply chain will function with predictable, repeated distribution patterns* (i.e., geographical locations, qualities, and quantities), thus operating in a lean and agile, effective, and efficient manner.

5. *Loyal customers provide useful and realistic feedback to the port.* Hence, the port can either diversify or retain its current service strategy, quality, and tariff range, knowing that there is a *specific target group that perceives the port's services as valuable, compared to the cost.*

6. If the port's service becomes obsolete or outdated, or if a similar port competes in the same market by providing competitive utility, quality, or tariffs, loyal customers are likely to express their complaints to the port's representatives, thus providing valuable feedback on the port's positioning in the current market.

7. This continuous feedback offers *primary market information* to the port, which, in turn, becomes aware of the volatile customer preferences and market trends, and *molds the port's corporate strategy* accordingly.

By implementing *selective relationship management*, numerous ports apply customer profitability analysis to weed out losing customers and target winning ones with special incentive packages.

7.5.3 Measuring Customer Loyalty

Measuring customer loyalty enables ports to assess the effectiveness of their brand message. It reveals consumers' perceptions on the brand image and utility.

When measuring customer loyalty, it should be considered that the comparative advantage of certain ports makes them more or less prone to attract loyalty. Also, certain services have longer purchase cycles than others.

Loyalty may be assessed in more than one way, that is:

- Contractual duration, that is, years of collaboration
- Volume of services required within multiple purchase cycles
- Preference of a specific port, despite competitors' offers or market share

As indicated in Table 7.1, loyalty may also be expressed as customers' tolerance or patience, despite time delays, port traffic, poor services, or higher tariffs.

7.5.4 Target Markets and Competitive Positioning

The plethora of emerging ports on the global arena contribute toward diminishing brand loyalty. This perception can be evaluated in depth by considering its causes and effects.

TABLE 7.1   Measurement of Customers' Loyalty

Indications of Customers' Loyalty	Measurement of Loyalty
Port delays, bottlenecks, traffic	Tolerance is measured by delay time.
Problems with cargo distribution or difficulties in berth accessibility	Tolerance is confirmed by degree of difficulty in port/berth accessibility.
Quality issues, negligence, human error	Loyalty may be measured by concluding long-term contracts and services despite negligence, errors made, and quality issues.
Choice among competing ports	Clients' loyalty for a certain port may be measured by the choice they have among competing ports. In fact, loyalty can be accurately measured in the presence of intense port competition. Comparison between similar ports may be made in terms of the following: • Technological advances, quality, tariffs, customer services • Logistics distribution; accessibility of port and supply chain • Market share, growth ratios, and competitive advantages

Source: M.G. Burns.

A port's inability or ineffectiveness to deliver services as promised minimizes brand market share and therefore encourages new ports to enter the specific market or market segment. Newcomers enter the market in a sense of urgency, often with competitive tariffs and incentive packages. This new market share frequently attracts investors, that is, firms with sufficient capital, and rarely innovators. Hence, new ports rarely have a clear brand image or elements of differentiation.

Their role is to cover specific, almost tailor-made needs for selected target groups and succeed where the previous port failed. In this respect, they enter the market as substitutes, yet with effective market targeting and competitive strategies, they may develop into leaders. This is due to the fact that a competitive package, efficient in terms of time and money, may be sufficient to attract and retain new clients.

Port service differentiation might be a risky endeavor, as any deviation from the previous service quality or utility contains the possibility of losing a share of the target group. Furthermore, limited experience in the specific market or target group does not allow ports to differentiate.

This explains why a very small amount of ports enter the market as leaders and differentiate, whereas the majority of ports pursue a low-differentiation and low-risk strategy.

It is worth noting that in the era of technology, rapid development, and easy access to industry information, followers tend to imitate the leading port in such a quick pace that consumers do not always distinguish which is the leading port and which are the followers. More likely, it looks like a sustained imitation process takes place, with low differentiation rates.

The aforementioned analysis explains the decreased loyalty levels in ports and maritime business in general. In fact, it suggests that in order to inspire and maintain brand loyalty, innovation and sustainable quality are required. It is also implied that loyalty should be mutual, that is, customers' loyalty should be retributed by the port's loyalty.

REFERENCES

Batumi Port. 2013. Available at http://www.batumiport.com/eng/. Accessed August 2, 2013.

Blanchard, R. 1999. Parting essay. Available at http://www.brandcoolmarketing.com/brand-quotes.html. Accessed August 1, 2013.

Georgia Ports. 2011. Panama Canal and Georgia Ports Authority renew and strengthen alliance. Available at http://www.gaports.com/corporate/tabid/379/xmmid/1097/.../2/default.aspx. Accessed August 2, 2013.

KazTransOil JSC. 2013. Available at http://www.kaztransoil.kz. Accessed August 2, 2013.

Massport, Massachusetts Port Authority. 2003. Panama Canal Authority forms strategic alliance with Massachusetts Port Authority; Alliance will boost trade between New England and Asia through the Panama Canal. Available at http://www.massport.com/news-room/News/PanamaCanalAuthorityFormsStrategicAlliance WithMassachusettsPortAuthority;AllianceWillBoostTradebetweenNewEnglandand Asiathro.aspx. Accessed August 2, 2013.

Machiavelli, Niccolo. The Prince. 1882. Translated by N.H. Thomson. Vol. XXXVI, Part 1. The Harvard Classics. New York: P.F. Collier & Son, 1909–14.

North Carolina State Ports Authority (NCSPA). 2013. Available at http://www.ncports.com/news/news-releases/. Accessed August 2, 2013.

Oxford University Press Dictionary. 2013. Available at http://www.oed.com. Accessed August 3, 2013.

Peel Holdings. 2013. Available at http://www.peel.co.uk/. Accessed August 2, 2013.

Philosophy Dictionary. 2013. Available at http://www.philosophy-dictionary.org/. Accessed August 3, 2013.

Plumer, B., Washington Post. 2012. Could a port strike really cripple the US economy? Available at http://www.washingtonpost.com/blogs/wonkblog/wp/2012/12/27/could-a-port-strike-really-cripple-the-u-s-economy/. Accessed May 14, 2013.

Port Manatee. 2009. Port Manatee signs strategic alliance with the Panama Canal. Available at http://www.portmanatee.com/userfiles/file/PDF/newsupdate.pdf. Accessed August 2, 2013.

Port of Houston. 2011. Port of Houston Authority and Panama Canal Authority renew strategic alliance. Available at http://www.portofhouston.com/inside-the-port-authority/communications/business-news/photo-release-port-of-houston-authority-and-panama-canal-authority-renew-st/. Accessed August 2, 2013.

Port of New Orleans. 2011. Panama Canal and the Port of New Orleans renew. Strategic alliance. Partnership reaffirms commitment to mutual growth and cooperation. Available at http://portno.com/CMS/Resources/press%20releases/prsrel080811.pdf. Accessed August 2, 2013.

South Carolina State Ports Authority (SCSPA). 2013. Available at http://www.scspa.com/About/news/pressroom/pressroom.asp. Accessed August 2, 2013.

The Port Authority of New York and New Jersey. 2013. PANYNJ. Available at http://www.panynj.gov. Accessed August 2, 2013.

The Port of Miami. 2013. Miami, Dade County. Available at http://www.miamidade.gov/portmiami/. Accessed August 2, 2013.

The Tampa Port Authority. 2013. Available at http://www.tampaport.com/. Accessed August 2, 2013.

UNCTAD. 2011. World Investment Report, 2011. Available at http://unctad.org/en/docs/wir2011_embargoed_en.pdf. Accessed August 2, 2013.

Virginia Port Authority (VPA) 2013. Available at http://www.portofvirginia.com/media/126619/cafr_web_2013___final.pdf. Accessed August 2, 2013.

World Bank. 2000. World Bank, Module 6, pp. 5–6: Designing a port reform toolkit, the World Bank (2000). Accessed June 12, 2013.

CHAPTER 8

Key Performance Indicators as Tools of Strategic Planning and Management

8.1 STRATEGIC PLANNING, DEVELOPMENT, AND MANAGEMENT: EXCEEDING THE CORPORATE OBJECTIVES

The best teamwork comes from men who are working independently toward one goal in unison.

James Cash Penney
Founder of J.C. Penney

The successful outcomes of strategic planning can have durations of multiple decades and become the basis for a port's future expansion. Therefore, port planning, sustainable growth, and profitability are the objectives and duties shared among nations, states, port authorities, and the maritime industry as well.

The primary goal of port planning is the strategic design of an efficient facility that serves as a functional component of the local, national, and global transportation network, through the following steps:

- Achieving optimum utilization and expansion of services and activities
- Attracting investment from reliable sources
- Forecasting the short-, medium-, and long-term industry trends and developing functional port systems and facilities that meet customers' demand over long periods
- Delivering overall efficiency and performance supervision, measurement and controlling of the port's operations performance—ongoing monitoring and evaluation of the port's input and output, for example, in operations, technologies, marketing, and so on
- Establishing the port's role as key logistics provider in the interface between sea, land, and air
- Developing efficient business/port intelligence research for the ports' competitors
- Consequently, building the port's niche markets through differentiation
- Establishing long-term partnerships with the region's most valuable customers

Strategic planning aims to accelerate commercial development as well as improve the utilization and overall performance of ports. The power shift from ports to terminal operators, charterers, and logistics conglomerates defines a port's objectives in the strategic planning process. This planning process focuses upon cost and time efficiency, operational excellence, safety, technological aptitude, regulatory compliance, and so on. The results of an efficient port planning strategy can be clearly seen in the cargo volumes transported through the logistics chains. Port success and efficiency are not only about sustainable growth: the elements of regulatory compliance, safety, security, occupational health, and respect for the environment are equally important paths that lead to success. A case study on the Port of Haifa, Israel, follows.

CASE STUDY: THE PORT OF HAIFA, ISRAEL

THE EFFICIENT PORT

Here's a brain teaser that only a few would be able to answer correctly: which is the fourth most efficient port in the world, out of the world's 41 leading ports? Perhaps the port of Los Angeles, the Port of Singapore that serves much of Asia, or perhaps even the picturesque Port of Rotterdam?

The answer is none of these. According to a special report published by OECD in 2012 on the world's most efficient ports, the fourth most efficient port in the world is Haifa's Container Port.

Haifa's extraordinary ranking as fourth among the world's 41 ports that won in the "efficiency race" was totally justified based on the principle that the size of a port does not necessarily reflect in its efficiency: smaller ports can still be extremely efficient.

Haifa Port (Figure 8.1) is the largest port of Israel, a country with a strategic geopolitical location abundant with oil and gas reserves. This modern port is located in a natural, protected bay, and its superstructure and infrastructure allow for the shipping and transportation of all ship and cargo types, including docking facilities for large passenger liners. The port is operated by the Haifa Port Company, a government-owned company that is committed to operating as an engine for the advancement of Israel's economy and growth as well as the source of income for

FIGURE 8.1   Haifa Port at dusk.

the tens of thousands of employees working at all levels of the port's activities. The port's enhanced infrastructure and continuous expansion are supported by an efficient rail and road distribution network.

In 2012, Haifa Port started to fulfill the potential in the Carmel Terminal, one of the biggest, most advanced container terminals in the Mediterranean. With its capacity that is around 1 million TEU a year, and with its great Kocks Gantry cranes (6) and Kone Cranes RMGs (12), it is something that decision makers in the shipping companies could not miss. Haifa's efficient terminals handle approximately 24 million tons of cargo each year, with sustainable growth anticipated, pursuant to the nation's oil and gas extraction operations.

HAIFA PORT ADVANTAGES

The port has many distinct advantages, namely:

- Strategic geographic location close to the busiest shipping route in the world—from and to the Suez Canal. The port is located in the heart of Haifa and enjoys a wide range of shipping services, overland and air transport to other parts of the country as well as commercial centers and places of entertainment.
- Technology, size, efficiency: Haifa's advanced, high-capacity, high-tech container terminals, and especially the Carmel Terminal, one of the biggest, most advanced container terminals in the Mediterranean.
- Service sustainability: 24/7 operations. Continuous service (even between shifts) for trucks and container terminals.
- Accessibility to and from the port is easy from multiple routes and the Israel Rail, with no need to travel through the city. An efficient network and investment in infrastructure make access even easier and faster.
- Cargo and service diversity: Haifa Port can handle most cargo types: container, general cargo, bulk cargo, chemicals, oil and gas, Ro-Ro, and so on, through a large number of approaches and quays.
- Use of the world's best and most advanced terminal management system—the NAVIS System.

Haifa Port and the Application of NAVIS and Terminal Operating System (TOS)

Haifa was among the first ports globally to employ a TOS, which is a major supply chain tool with the principal purpose of monitoring and controlling the flow, warehousing, and distribution of cargoes in the vicinity of container terminals and the port limits. To secure efficient and accurate cargo tracking options in the port's vicinity, TOSs frequently employ alternative information technology (IT) solutions, for example, web-based and mobile systems, wireless LANs, EDI processing, and radiofrequency identification (RFID). The software optimizes and manages the available port assets, labor and equipment, and workload planning, and this continuously updated database enables the port's efficient and cost-effective decision making. This efficient system is used on desktops and cell phones (Figure 8.2a and b) to show how every worker and every crane works

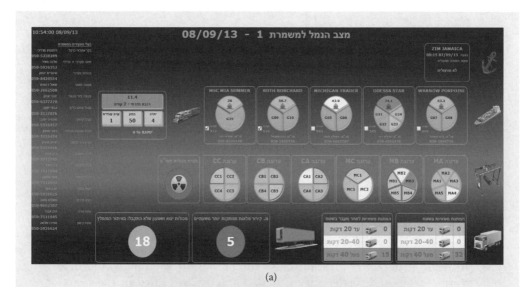

(a)

(b)

FIGURE 8.2   (a) Desktop and (b) mobile versions of TOS.

each second, in the quays and in the storage areas, as signified with the green, yellow, and red colors, wherein the red color signifies an emergency and immediate action required. The screens are connected with closed-circuit cameras within the port and on big LCDs as well.

HAIFA'S LEADERSHIP

FIGURE 8.3   Mendi Zaltzman, CEO Haifa Port.

Haifa Port's CEO, Mendi Zaltzman (Figure 8.3), is considered as one of the most experienced port executives in Israel. With no less than four terms as a CEO in the ports of Eilat and Haifa, he completes four decades in Israeli ports and in the Ministry of Transportation. Since his return to Haifa Port in 2008 as CEO, Mr. Zaltzman has led an impressive and unprecedented revolution in the port's development and productivity, placing Haifa Port today as a top-level container terminal. He initiated the tailor-made upgrade of the port's TOS with the latest Navis TOS, as well as new ultramodern cranes and cargo handling equipment (CHE) in the new Carmel container terminal with different, state-of-the-art STSs and RMGs.

In less than two years, his hard and brave decisions enabled the 30% increase in Haifa's productivity. In 2012, after long negotiations, his lasting dream to bring a major global line to Israel came true with the MSC Golden Gate Service to Haifa. Since then, the port's container handling records has been rising at a steady pace.

Zohar Rom is in charge of all marketing communications, advertisement, public relations, and spokesmanship in Haifa Port. Since his arrival, the Port of Haifa's image in media improved dramatically both in Israel and around the world. He also designed and implemented the new web site of the port as well as the first mobile site in Israeli ports. In 18 months spanning between 2012 and 2013, the site page views were up 225%.

Accurate forecasts and market analyses are necessary for an effective strategic planning process. For this reason, feedback and direction should be obtained from a diversity of sources, both internal and external, such as the following:

- Key industry players
- Local and national markets
- Political authorities
- Policy makers
- Staff members from different departments and levels of authority
- Local communities

A port's plans should be both measurable and tangible. They should explicitly contain past achievements, proven track record, and resources gained, on the basis of which the future forecasts and goals will be founded, including the following:

- Increasing participation in joint ventures and strategic alliances with industry leaders
- Becoming a dynamic component of the regional socioeconomic decision-making process
- Developing efficient business development and asset management
- Maximizing profit and operating margin
- Boosting regional employability through direct and indirect jobs
- Improving regulatory compliance (especially addressing issues of safety, security, and the environment)
- Achieving optimum utilization and revenue of sea–land infrastructure, by reshaping tariff policies
- Keeping abreast with technological advancements and wisely investing in equipment that will improve the port's productivity and efficiency
- Organizing, monitoring, and controlling ships' traffic
- Increasing revenue through freight levels
- Increasing import and export volume
- Achieving optimum rates on return on investments, return on assets, and pertinent funds obtained

8.2 PORT PRICING STRATEGIES: TARIFF CHANGING AND COMPETITIVENESS

8.2.1 The Shipping Demand Paradox

According to the basic principles of derived demand, the consumers' demand for the final product is directly dependent on the demand of a factor of production (e.g., port and maritime services) and therefore will determine its price.

However, in real-world shipping trade, this is not the case: despite the high demand for shipping over the past 50 years, retail prices in the developed and developing countries have risen by at least 300%. Crude oil price in particular has increased from $19 in 1949 to $110 in 2011 and $120 in 2012 (ESA 2012). At the same time, freight rates and supply chain and inventory costs have been reduced.

Commodity pricing factors include production cost, location, competition, economy, and product value (quality, utility, and scarcity). Sea transport pricing entails three key components: commodity pricing, freight rates (carrier costs), and port tariffs. While freight rates and raw material prices seem to be quite homogenous, there seems to be a disproportion in global manufactured and value-added commodity pricing, as well as wages. Furthermore, port tariffs are frequently controlled by geographical pricing, that is, determined by factors such as competition from other regional ports, port traffic, and trade agreements in the area. In this respect, port tariffs do not reflect the actual transportation and port costs in the area or the time efficiency, productivity, or advanced technological designs that contribute to efficiency and effectiveness.

As an example, South African *port tariffs* were 874% above the *global* average for containers and 744% above the global average for automotive cargo (Crotty 2013). On the other hand, South African port tariffs on coal exports were 50% below the global average and tariffs on iron ore exports were 10% below the global average.

This disproportionate profit margin of the ports could be named as *the shipping demand paradox*, as an industry's oversupply reduces profit margins in disproportion to its vast contribution as a factor of production to a profitable industry and economy.

8.2.2 Port Pricing

Port pricing and the tariff setting mechanism are dependent to a great extent on the following:

a. *Expenditures*, which include all expenses undertaken by the port entity, terminal operators, and logistics and transportation companies that provide services within/for the port. Expenditures or costs are distinguished into accounting, opportunity, and economic cost.
   - *Accounting cost* reflects the overall expenditures of capital or resources, invested in a specific venture, for example, port operations. This expenditure was spent in the past and the pertinent accounting transactions are recorded as journal entries.
   - *Opportunity cost* pertains to the unmaterialized rewards that have been rejected or declined, in order to pursue a more attractive investment. The opportunity cost demonstrates the advantages a port could have gained should an alternative investment be selected.
   - *Economic cost*, which is the sum of the accounting cost plus the opportunity cost.
b. *Income* is a port's portion of the revenue generated for shipowners and cargo owners that the port authority can receive in exchange of services provided. Income should be closely related to the value of these services offered. In the case of overcharging their clients, ports may see their clientele conclude business with competitive ports and even change their original trade routes in search for a more cost-effective alternative.
c. *Profit* is estimated by the difference between input (factors of production, overall service costs) and output (cost of services provided, and variable and fixed expenses) costs. It can be expressed by the following equation:

$$\text{Profit} = (S - \text{VS}_c) \times Q - \text{FS}_c \tag{8.1}$$

where S is the unit price of a service, $\text{VS}c$ is the variable cost of a service unit, Q is measured service quantity, and FS_c is fixed service costs.

8.2.3 Pricing Systems and Price-Setting Considerations

Pricing, Investments, and Market Differentiation

Pricing is a powerful port growth strategy: it is a fund-generating tool that enables ports to produce the resources necessary for new investments and expansion. A larger port with increased market share and limited port congestion is an attractive alternative for port managers. In addition, these additional funds may enable the port to differentiate, by investing in customized, innovative technology that will attract a new segment of customers.

a. *Pricing and Port Competition*

Increasing competition among ports compels them to become more efficient and economically supportable. Pricing is a valuable instrument that assists ports in attaining the aims set within the port development plan(s).

Strategic planning enables ports to endure cutthroat market competition by exercising strategic port pricing, market differentiation, and cost leadership practices. Increasingly, ports pursue customer loyalty, as the true winners in a competitive market are the ports with established long-term partnerships (UNCTAD 1995).

A competitive price setting strategy may also deter competitors from replacement ports: ports that offer a high "value for money," that is, align their tariffs to the benefits they offer to their customers, are more likely to increase their market share and hinder the market entry of replacement ports.

When ports are facing intense competition from replacement ports, certain marketing schools strongly advise ports to retain competitive, yet not cheap tariffs, as clients perceive low cost with low value. In these cases, low tariffs may be perceived by the market as a sign of weakness, reduced market share, and diminished competitive edge.

b. *Undercharge versus Overcharge*

The basic principle of port pricing suggests that the tariffs should reflect the supply–demand relation at any given time within the market cycles. Since port pricing is a product of market equilibrium, ideally, port managers should refrain from *overcharging*, that is, a condition when the charge of port services is much higher than both the original cost and the value or benefits that the customers enjoy. In this case, the port is likely to lose clients, as they will eventually seek for a replacement port. At the same time, port managers should avoid *undercharging* when possible, that is, the situation where the actual cost and value of services are higher than the tariffs and the port's overall pricing system. In the real market, intense competition frequently drives down the tariffs in neighboring ports, ports that serve the same supply chain or similar market segments.

c. *Pricing in Centralized versus Decentralized Port Systems*

For the pricing and tariff mechanism to be successful, a certain degree of flexibility and autonomy is required. Centralized seaport systems are typically less flexible and require more time in data processing and approvals. Hence, tariffs in centralized systems may be outdated, owing to the additional time required in the decision-making process.

Port managers need to conduct a market analysis and forecasts for specific geographical and market segments (ship types, commodity markets, buyers, etc.). A clear picture on the port's expenses versus potential income based on the market conditions will enable port managers to establish the optimum price setting.

d. *Port Pricing and Cost Coverage*

A port's primary objective is to cover the port's overall fixed and variable costs; to cover the expenses for the services offered, it is required to secure income that is equal to, or preferably exceeds, the costs.

Port economic policies may vary; hence, certain ports may adopt a flexible policy, wherein the port's overall income should cover or exceed the port's overall expenses. In this manner, it is easier for port managers to reach the target set. However, the other extreme of port economic policies stipulates that the cost versus income balance should be achieved by each port segment, that is, each port terminal, department, contract or service provided, and so on. This target is more difficult to achieve, for the simple fact that port occupancy and demand for specific services constantly fluctuate, as the demand for different market segments (crude oil tankers, bulk carriers, liquefied natural gas [LNG], passenger ships, etc.) varies and is never of similar or identical demand. The acceptance of this market reality should encourage more port authorities to estimate costs and earnings in a holistic manner. This decision will make it easier for port managers and accountants to reach their periodic financial goals and will take away from the organization unnecessary pressure.

e. *Port Services and Tariffs*

a. Port dues on ships (paid by shipowners)
b. Port dues on the cargo (paid by charterers—cargo forwarders or receivers)
c. Aids to navigation (stipulated by the charter party)
d. Pilotage (stipulated by the charter party)
e. Towage (stipulated by the charter party)
f. Berthing and unberthing (stipulated by the charter party)
g. Berth occupancy—time (stipulated by the charter party)
h. Cargo handling at port and stevedore expenses (stipulated by the charter party)
i. Lease of terminal's CHE
j. Ship services: bunkering (stipulated by the charter party)
k. Ship services: water, garbage removal, ballast water management, power generators, communication expenses)
l. Ship's stowage planning (paid by shipowners)
m. Cargo inspections; tally of cargo (stipulated by the charter party)
n. Cargo storage, warehousing (paid by charterers—cargo forwarders or receivers)
o. Logistics: intermodal transportation and cargo distribution (paid by charterers—cargo forwarders or receivers)

Table 8.1 shows the function of different tariff categories.

8.3 KPIs: MEASURING FINANCIAL AND OPERATIONAL PERFORMANCE

As modern port management and the shipping industry grow, key performance indicators (KPIs) or key success indicators are becoming a principal port management tool utilized to measure productivity and efficiency. KPIs enable port authorities, among other business entities, to outline and measure their growth level between setting and achieving their corporate goals.

TABLE 8.1   Function of Different Tariff Categories

A. General Tariffs

Conservancy, port dues

- Value based, corresponding to the value of the vessel (type and size)
- Produce revenues to pay for waterside infrastructure, floating equipment, and administration

Wharfage

- Value based, corresponding to the value of the cargo
- Produce revenues to pay for wharves and landside infrastructure, equipment, and administration

B. Facilities Tariffs

Berth hire

- Performance based to encourage vessels to reduce time at berths, especially during periods of congestion, or at underutilized berths, to attract vessels

Transit storage

- Performance based to encourage consignees to transfer in-bound cleared cargo to other storage areas and to encourage shippers to store cargo in preparation for loading

C. Service Tariffs

Pilotage

- Cost based to cover the variable costs of pilots and the pilot boats

Towage

- Cost based to cover the variable costs of tugboats and crew

Berthing/unberthing, mooring

- Cost based to cover the variable cost of the gangs

Stevedoring, wharf-handling, receiving/delivery

- Cost-based tariff to cover the variable costs for the cargo-handling labor and equipment

Equipment hire

- Cost based to cover the fixed and variable costs for the equipment and its operators

Cargo processing (including consolidation/deconsolidation)

- Cost based to cover the variable costs for the cargo-handling labor and equipment

Warehousing

- Value based corresponding to price of private service warehousing located outside the port

Fuel, utilities

- Cost based to cover the direct cost for the amount consumed

Source: UNCTAD 1995. World Investment Report. Available at http://unctad.org/en/Docs/wir1995 overview_en.pdf (accessed on August 2, 2013).

Port managers measure KPIs as either considering the port as a separate, autonomous corporate entity or in an integrative manner, where the port is integrated in its logistics chain.

Goal setting is a part of performance management. KPIs also assess a port's optimum utilization of resources, by measuring and comparing all input/output variables and functions. Therefore, KPIs are deeply connected with the goal-setting process, as they determine the likelihood and pace that a port will meet its goals.

The first known goal-setting concept was generated by *Aristotle*, the ancient Greek scientist and philosopher (384–322 BC). In his theory of "Final Causality among the Four Causes" (Τέλος, τα τέσσερα αίτια), he supports that action is only generated when an ultimate goal exists; hence, the goal-setting process entails reaching our highest possible potential (Aristotle 1924, 2013). The theory also suggests that throughout the process of setting and achieving our goals, practicing excellence is necessary, and this can only be accomplished with two major tools: intelligence and ethics.

Two thousand years later, Locke et al. (1981) originated their own theory as a continuation of Aristotle's findings: from 1969 to 1980, they screened the attitudinal outcomes of goal setting and verified that the higher and more specific the goals, the greater the performance, as compared to minimal effort or lack of goals.

A port's goal-setting process pertains to defining its long-term development and growth plans; the element of evolution is strongly present, as the port authority decides that its current position has not achieved its full growth, and a new positioning and aims can bring greater success in the future. Consequently, KPIs measure a port's production and business drivers. Ports can achieve their corporate goal only when they have tangible measurements of where they currently stand. Defining the port's current position can be measured by KPIs, in four distinctive areas, that is, Economics, Marketing, Operational and HR metrics:

a. *Economics metrics*: Successful ports strive to achieve cost efficiency, economies of scale, and profit maximization in critical corporate sectors. The following measurable elements help establish a system of financial goals, motivate the human force, and energize the goal-setting, goal-attaining process:
   - Budget allocation planning; spending and results
   - Costs and profit, profit and loss account
   - *Operating income* (revenue—cost of goods sold—depreciation)
   - *Return on capital* (operating profit/equity shareholders' fund)
   - *Return on investment* (net operating income/average operating assets)
   - *Return on assets* (net income/total assets)
   - *Cash flow return on investment* (cash flow/market value of capital employed)
   - *Economic value added (EVA)* (net operating profit after taxes—(capital × cost of capital)
   - Residual cash flow (net adjusted cash flow—cost of capital)
   - Market value added (equal to value less capital)
   - Residual income valuation:

$$P_0 = B_0 + \sum_{t-1}^{\infty} \frac{RI_t}{(1+r)^t} = B_0 + \sum_{t-1}^{\infty} \frac{E_t - rB_{t-1}}{(1+r)^t} \qquad (8.2)$$

where B_0 is the present book valuation of equity, B_t is book valuation of equity during set time (t), E_t is the net income over set time (t), r is ROR on equity, and RI_t is future time residual income.

b. *Marketing and customer service metrics*: A port's corporate survival and growth revolve around meeting and exceeding clients' expectations through delivering value-added services:
   - Service satisfaction perceived by customers, that is, terminal operators, liner and tramp shipowning companies, ship managers, charterers, industries, the community
   - Existing clients' retention
   - Market share acquisition/loss
   - New or niche market penetration

c. *Operational metrics*: This entails quantifying operational productivity through KPIs in order to estimate the output-to-input ratio.
- Input versus output, economies of scale, expenditure, cost minimization
- Performance quality, quantity, efficiency, regulatory compliance
- Time measurement, that is, turnaround time, port traffic, eliminating bottlenecks
- Optimum utilization of facilities

Certain operational performance indicators as recommended by UNCTAD (1976) include the following:
- Tonnage serviced at port
- Berth occupancy earnings per ship/cargo tonnage
- Cargo handling revenue per ship/cargo tonnage
- Labor-related expenses
- Capital equipment expenditure per cargo tonnage
- Participation per cargo tonnage
- Overall participation
- Late arrivals
- Ships' or port workers' idle time
- Service time
- Ships' turnaround time
- Tonnage per vessel
- Percentage of time berthed ships labored
- Gang utilization per ship per shift
- CHE utilization per ship per shift
- Tons for each ship hour while in port
- Tons for each ship hour at berth
- Tons for every gang hour
- Time gangs were idle

d. *Entrepreneurship and human resources metrics*: The human factor element is measured herewith, including talent, innovation, efficiency, compliance, and human error. Namely, the strategic and tactical decisions are set by the company's leadership, wherein employees' skills utilize the port's assets and other resources, in order to ensure that corporate goals will be met, through factors such as the following:
- Entrepreneurship, innovation
- Talent recruitment and retention
- Motivation and performance
- Training, development, drills, life education
- Avoiding brain drain
- Matching skills with tasks
- Productivity, time versus money

Labor input is typically measured as the number of hours worked, quantity of employees assigned to a specific task or a specific department, and an evaluation of the conversion of part-time jobs into full-time jobs. For efficiency and GDP development analysis, labor input is estimated by the total amount of hours worked. The measurement of labor productivity is increasingly perceived as inaccurate, for numerous reasons, such as absenteeism, output quantity versus quality, time worked versus actual productivity, and the correlation between salaries and productivity. Total factor productivity seems to drill down and encompass all factors of production as opposed to just one of them. Hence, total factor productivity increasingly becomes an important element of measuring the average productivity of labor (OECD 2001).

The overall measurement of these four key areas should bring direct benefit to the port, as well as direct and indirect benefits to the community, the state, and the nation in terms of regional economic and trade growth, supporting national goals, supporting the employment markets, contribution through taxation, and so on.

A case study on the Port of Antwerp, Belgium, follows.

CASE STUDY ON THE PORT OF ANTWERP, BELGIUM

The Port of Antwerp, Belgium (Figure 8.4), is Europe's second largest port after Rotterdam (Netherlands) and ranks thirteenth among the top 20 global ports. The port of Antwerp is Europe's significant model of excellence in financial, commercial, and industrial growth. A radiant sea–land interface, Antwerp is the greatest built-in chemical industry cluster in Europe. Sixty percent of Europe's buying power is positioned inside a radius of 310 miles (500 kilometers) from the port.

In 2012, the Port of Antwerp transferred over 184 million tons of maritime cargo, with sharp annual increases therefrom, especially in the oil and gas and the container sector.

PORT OF ANTWERP COMPARATIVE ADVANTAGES

The port's key comparative advantages include the following:

- Strategic location in the heartland of Europe secures the port's accessibility and connectivity with Europe's major rail, highway, and maritime networks.
- The strategic inland location of the Scheldt estuary, connecting the Flanders (Belgium) with France and the Netherlands, in a deep-draft (100,000 DWT), yet tidal navigational passage of 50 miles (80,000 kilometers).

FIGURE 8.4   Aerial photo of the Port of Antwerp. (Courtesy of the Port of Antwerp.)

- The port's increased draft, capable of accommodating the largest ocean-going vessels.
- Outstanding navigational accessibility, with a broad variety of sustainable services with strategic alliances throughout the supply chain (navigation, logistics, industrial trade zones).
- Freight-generating potential for the hinterland and the port market.
- Container handling facilities, an incomparable pipeline system, and the strongest concentration of supply-chain warehousing and distribution area supported by optimum labor efficiency.
- Innovative, state-of-the-art technology: for example, the port has obtained subsidies from the European Union, in order to design and construct an LNG bunkering station for barges in the port of Antwerp.

STRATEGIC PLANNING

The port's strategic planning and development, now and in the future, includes the following:

- Reinvigorating the port's leadership status worldwide, in a sustainable manner
- Optimum utilization of the port's navigational resources, infrastructure, and superstructure; designing a long-term perspective for the port's accessibility throughout the supply chains
- Long-term improvements of the port's high added value at a nationwide and continental level
- Ameliorating hinterland access to expedite the intermodal growth of commodities' trade; integration of all transportation modes with Belgium's national transportation Mobility Master Plan
- Applying powerful innovation plans to maintain the leading role of the port's social standing and competitive advantage
- Enhancing functional performance by means of increased storage efficiency. Architecture and utilization plans of vacant land space
- Regulatory compliance: establishment of the port of Antwerp as a safe and secure port
- Environmental response and proactiveness to environmental emergencies
- Establishing a research information center that accumulates all the pertinent data to implement synchronized ecological and environmental port and regional policies
- Managing environmental issues from a perspective of social responsibility and community engagement; adding value to the existing environmental plans and practices
- Service to the society: participating in community initiatives for an environmentally friendly culture with particular focus on regional residents and employees
- Fortifying community support for the port between all national and international stakeholders. Being conscious of soft values on top of the development of jobs and welfare

- Enhancing understanding of the significance of the port of Antwerp, endorsing community integration at a local and regional level, while securing optimum participation of the key players
- Enhancing the port's social status as a desirable employer
- Securing a healthy public climate

Source: EC Europa 2013; MSC, Belgium 2013; Port of Antwerp 2012; Sustainable Port of Antwerp 2013.

8.4 PORT EQUIPMENT AND BERTH FACILITIES: OPERATIONS, MAINTENANCE, AND DEPRECIATION

As the modern era for ports and the maritime industry are increasingly concerned about strengthening the port's competitive edge, the industry seeks to mitigate contemporary challenges in a highly intricate, multienterprise market configuration. Port authorities and managers seek for solutions and corporate growth through people- and technology-driven alternatives.

First, the technology factor is addressed through a plethora of innovations and heavy investment in the port's IT, superstructure, and infrastructure. Second, the human factor is associated with leadership and teamwork issues, HR talent and productivity development, labor unions' issues, strikes, working conditions, and so on. New or reinvented management and HR development strategies are applied to address the modern seaports' challenges. Section 8.4 deals with technologies, and Section 8.5 addresses the human force element.

New technologies are increasingly being introduced in port infrastructure and superstructure, perceived as remedies to growth of trade volume, changes in trade routes, intense market competition both locally and globally, and overall market fluctuations.

8.4.1 Port Cargo Handling Equipment

Port technology is a priority within a port management strategy: as intraport competition and the increasing industry demands increase, they literally dictate to ports the type and quantity of investment required for technology, for example, for specialized terminals, berths and docks, advanced pavement designs, warehouses and sheds, as well as cargo handling equipment (CHE).

The design and type of a port's cargo handling gear may vary, in accordance with the following:

a. The port's investment portfolio and budget restrictions
b. Energy efficiency and emissions dictated as per port and state environmental restrictions
c. Energy cost dictated as per port's financial restrictions
d. Ship and cargo type, that is, tanker ships, LNGs, containers, break bulk, reefer ships, bulk carriers, and so on
e. Ship size, cargo volume, and port draft restrictions, for example, compare a deep-sea mega-port capable of handling ULCCs, as opposed to a smaller port with draft restrictions, which may need feeder ship and other cargo handling arrangements

TABLE 8.2   CHE Type per Cargo Type

Liquid Bulk Carrier Terminals	Container Terminals	Dry Bulk Carrier Terminals	Break Bulk Terminals	Ro-Ro's, Vehicle Terminals	Passenger Ship Terminals
Forklifts	Top handlers	Forklifts	Forklifts	Forklifts	Forklifts
Trucks	Yard tractors	Loaders	Loaders	Electric pallet jack	Tractors
Cranes	Rubber tired gantry cranes (RTG)	Sweepers	Hydraulic excavators	Trucks	
Loaders	Forklifts	Trucks	Trucks		
	Side picks	Bulldozers	Bulldozers		
	Man lifts	Skid steer loaders	Skid steer loaders		
	Trucks	Man lifts	Sweepers		
	Sweepers	Cranes	Cranes		
	Rail pushers	Various	Man lifts		
	Reach stackers		Yard tractors		
			Various		

Source: M.G. Burns.

f. Intermodal arrangements for the cargo, that is, transport via oceangoing shipping, short-sea shipping, pipelines, rail, 18-wheeler trucks, and so on
g. Concrete structure of berths, seismically designed or upgraded to handle heavier cargoes, for example, a berth consisting of hybrid concrete–steel superstructure, reinforced with gravity walls and sizeable concrete shafts

Table 8.2 shows the CHE types that a typical port equipment list contains. The table distinguishes the CHEs that berthing facilities utilize, segmented as per cargo type.

Detailed information on the purchase, operation, and maintenance of CHE is developed and retained by the port authorities and the respective terminal operators.

The following case study illustrates some of the successes and "firsts" to come out of the Port of Houston Authority.

CASE STUDY: PORT OF HOUSTON, TEXAS (USA)

POHA: AN EQUATION FOR SUCCESS

The port celebrated its 100th birthday in 2014, making it one of the youngest global mega-ports. It boasts providing over 1 million jobs and $178.5 billion in state economic activity. Its remarkable list of firsts verifies the old saying that "Everything is bigger in Texas!"

- **1914:** First port to be constructed with federal capital and complementing state capital support, paving the way toward growth for most US ports!
- **1919:** First direct cotton distribution to Europe carried onboard *M/V Merry Mount.*

- **1956**: First container ship, *M/V Ideal X.*
- **1981**: First double-stack container train.
- **1997**: Baytown Tunnel (1041′ length × 35′ diameter) was the largest tunnel removed as part of a port's dredging and development operations, without shutting down the Houston Ship Channel: the tunnel was subsequently replaced by the Fred Hartman Bridge. The overall logistics of redirecting the cargoes was most efficient, with zero accidents and zero effects on the port's navigational safety.
- **2000**: First port to conduct air emissions testing with its outdoor devices.
- **2002**: First port to conform to ISO 14001 requirements for environmental excellence.
- **2004**: First port to be reaccredited to ISO 14001 standards.

Source: POHA Interview 2013 and Port of Houston Authorities 2013.

8.4.2 The "Port Equipment List"

The *"Port Equipment List,"* typically developed through the port's software, contains information pertaining to all of the port's departments and functions: from a top management, port planning, and accounting perspective, an efficient list of the port's assets is necessary. The list is also used for commercial purposes by the cargo operations department. In addition, the port's technical and maintenance departments ensure that the list is in line with their scheduled maintenance, inspection, auditing, repairs, sale, and purchase plans.

Table 8.3 contains a concise version of the data typically found in the "Port Equipment List."

8.4.3 IT, Logistics, and Operational Port Equipment

Modern-day ports are equipped with a wide range of technologically advanced machinery, IT, systemic control, and cargo-handling software, with the principal focal points being the following:

a. *Global digital trading network*: In the 1990s, US ports, shipping lines, and charterers/shippers have developed a global paperless trading network with the purpose of swift and efficient data exchange.
b. *IT-based alarm and notification systems; visitor-tracking software*: To mitigate seaport security, IT systems now offer wide-area satellite monitoring, which ensures the safe and secure movement of cargoes, employees, and passengers alike. Groundbreaking technologies generate improved versions of IT-based alarm and notification systems that employ consolidated information based on multiple sources. Satellite imagery is used for wide-area monitoring and circumference coverage, offered through intelligent object detection and closed-circuit cameras.
c. *Vessel traffic services (VTSs)* are designed to deliver live supervision and navigational assistance for ships in ports, as well as in enclosed and high-traffic seaways. VTS systems are distinguished into two categories:
  i. Surveilled VTS systems, which contain at least one ashore sensor, such as AIS, radar, or closed-circuit TV systems in order for operators to supervise and manage ship traffic activity.

TABLE 8.3   The "Port Equipment List"—a Concise Version

Port Equipment List—CHE			
Port: Terminal:		**Fiscal Year:**	
Part A: Equipment Life Cycle Management			
	Purchase Data	**Planned Maintenance Status**	**Repair Status**	**Certification Status**
Equipment data input	• Purchase date: • Port ID #: • Manufacturer: • Type: • Expected life: • Warranty: • Price: • Location:	• Maintenance date(s): • Inspection date(s): (internal and external) • Inspection findings:	• Breakdown date(s): • Repair date(s): • Cost: • Root cause analysis: • Repair documentation:	• Certificates issued: • Compliance status: • Audit dates: • Audit findings:

Part B: Asset's Current Status

(Select one or more options, as appropriate)

a. Good working conditions
b. Repairs required
c. Inoperable
d. Obsolete
e. Withdrawn from cargo operations
f. Disassembled for the value of its spare parts
g. Secondhand sale to third parties
h. Sold for scrap/recycling

Comments:

Part C: CHE

Technical Data and Specifications

CHE type:

CHE ID number:

CHE manufacturers, year and model:

Engine manufacturers, year and model:

Horsepower (kW):

CHE and engine manufacturers, year, model:

Energy type of fuel used (ULSD, gasoline or propane):

Annual operating hours:

Engine performance:

Emission Control Systems (ECS):

(Select as appropriate)
- Diesel oxidation catalyst (DOC)
- ULSD fuel
- Onroad engines
- Emulsified fuel
- Diesel particulate filter (DPF)
- Diatomic oxygen gas (O_2)

Emission Control System (ECS) year built/fitted:

Source: M.G. Burns.

ii. Nonsurveilled VTS systems comprise at least one control area where vessels must report their ID, flag, trade route, load/discharging ports, speed, and so on, to the authorities. Their functions and processes enable them to facilitate smooth ship traffic and ameliorate navigation even during extreme weather conditions. Most important, they are designed to control safety at port and in inland waterways, by mitigating groundings, collisions, and so on (USCG 2013).

d. *Innovative logistics technologies*: Modern logistics systems seek to increase overall efficiency while minimizing operating and maintenance costs.

A *warehouse management system (WMS)* is a critical component of ports and their logistics networks as they manage, control, and monitor cargoes in the process of receiving, putting away, picking and scanning, packing, and shipping.

Fixed logistics equipment:

e. *Belt weighers or in-motion checkweighers (IMC)* are designed to check the weight of shipments while in motion and reject them if they are underweight or overweight, that is, if their weight is not in line with the bills of lading or contract of affreightment and other documentation available at the port. IMCs frequently combine extras such as the following:

- Intelligent segmentation or rejection dependent on cargo weight
- USDA/FDA compliance with reporting standards
- Specific weight warehousing with statistical reporting
- Freight tracking mechanisms
- Metal detectors and x-ray devices

f. *Automatic storage and retrieval systems (ASRS)* are becoming increasingly essential in today's modern warehousing environment where time is equally important to high-density storage for containers, pallets, and so on. Intelligent ASRS units function with forklift trucks that deliver containers or pallets for handling to the system and remove empty containers or pallets to other designated areas. Cargo data and ID number for tracking will be input to the main (host) computer for network monitoring, control, and tracking. Cargo automatic retrieval is made possible as cargo location is registered in the local computer system.

g. *Sorting systems technologies (SST)* represent a state-of-the-art solution that can swiftly and precisely merge high volumes of diverse cargo streams into numerous predetermined shipment segments for transport and distribution. Depending on the port's WMS, the sorting process can be designed either as an integrated part of the logistics network or as an independent entity within a port or terminal.

h. *Automated conveyor systems* are not as sophisticated or with equally heavy-lift capacity as SSTs above, yet they ensure high-speed, flawless transfer of containers, boxes, and so on, to I/O and other logistics systems over long distances within the port's warehouse(s). As computer input designates the cargo transfer data (cargo ID, time and place of transport), cargo units are moved at the chosen location just in time!

i. Industrial robots are invaluable logistics tools for WMSs, which, based on barcode and RFID labels, are capable of identifying cargoes, order picking, palletizing and depalletizing, product packing, labeling, and so on.

j. Improved freight handling systems for increased operational performance. Modern wireless solutions monitor freight activity at port and gantry cranes,

and onboard ships, rail, and trucks. To ensure accurate positioning, they operate through global positioning systems and RFID. Port technologies can swiftly handle cargoes through the process of check-in, scanning, registration, and handling.
k. Logistics IT systems include the following:
  • Mobile technologies with built-in bar code scanners to allow identification of containers.
  • WMS software is especially designed to serve the entire logistics network and is distinguished into (a) business monitoring, (b) operational monitoring, and (c) cargo handling integration monitoring.

8.4.4 Depreciation Methods

Depreciation is an accounting principle where an asset's capital cost is dispersed throughout its life. The asset's actual value is reduced over time because of natural wear and tear. Salvage value or residual value is the estimated asset's value at the end of its useful life.

Table 8.4 includes a concise version of the prevailing depreciation methods.

1. **Straight-Line Depreciation Method**
   This is a straightforward and the most frequently used method, where the asset's value is equally depreciated each year throughout its useful life. It is measured by estimating an asset's purchase price, subtracting its residual or scrap value, and dividing it by the asset's anticipated useful life in years. At the end of the asset's commercial life, what remains is its salvage or scrap value.
   Accelerated depreciation methods: Sum of the Years Digits Depreciation (SYD), and Double Decline Balance (DDB)
   When using these methods, companies typically declare less revenue in the starting years, and their revenue will increase throughout the asset's commercial life. The most frequently encountered options of accelerated depreciation methods include the double-declining-balance and the sum-of-year methods.
2. **Double-Declining-Balance Depreciation Method**
   This method multiplies by 2 the SL depreciation sum of the first year and uses this fraction to estimate the depreciation over the next years of the asset's useful life.
3. **Sum-of-the-Years Digits Depreciation (SYD)**
   The sum-of-year depreciation method generates a variable depreciation cost, although the last year of the asset's useful life, its accumulated depreciation, is equivalent to the straight-line depreciation amount.
4. **Unit-of-Production Depreciation**
   This method estimates the asset's depreciation through a set rate for every unit of production. For this estimation, the expenditure per single production unit is estimated and subsequently multiplied by the final number of units generated over a specified period.
5. **Hours-of-Service Depreciation**
   This depreciation method is similar to the unit of production technique, the difference being that the depreciation amount is estimated by the sum of service hours throughout a specified time.

TABLE 8.4   The Asset Depreciation Methods

Depreciation Methods	Formulas	Time-Based Depreciation	Accelerated Depreciation	Activity-Based Depreciation
Straight-line method (SL)	SL = (Cost—residual value)/useful life	✓		
Hours-of-service depreciation	HSD = Purchase cost—residual value/useful life in terms of service hours = Depreciated amount/useful life			✓
Double-declining-balance method (DDB)	DDB in year 1 = Depreciation base * (2 * 100%/useful life of asset in years)	✓	✓	✓
Sum-of-the-years method	$SYD = Depreciable\ base \times \dfrac{Remaining\ useful\ life}{SYD\ numbers}$	✓	✓	
Unit-of-production (machine hours) depreciation	$\dfrac{Number\ of\ units\ produced}{Life\ in\ number\ of\ units} \times (Cost—salvage\ value)$			✓

8.5 PERFORMANCE MANAGEMENT AND THE HUMAN FACTOR

> One machine can do the work of fifty ordinary men.
> No machine can do the work of one extraordinary man.
>
> **Elbert G. Hubbard**
> *1856–1915*

8.5.1 Performance Management

Performance management is a corporate method of combining factors of production, techniques, and human resources for the company to achieve its strategic goals and competitive edge. Although corporate performance is both anthropocentric and technology driven, the technology factor is either dependent upon a port's financial powerbase, its earnings, investors and stakeholders' relations, and so on. On the other hand, the human factor is much more significant, as it also determines technological performance and overall productivity.

The powerful port workers' union is an example of how the human factor prevails over machinery: $1 billion a day is the cost for a strike in a major global port, encompassing the cargoes that are directed to nearby seaports, contracts' cancellation, and income loss for the regional business and transportation companies.

The performance management process commences by

a. Motivating our teams
b. Pinpointing strengths, weaknesses, and need for progress
c. Appreciating talent
d. Joining forces and resources and transforming personal goals into ambitious, large-scale corporate goals
e. Analyzing and understanding past performances, which will allow the port enterprise to define its future performance

Performance management is broad enough to encompass the entire port business entity and flexible enough to concentrate on the performance of a company segment such as a terminal, a single department, a group of workers, or a single member of staff. This method of monitoring performance verifies that the company moves toward its preset goals and objectives.

8.5.2 Human Factor in Port Performance

The human factor is becoming increasingly important in modern port management. And yet, this was an element that has been overlooked during the past decades of globalization and innovative technologies. However, the 2008 economic crisis acted as a wake-up call for the industry, as mega-ports with state-of-the-art technologies suffered great commercial and financial losses because they overlooked the human element. On the other hand, smaller ports with limited resources flourished and discovered their global competitive edge, simply by being enthusiastically driven by people.

Port recruiters frequently state that in order for modern ports and shipping businesses to remain competitive, they need to hire forward-thinking professionals with the ability to critically evaluate the market conditions and solve problems from a practical, rather than an abstract, theoretical approach. On the other hand, the solution possibly lies in the failure of their leaders to motivate employees in such a way that it brings out all the much desired good ideas, values, and abilities.

Port managers are increasingly being asked to enhance the corporate "human capital" through a plethora of HR methods, such as the following:

a. Diligent and efficient recruitment processes where the right persons are hired for the right job
b. Motivating workers through attractive employment benefits and remuneration
c. Enhancing employees' training and development, as well as life-education programs
d. Acknowledging and rewarding innovation, talent, performance, loyalty, tenacity, and so on
e. Encouraging internal promotions as opposed to external hires

Today's businesses require rapid response and adaptation to ever-changing and unpredictable circumstances. Leaders are continually dealing with the issue of how to redesign their organizations to ensure that they remain competitive. Innovation and transformation are an integral part of this process; hence, ports and other global market segments should

• Be more innovative in thinking
• Develop different and more productive ways of observing
• Generate creative solutions
• Observe key forces and trends influencing the business world

8.5.3 Human Factor—Survey Analysis

An original survey was conducted by the author in order to assess modern port and shipping employees at a global level. One hundred fifty questionnaires were distributed in ports' and shipping organizations' management and employees.

Table 8.5 reveals the results of the survey and identifies significant elements pertaining to employees' productivity, loyalty, motivation, information exchange, and so on.

While the questionnaire addressed individual roles and how these affect performance trends, teamwork and group performance are another dimension of corporate productivity, strongly driven by unity, a by-product of mutual trust and loyalty, quality of communication, corporate structure, ethics, paradigms, and culture.

Group level:

• The *"halo error"* or *"halo effect"* has to do with the cognitive judgment or bias (positive or negative opinion) we tend to hold against colleagues depending on first impressions, physical appearance, ethnicity, race, and so on.
• The *"cannibalization effect"* is found in teams where productivity is poorly monitored; hence, credit for the team's achievements is distributed unevenly. As teams operate based on inequality and lack of fair reward systems, the hardworking team members may eventually become demotivated, to the detriment of the organization and the team.

TABLE 8.5   The Human Factor: Questionnaire Analysis

1. The vast majority of white-collar workers and blue-collar workers confirmed that their productivity depends on job satisfaction and working conditions.
Motivation is a paramount factor in productivity, while motivators in order of importance are as follows:

   • Opportunities for professional advancement
   • Sincere appreciation of employees' unique skills, talent, experience, and contribution
   • Financial incentives
   • Lifetime skills development, drills, certified training

2. 91% of white-collar workers and 76% of blue-collar workers feel more productive when they are allowed to take professional initiatives, which are in line with the corporate processes and protocols and regulatory compliance.

3. 83% of employees' productivity and their willingness to take initiatives for the port/shipping company's benefit are closely related to professional confidence and familiarity. These derive by training, education, and hands-on experience.

4. 76% of the respondents consider that their loyalty to the port/company increases as they are being increasingly involved with the port's/company's strategy and long-term development plans.

5. 63% of the respondents answered that their loyalty to the port/company is closely related to their immediate supervisor, rather than the company itself.

6. 94% of the respondents verified that the employees' willingness to significantly increase their productivity (e.g., through working longer hours, reaching tight deadlines, or being involved in strenuous or even risky endeavors) increases as they become increasingly informed or actively involved with the port's/company's strategy and long-term development plans.

7. 58% of the respondents consider that employees' productivity is directly proportional to their supervisors' alignment with the corporate goals and protocols.

8. 73% of the respondents need to feel corporate loyalty in order to share more of their ideas and suggestions on improving corporate performance.

9. 66% of the respondents would prefer the company meetings to be more interactive. They would welcome problem-solving tools such as brainstorming techniques or root cause analysis sessions.

• Similarly, *"dispensability of effort"* and the *"counterproductive team"* occur when certain team members are not sufficiently productive, which in the long run will reduce the entire team's motivation levels, output, and efficiency.

• *"Responsibility diffusion"* describes certain team members' detachment from personal achievement and the loss of personal responsibility contributing to bare minimum working input and abstinence from team goals.

• *"The invisible team,"* frequently present in larger ports, business entities, and teams, pertains to the impersonal employment methods and the lack of methodic and unbiased employees' appraisal, which leads workers to absenteeism as a result of inadequate team connection and constrained expectations of recognition.

• *"The free rider effect"* within a *"loafing team"* relates to the phenomenon of the employees who reap the benefits of free corporate events and benefits, for example, networking, economic advantages and benefits, commissions, bonuses, and so on, yet they are not inclined to deliver the desired results and earn these benefits like other team members. This will quite likely occur in ports and businesses where work productivity is inadequately assessed, missing sustainability, consistency, and impartiality.

- *"Collocated versus dispersed team"* relates to Latane's social impact theory, which appreciates that smaller sized teams are characterized by visibility; therefore, individual effort and efficiency can be monitored and compensated. Through an award system of individual effort, business associates will be encouraged to operate in a dynamic manner and optimum, sustainable levels of productivity. On the contrary, visibility may not be achievable in larger teams, where personal productivity cannot be evaluated in a sustainable, impartial, or consistent manner. Hence, team efforts tend to slack as personal contribution cannot be measured. Very similar effects take place when teamwork is characterized by transparency and public exposure. When this happens, monitored and accurately assessed productivity makes employees work harder. In contrast, when personal and collective efforts remain private and contribution is not recognized, productivity diminishes.

REFERENCES

Aristotle, Ancient Greek Scientist and Philosopher. Third century BC. Physics II 3 and *Metaphysics* V 2, Aristotle offers his general account of the four causes. Accessed August 4, 2013.

Aristotle. 1924. Aristotle's Metaphysics: A Revised Text with Introduction and Commentary. OUP, Academic Monograph Reprints. 2 Volumes. Oxford University Press, USA. Revised edition (December 31, 1924).

Crotty, A. 2013. Drop in port tariffs to boost exports: Business Report, South Africa. Available at http://www.iol.co.za.

ESA, Economics and Statistics Administration, US Department of Commerce. 2012. Available at http://www.esa.doc.gov/Blog/2012/02/22/crude-oil-and-gasoline-prices-2012.

European Commission, Europa, Innovation and Networks Executive Agency, (EC INEA Europa). 2014. Renovation studies for Antwerp's Royers Lock to benefit from EU funding. Available at http://inea.ec.europa.eu/en/news__events/newsroom/renovation_studies_for_antwerps_royers_lock_to_benefit_from_eu_funding.htm. Accessed January 21, 2014.

Locke, E., Shaw, K., Saari, L. and Latham, G. 1981. Goal setting and task performance: 1969–1980. *Psychological Bulletin (American Psychological Association)* 90(1): 125–152. Accessed August 4, 2013.

Mediterranean Shipping Company, Belgium (MSC). 2013. Operations. Available at http://www.mscbelgium.com. Accessed August 12, 2013.

OECD. 2001. Measurement of Aggregate and Industry-Level Productivity Growth. OECD Manual. Available at http://www.oecd.org/std/productivity-stats/2352458.pdf.

Port of Houston Authorities (POHA). 2013. Interview.

Port of Houston Authorities (POHA). 2013. About the Port: Overview; History; Economic Impact. Available at http://www.portofhouston.com/. Accessed August 12, 2013.

UNCTAD. 1976. Port performance indicators. Available at http://unctad.org/en/PublicationsLibrary/tdbc4d131sup1rev1_en.pdf. Accessed August 4, 2013.

UNCTAD. 1995. World investment report. Available at http://unctad.org/en/Docs/wir1995overview_en.pdf. Accessed August 2, 2013.

USCG. 2013. Vessel traffic services. Available at http://www.navcen.uscg.gov/?pageName=vtsMain. Accessed August 15, 2013.

CHAPTER **9**

Leadership and Teambuilding

All of the great leaders have had one characteristic in common:
it was the willingness to confront unequivocally
the major anxiety of their people in their time.
This, and not much else, is the essence of leadership.

John Kenneth Galbraith
Canadian–US Economist and Diplomat, 1908–2006

9.1 LEADERSHIP AND TEAMBUILDING COMPLIANCE: STCW, THE MANILA AMENDMENTS

For a contemporary port to face today's complex and ever-transforming global trade patterns, new methods for thinking and acting are needed. Although capital investments and state-of-the-art technology do make the difference in terms of growth and development, the availability of resources alone does not guarantee commercial success, neither a profitable return on investment. Inconsistent leadership, teambuilding getting out of sync, inaccurate forecasting, unsuccessful partnerships, or expansion in the wrong market segment, all show that leadership, teambuilding, and strategic decision making really do make a difference.

Indeed, the core material where port growth is based upon is entrepreneurship and human talent. Hence, top port management must change its role through reshaping their strategic corporate message, that is, "what should we do next?" toward a new behavioral context, that is, "what is the corporate structure in which we work and make decisions?"

STCW, the International Convention on Standards of Training, Certification, and Watchkeeping for Seafarers, was established in 1978 by IMO's convention, aiming to improve the industry's standards. Pursuant to a number of revisions, the 2010 Manila Amendments effective as of January 1, 2012, with full implementation as of January 1, 2017, adopted new specifications for leadership and teamwork, highlighting the necessity to improve the industry's pertinent management tasks—such as resource management, time management, project management, strategic planning, and so on (IMO 2011). Additional skills that need to be developed bring up issues of effective communication, eloquence, motivation, generation of new ideas based on brainstorming, and so on (ITF 2010). The STCW Convention and the 2010 Manila Amendments consist of provisions for deck and engine officers and crew to comply with increased proficiency and training requirements in the aforementioned skills and competencies.

Within the amendments put into practice, many significant modifications entail mandatory training in leadership and teamwork, resource management, environmental awareness, and so on. The amendments signify the first time that mandatory IMO train-ing involves the development of soft skills and signifies a new awareness of how the human factor alone may affect the industry's performance, despite the high automation and capital-intensive practices that characterize the industry. Some of the topics that could be included in the mandatory training include human error, time management, planning, decision making, leadership and teamwork during crisis management and emergency response, and so on.

Effective leadership and teambuilding find practical applications and usefulness at an administrative, operational, and emergency response level. While evaluating the element of human error in maritime accidents and corporate leadership insufficiencies, it becomes obvious that the coexistence of leadership and teambuilding is rather significant for ports and the maritime industry as a whole.

9.1.1 Leadership and Teamwork

The corporate survival of a port entity in the twenty-first century requires a corporate architecture in which leadership and teamwork go hand in hand, that is, in the forms of leadership teams and leading team members. In the era of global change and unpredict-ability, successful ports should form their teams in adoptable, pliant working schemes, rather than rigid, unchangeable patterns.

The formation of a solid leadership and teamwork entity will provide the port with stability needed in order to overcome any corporate challenges and bring fresh ideas and out-of-the-box alternatives in an ever-changing global market. In fact, two of the most challenging issues that global businesses face today are the corporate inability to real change and the gradual shift into old, obsolete working patterns.

And yet, modern ports should be open to radical corporate changes, if the demands of the market deem this necessary. A recent study by the Oxford University revealed that among 100 large corporations that had to implement radical IT changes, only 16% of the companies were able to make a clear cut from their past working practices and fully take advantage of their new, modern IT systems (CNN 2013).

Port teams should consist of capable members with complementary talents and experiences. Leaders should hold a clear vision of growth, tangible strategies, and opti-mism supported by realistic plans and expectations. Finally, port leaders and their teams should be designed with powerful conflict-resolution mechanisms that will help the ports strive to achieve goals of expansion and mutual growth, instead of being involved in internal.

Teambuilding should be designed in a manner where complementary qualifica-tions and capabilities will enable the port's leaders to materialize their strategies and business targets. Modern ports, just like any other global business sector, typically handle larger projects that necessitate multiple skills and a wide variety of profes-sionals to work together. Long working hours are the norm in the shipping indus-try. Tensions and anxiety may arise when teams work for long hours in projects that are urgent, demanding, and complex. Under such stressful working conditions, conflicts and personality clashes are likely—especially when teams consist of hetero-geneous personalities. Recent studies reveal that conflict at the workplace and persist-ing clashes over the same issues are likely caused by radical differences in personal values. Capable leaders should be able to set the port's values in an upfront manner and inspire its employees to honor and share them. The port's HR department should

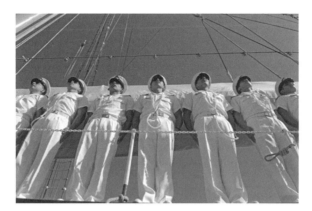

FIGURE 9.1 Leadership, teamwork, and a unity of purpose are vital in the port industry. (Courtesy of the Port of Haifa, Israel.)

focus on hiring professionals that share the same corporate values and culture and enjoy the same working environment. The hiring of the most skillful personnel may not be enough, if these employees have conflicting core values or working cultures (Figure 9.1).

9.1.2 Motivation

Successful, effective team leaders have the character and the power to translate the port's corporate vision into a manner that team members can perceive as their own, personal goal and vision.

Leaders and their teams should set the port's objectives by using brainstorming techniques and SWOT (i.e., Strengths, Weaknesses, Opportunities, and Threats) analyses. Ideally, the organizational goals should be aligned with the team members' professional goals. The implementation of participatory management and the team members' encouragement in expressing their views, knowledge, ideas, and recommendations will only enrich the port's decision-making process. The usefulness of participatory management can be verified during difficult times when crucial decisions have to be made. Lack of knowledge, lack of preparation, and the employees' inability to be an active part of a port's radical transformation (for better or for worse) have frequently been detrimental to the port and beneficiary to the port's competitors.

Professional skills and talents can be useful when they are parts of well-designed, well-composed teams. It is all about each member understanding his or her own professional dynamic, and bringing to the table the skills that combined with others can bring the results that the company needs.

9.1.3 Redefining Leadership and Teamwork through Leadership Styles

A significant yet overlooked area that will define a port's or a company's leadership and teamwork possibilities is leadership styles.

An evaluation of the two major leadership styles together with their advantages and disadvantages will help assess which leadership style better promotes teamwork:

a. Autocratic Leadership—Main Advantages and Disadvantages

Advantages

- The organization's most critical decisions are being taken or approved by the company's CEO. In case decision making or implementation of ideas has caused damage to the organization, the leader alone will suffer all damages.
- The risks (financial, expansion, strategic) are being taken by the leader(s) only. The team members are only obliged to follow the set rules.
- In case of fault, it is easy to identify where the responsibility should be attributed. The part at fault can be easily isolated, deprived of power, or even discharged.
- Team members need to strictly comply with the corporate rules, protocols, and job instructions, with minimum or no margin for innovation, improvements, or risk in the decision-making processes.
- The company's focus is toward retaining the status quo.

Disadvantages

- Lack of corporate flexibility may not allow the company to adapt to a volatile, ever-changing global market.
- The company suffers from loss of talent and new ideas, as the rigid corporate structure is more focused on retaining the status quo, rather than moving forward.
- The company gradually becomes unable to survive during market fluctuations, as its static rules are unable to offer new solutions to newly developed problems.

b. Virtual Leadership—Main Advantages and Disadvantages

Advantages

- Flexibility can be perceived as the company's ability to survive under any market changes.
- Participative decision making promotes a culture of teamwork, where different skill sets, talents, and professional backgrounds form an efficient puzzle of complementary skills and possibilities.
- Corporate visibility and open communication enable employees to make collective suggestions and share their personal views. Although the final decision is taken by the top management, the employees feel more gratified as their opinion was both heard and taken into consideration. Management's decisions were justified.
- Team members may become more empowered as they participate both in risk and credits.
- The risks (financial, commercial, and strategic) are being proportionally taken by all team members. Team members are expected to take initiative and find solutions to problems, and their performance will be evaluated depending on their ability to seek solutions. The credits are attributed accordingly.
- In case of fault, all members will try to rectify the situation. The ability for timely corrective action will establish the team's potential for initiating and undertaking new projects.
- It is an "all for one and one for all" endeavor, where the overall team performance will determine the final success.

- Unanimous and egalitarian procedures are followed during the decision-making processes.

Disadvantages

- Leaders must be willing to empower their teams and trust the company's future in the hands of their employees. This "exposition" might be considered as a high-risk business, which many companies are not prepared to take.
- Working and contributing in teams might be a frightening experience for individuals who are used to working alone or under an authoritative influence. They might be unwilling to take a risk or expose their ideas, under the fear of failure.
- In the case of error, the responsibility is attributed to both the leaders and their teams, making the root cause analysis more complex, as the root of error may not be easily identified. For this reason, the corporate culture emphasizes corrective action, rather than a blame culture (Burns-Kokkinaki 2002).

Over the past three decades, the pendulum swings from hierarchies, traditional management, spans of control, to more democratic formulas of empowerment, reengineering, and participative management. In order to reap the benefits of leadership and teamwork, a balance must be achieved. What is certain is the necessity to deal effectively and efficiently with the accelerating changes of the new era, in a climate where *Leading by Example* and *Leading by Teamwork* are implemented.

The following illustrates the Maersk Group as a case study of a leading global conglomerate.

CASE STUDY: THE MAERSK GROUP—APM

A Conglomerate with Style

Born on July 13, 1913, Maersk Mc-Kinney Møller (Figure 9.2) reengineered his family's maritime agency into a global conglomerate, handling over 15% of the world's merchandise carried by sea. His entrepreneurial disposition motivated him to enter diverse markets and at times own an airline, an IT company, industrial products, supermarket chains, and so on. Originating from Danish concessions in the North Sea, he established the Danish Underground Consortium and the concession to survey, drill, and produce oil. This monopolistic advantage motivated him to establish a global oil drilling enterprise and a fleet of 140 ships to carry this oil.

Today, Maersk is a leading-edge corporation, persistently innovating, growing, generating revenue, drilling oil, moving cargoes, procuring, training, and much more. As a corporation's global borders expand, it is easy to lose its vision, sense of control, and business personality, that is, the qualities and traits that make it distinguishable within the industry. With Maersk, this has never been an issue.

A.P. MOLLER–MAERSK GROUP

The A.P. Moller–Maersk Group is a global organization with 110,000 employees and offices in 125 countries, headquartered in Copenhagen, Denmark. Besides

FIGURE 9.2   Arnold P. Maersk Mc-Kinney Møller (1913–2012).

owning one of the world's largest shipping companies, both containers and tankers, the A.P. Moller–Maersk Group is linked to a number of activities within the energy, manufacturing, shipbuilding, and retail market sectors.

The Maersk group entails the following business units:

Maersk Fleet: with a fleet of over 1000 oceangoing, modern, diversified, and innovative ships:

Maersk Line (approximately 600 ships)
Maersk Tankers (approximately 400 ships)

Maersk Oil, which operates production of approximately 600,000 barrels of oil equivalent per day under difficult operating conditions, such as deepwater, high temperature, and high pressure.

APM Terminals, which is an efficient global terminal network of 20,000 person-nel in 68 international locations with operations in 69 port and terminal facilities as well as over 160 inland services operations. The company was named both "Port Operator of the Year" at the Lloyd's List 2012 Global Awards and "International Terminal Operator of the Year" for 2012 by *Containerisation International* (APM Terminals 2013).

Maersk Drilling, which is the ninth biggest drilling contractor globally and a growing leader in the deepwater market. Maersk Drilling's fleet includes some of the world's most advanced harsh environment jack-up rigs, deepwater semi-submersibles, drilling barges, and workover barges.

Maersk Drilling supports global oil and gas production by providing high-efficiency drilling services to oil companies around the world (Maersk Drilling 2013).

In addition, the group controls Maersk FPSOs, Maersk Supply Service, Damco, Svitzer, Maersk Container Industry, Maersk Training, Dansk Supermarked, MCC Transport, Safmarine, and Seago Lines.

LEADERSHIP THROUGH INNOVATION

In modern fast-paced global business, innovation is the only way for companies to assume a leadership role. On the basis of this principle, Maersk invests heavily in innovation throughout all corporate levels. The following section demonstrates

Maersk's selected innovations and verifies the company's dedication to lead through innovation:

"Triple-E Class" Containers—The World's Biggest Ships

Maersk Line has designed and built the "Triple-E class" containers (Figure 9.3), which are the world's largest container vessels, having a capacity of 18,000 TEU. The "Triple-E" title indicates its three primary purposes: (i) Economy of scale, (ii) Energy efficiency, and (iii) Environmental improvements. Maersk granted Daewoo Shipbuilding two $1.9 billion agreements ($3.8 billion total) to construct 20 such vessels, which are scheduled for delivery from 2013 to 2015. Some of the principal characteristics of the ships are as follows:

- Twenty percent less CO_2 emissions per container moved, as compared to the *Emma Maersk*, previously the world's largest container vessel, and 50% lower than the industry average on the Asia–Europe trade lane.
- The ships will be built with a waste heat recovery system, reducing up to 10% of main engine power, which equates to the typical annual electricity usage of 5000 European households.
- The fleet is capable of navigating for 114 miles (184 kilometers) using 1 kWh of energy per ton of freight, while a jumbo jet travels 0.31 miles (half a kilometer) consuming the same amount of energy per ton of freight.

APM Fast Crane Concept: APM Terminals

APM Terminals' FastNet Cranes signify a massive plunge in terminal productivity, by enabling STS gantry cranes to work adjoining bays of larger-sized container ships and thus doubling crane productivity (Figure 9.4).

During present cargo operations at port, crane legs determine the lowest space of one bay but limit the accessibility of adjacent bays, leading to lost possibilities for

FIGURE 9.3   Maersk Line Triple-E Class Containers. (Courtesy of A.P. Moller–Maersk Group.)

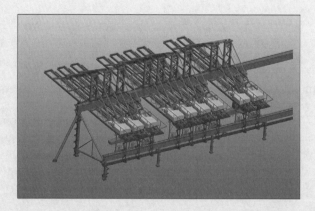

FIGURE 9.4   APM fast crane concept: APM Terminals. (Courtesy of A.P. Moller–Maersk Group.)

optimum production. Maersk's innovative technology allows the number of cranes used on a vessel to be doubled, thus minimizing turnaround time and port stay, with no increase in the handling rate of separate cranes (APM Terminals 2013).

Maersk 2011: Maersk Hybrid Crane to Reduce the Carbon Footprint

As modern technological innovations increasingly focus on low-CO_2 emissions and fuel-saving improvements, APM Terminals will save 20% per TEU handled. After the auto manufacturers' innovation in using two or more energy sources for propulsion, APM Terminals has launched a new technology enabling its rubber tired gantry cranes at the terminal to use diesel and electric power in order to move, load, and discharge/unload containers (Figure 9.5). APM Terminals has commenced refitting over 400 cranes in five continents with pioneering technology, having

FIGURE 9.5   Maersk hybrid crane. (Courtesy of A.P. Moller–Maersk Group.)

environmental, economic, and commercial advantages. According to Ross Clarke, head of Design and Operations for New Terminals at APM Terminals: "The refitted cranes will lead to energy, maintenance and cost savings, and APM will deliver cost-competitive, sustainable services."

APM Terminals, New Nigerian Mega-Port

APM Terminals is presently one of the greatest port and terminal operators in Africa, and in West Africa in particular, where APM Terminals Global Terminal Network include nine facilities, including Apapa Container Terminal and West Africa Container Terminal in Onne, Nigeria.

APM Terminals and its consortium associates aim to develop a new greenfield mega-port project and free trade zone at Badagry in Nigeria's Lagos State, ready for operations in 2016. The Nigerian Minister of Transport, Senator Idris Umar, considers the project as a demonstration of a public–private partnership development that would tackle congestion and establish Nigeria as a leading African maritime hub.

The deepwater port will be one of the biggest in Africa and will include groundbreaking facilities for oil and gas, container, bulk, Ro-Ro, and general cargo. The bordering Badagry Free Trade Zone will consist of oil refineries, an industrial park, a power plant, and warehousing and inland container deport capabilities (http://www.apmterminals.com).

9.2 EMPLOYEE MOTIVATION, TRAINING, AND DEVELOPMENT

The growth and development of people is the highest calling of leadership.

Harvey S. Firestone

Over the past years, the shipping world has become more complex, with increasing global competition, pressure for increased profits, market share expansion, and at the same time a need for quality and customer satisfaction. There is an urge to improve quality and increase productivity, whereas costs have to be minimized. To achieve this, ports need to utilize their comparative advantage(s), become flexible, and able to withstand market fluctuations. Human resources are a powerful weapon enabling ports and ships to meet this target.

Motivation is defined as an internal drive that activates behavior and gives it direction. The term *motivation theory* is concerned with the processes that describe why and how human behavior is activated and directed. It is regarded as one of the most important areas of study in the field of organizational behavior.

The following illustrates NYK Group as a case study of a leading global shipping magnate.

CASE STUDY: NYK GROUP (NIPPON YUSEN KABUSHIKI KAISHA), JAPAN

"Continuance Is Power" 継続は力なり。

NYK Group is one of the world's largest and oldest shipping companies and a core member of the Mitsubishi family, headquartered in Tokyo, Japan. After 120 years of global leadership, the company retains its traditional Japanese nobility while undergoing a dynamic transition from steamships, to fuel oil, to state-of-the-art solar-powered ships, to innovative "Super-Eco Ship 2030" designs (Figure 9.6), to entry in the offshore shuttle tanker business and ultra-deepwater drill-ship business in partnership with Petrobras. The group's rapid, yet steady expansion was achieved as the company remained loyal to its core values of "Integrity, Innovation, and Intensity" (3 I's) (NYK Report 2013). NYK actively contributes to the betterment of societies through safe and dependable monohakobi (transport).

The group's philosophy of corporate social responsibility (CSR) is the idea that each employee can earn the respect and trust of society through the three pillars of "sound and highly transparent management," "safe, environment-friendly operations," and "workplaces that instill pride" (Yusen Logistics 2013).

NYK GROUP: HISTORY

1870
Tosa Clan establishes Tsukumo Shokai Shipping company.
1872
Tsukumo Shokai renamed Mitsukawa Shokai.
1873
Mitsukawa Shokai renamed Mitsubishi Shokai.
1875
Mitsubishi Shokai starts Japan's first overseas liner service between Yokohama and Shanghai. Mitsubishi Shokai changes name to Mitsubishi Kisen and then to Mitsubishi Mail Steamship Company.

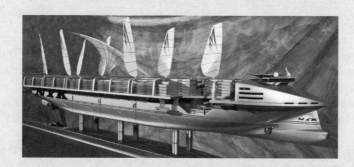

FIGURE 9.6   Super Eco Ship 2030, NYK's ecofriendly ship with innovative power generation, that is, having wind, solar, and traditional fuel options. (Courtesy of NYK Group.)

2008

"New Horizon 2010," the company's new medium-term management plan, was released.

NYK Cool Earth Project was launched.

Auriga Leader, a solar power-assisted vessel, was completed.

2009

The Emergency Structural Reform Project "Yosoro" was implemented.

Exploratory design for NYK Super Eco Ship 2030 was released (NYK Group 2013).

Good Design Awards received for two of NYKs environment-friendly ships, Auriga Leader and NYK Super Eco Ship 2030.

Participation in a project for ultra-deepwater drillship, to be chartered by Petrobras, begins.

2010

Yusen Logistics was established to integrate the NYK Group's logistics.

Two module carriers equipped with an innovative air-lubrication system were delivered.

Offshore shuttle tanker business was entered.

2011

A new medium-term management plan, "More Than Shipping 2013," was launched.

NYK sends support teams of volunteers and offers aid to the area devastated by the Great East Japan Earthquake.

NYK-TDG Maritime Academy graduates its first class, enhancing the NYK Group's measures to proactively employ seafarers.

2012

NYK's Ship Energy Efficiency Management Plan was the first in the world to be certified by ClassNK.

A plan was announced for the building of four post-Panamax pure car and truck carriers equipped with the latest energy-saving technology.

NYK hosted a cruise on *Asuka II* to aid reconstruction of the area damaged by the Great East Japan Earthquake and ensuing tsunami.

NYK Fleet: Continuance Is Power

Energy

Petroleo Brasileiro S.A. (Petrobras), of Brazil, begins operation of ultra-deepwater drillship for presalt fields.

Wheatstone LNG project, Australia: the group participates in this joint project between private and public sectors in Australia.

Bulk carriers

M/V Soyo, a coal carrier, is launched, which is equipped with an air-lubrication system.

M/V "Shagang Sunshine" is a 250,000 DWT iron ore carrier with the maximum loading capacity permitted at ports in Western Australia.

Car carriers

Four "Next-Generation Car Carriers" will be built and delivered in 2014 and 2015.

Containerships

The Innovative Bunker and Idle-time Saving (IBIS) Project begins in earnest, with the purpose of realizing optimized and extremely economic vessel operations.

Cruise ships

Crystal Cruises Inc. was voted as World's Best Large-Ship Cruise Line, by readers of US travel magazine *Travel + Leisure*, for the eighteenth consecutive year. In addition, the company was voted as the best cruise line (mid-size) by the magazine *Condé Nast Traveller*, for the sixth year in a row.

Tanker ships

VLCC "Tateyama" establishes long-term charter with the Thai Oil Group.

REFERENCES

NYK Group. 2013. Available at http://www.nyk.com/ENGLISH/csr/envi/eco ship/ecoship_interview.htm. Accessed July 24, 2013.

NYK Report. 2013. Available at http://www.nyk.com/english/ir/library/nyk/ pdf/2013.pdf. Accessed July 27, 2013.

Yusen Logistics. 2013. Available at http://www.eur.yusen-logistics.com/about-us/company-information/corporate-social-responsibility. Accessed August 21, 2013.

Motivated team members become more energized and hardworking, positively contributing in a port plan that appreciates their individual talents, background, and contribution. Powerful teams are based on trust, and the certainty that the port's success will also become their own professional success.

- Leaders should clearly apprehend their team members' qualifications, talents, and background, for the appropriate allocation of responsibilities.
- Although leaders should be open to new ideas, suggestions, and "out-of-the-box" perceptions, the decision-making process should be characterized by decisiveness and assertiveness.
- Leading by example and servant leadership principles seem to motivate their teams. Hence, leaders should be prepared to practically show the way to their employees by working harder and showing the way, as opposed to theoretically describing the company's objectives.
- Empathy is the key skill that twenty-first century ports and shipping companies require. By developing empathy skills and the awareness of what our team members think, believe, and feel, we can become more effective in motivating them and achieving the corporate goals.
- The future belongs to the ones who adapt to change. Leaders should be versatile and intelligently transform to the new market realities.
- Arguably, emotional intelligence can help leaders and team members to communicate effectively and understand motives, conceptions, and principles. At the same time, business decisions should not be governed by emotions.
- Self-awareness always leads to self-actualization. By understanding our own skills, we will find the path to success both as team members and as leaders.

- Teambuilding is all about building high spirits, developing talents, and motivating our people.
- The team's objectives should be aligned with the port's or company's objectives. At the same time, team members should use their complementary skills in order to find their own talent niche.
- Employee appraisals should be objective, consistent with the ultimate intention of overcoming obstacles and misunderstandings. Judgment should never be personal. Communication should be an opportunity for growth.
- Teams should retain their high spirits, strength, and confidence through contingency plans and alternative paths to overcome obstacles.

As modern ports and shipping organizations are increasingly becoming diverse and multicultural, it has become crucial to identify the core values that surpass national boundaries and establish homogeneous and homologous standards and competencies.

In today's business, excessive competition and economic pressures require employees to take initiatives, be inventive, and offer a great deal of their time, skills, and imagination. To meet these new demands, employee motivation is necessary. As the pyramid-like structure tends to fade, together with the distinction between "manager" and "employee," empowerment is considered as a key motivator and a vital tool for all organization team members.

Training for leadership and teamwork development requires both a theoretic approach through in-class training and practical experience through drills and hands-on participation, both onboard and onshore. Over the past two decades or so, a number of experts from many fields have produced literature on the types, manifestations, and results of leadership. During the past few years, poor business performance has taught us that a new understanding of leadership must emerge. The question raised is whether "distributed power" must replace the "command-and-control" policy.

The following illustrates how West Gulf Maritime Association contributes to regional maritime leadership, teamwork, and training.

CASE STUDY: WEST GULF MARITIME ASSOCIATION (WGMA)

THE ROLE OF WGMA IN THE SHIPPING INDUSTRY

Organized in 1968, the WGMA has a rich tradition of serving the maritime industry as is a regional, not-for-profit, trade association serving as an advocate, training, and educational group. WGMA is proactive in promoting maritime commerce on behalf of our members. WGMA members are the leading commercial operators in every Texas port plus the growing Port of Lake Charles, Louisiana. Members can benefit from the networking opportunities with other WGMA members who work in the shipping industry and experience many of the same challenges and opportunities. Their meetings bring together the "Who's Who" of the maritime industry in a relaxed and informal setting.

WGMA's membership's diversity and the *distinctively proactive advocacy role they play on their behalf are unlike ordinary trade associations* serving the West US Gulf region. In addition to their advocacy role, they are uniquely focused on promoting the well-being of our industry through the dissemination of factual information and useful communication on a timely basis. Finally, but most importantly,

WGMA has close working partnerships with relevant government agencies that influence maritime interests. WGMA fosters, where appropriate, dialog between industry stakeholders and government agencies.

The WGMA communicates and coordinates with various governmental entities throughout the West Gulf. These agencies view the WGMA as an important conduit for ongoing dialog with the maritime industry. Standing WGMA committees work with the association's staff to project a unified voice concerning industry issues in the West Gulf area. The WGMA member committees currently address industry education, technology changes, environmental issues, government affairs, safety and security issues, and a full range of maritime operations.

Through its Maritime Affairs department, the WGMA provides a forum for discussions and exchanges of information between member companies pertaining to a wide variety of industry challenges. The WGMA provides a Daily Industry Update Report useful in providing our members with events and forums affecting navigational, environmental, and safety concerns.

The WGMA conducts periodic member meetings in a variety of port cities. The WGMA staff brief members on news and developments affecting their businesses. Acting through committees appointed from the membership, the WGMA negotiates and administers various multiemployer collective bargaining agreements with the International Longshoremen's Association in West Gulf ports.

Forecasting the Future of Shipping

Houston and Texas have seen phenomenal cargo and maritime infrastructure growth over the decades. That trend will continue, demonstrating impressive expansion and many firsts at a national and global level. In addition to the cargo and commercial growth, we now see a dramatic shift in influence and commercial control with Houston's emergence as the place to be in maritime circles. The shift will eventually establish Houston as a leading global and national shipping center.

As for the future of the global shipping industry, the carrier alliances and vessel consortium of the mega-container operators will affect the container trades. Smaller carriers will need to band together to leverage the same economies to compete. Cargo will move in the most cost-efficient and time-consistent route.

Foreseeing the Future Trends of Gulf Ports

The outlook for gulf ports is very favorable. A number of key factors will greatly affect this region in the coming years:

- The deepening and widening of the Panama Canal. Cargo will continue to be diverted from the US West Coast to the US Gulf and US East Coast ports. Companies such Wal-Mart, Home Depot, and others now utilize multimillion square foot warehousing and distribution centers in the Houston area. This impact will be felt not just in the container trade but also with agriculture bulk products and building materials and in the high-value chemical sector.
- Trade with Cuba will grow in the coming years. Although Miami would appear to be a logical port for consumer goods traded with Cuba, Texas

will still garner a large market share of agriculture products, machinery, engineering/construction project sourcing with Houston's large EP&C base, and of course the huge chemical manufacturing infrastructure found there.

- The LNG market has already started to change the face of maritime shipping in the US Gulf, Texas and Louisiana in particular. This trend will continue. *LNG as a feedstock combined with the competitive edge of lower energy costs* will create a boon of new business, found in the petrochemical and manufacturing sectors. Some experts indicate that Texas could become the "Saudi Arabia of the world LNG market." The future for LNG is very bright and will be favorably felt within the maritime circles.

- Texas is a pro-business and pro-maritime state. Industry has found a user-friendly environment to relocate or expand. WGMA is approached by all types of maritime-related companies looking at expanding or relocating to Texas. Four out of five companies end up coming here or enlarging their operations. There are huge capital expansion projects planned with the maritime sector up and down the US Gulf. Louisiana alone may see over $40 billion in new projects in the next 10 years. This can be multiplied 10-fold for the Texas ports.

Interview with Niels Aalund
Officer and Senior Vice-President, WGMA

9.3 LEADERSHIP AND TEAMBUILDING ASSESSMENT, DRILLS, AND BRAINSTORMING EXERCISES

Never tell people how to do things. Tell them what to do and
they will surprise you with their ingenuity.

General George S. Patton
American General in World War I and II, 1885–1945

Aristotle, the great ancient Greek philosopher, wrote that to be a good leader, one must be a good follower. Therefore, a valuable team member who works in an efficient and supporting manner is most likely a potential leader for the organization. Teamwork and cooperation ensure unity, and unity generates strength in any business or social entity. Successful teamwork is built on clear, consistent corporate goals and values, established by great leaders. Teamwork is defined as the effective and efficient collaboration of employees, toward the same purpose. Successful teams are characterized by principles of equality, mutual support and esteem, and sharing of knowledge. Most important, successful teams are based on the certainty that their performance is measured and evaluated in a consistent, impartial, and equitable manner. Figure 9.7 shows the four principal leadership styles.

What creates effective leaders is a subject of frequent study and discussion, yet the definitions of leadership are as varied as the explanations. The shipping industry and port management in particular are very demanding industrial sectors, where leadership errors can have severe impact not only in financial and commercial terms but also in the

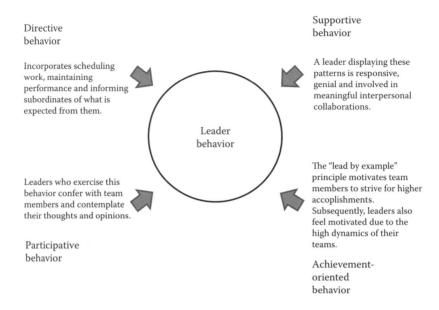

Directive
behavior

Incorporates scheduling
work, maintaining
performance and informing
subordinates of what is
expected from them.

Supportive
behavior

A leader displaying these
patterns is responsive,
genial and involved in
meaningful interpersonal
collaborations.

Leader
behavior

Leaders who exercise this
behavior confer with team
members and contemplate
their thoughts and opinions.

Participative
behavior

The "lead by example"
principle motivates team
members to strive for higher
accoplishments.
Subsequently, leaders also
feel motivated due to the
high dynamics of their
teams.

Achievement-
oriented
behavior

FIGURE 9.7   Four different types of leader behavior.

ports' or ships' safety, security, environmental pollution, social responsibilities, and so
on. Modern headlines have frequently covered stories where leading ports and leading
global corporations have collapsed overnight owing to a single strategic misjudgment—
which usually reveals a broader systemic failure. Ports are involved in large-scale business
and operations. Because of the wide range of activities in a global network of partners
and markets, leadership errors or omissions have a significant impact in the port's future.

Brainstorming is a team solution-seeking procedure that requires the creative generation of remedies through suggestions and recommendations from each team member on
an equal basis.

Within a casual, comfortable environment, team members contribute through an
"out-of-the-box" approach, suggesting that the feedback obtained can be unconventional, unorthodox, or extreme.

The benefit here is that the collective contribution can be assessed and a final solution
can be pursued through combining multiple views. A think tank of ideas, solutions, and
growth tactics should be established, for the feedback of the consecutive brainstorming
sessions to be registered.

After the initial triggering of ideas, leaders should consider the pertinent practical
elements such as resources, feasibility studies, and so on.

The steps of brainstorming should have the following structure:

- Identifying the problem
- Setting a feasible and yet ambitious goal
- Establishing measurable, tangible objectives that should be met within a specific
  deadline
- Pinpointing the available tools, assets, resources, and limitations—both tangible
  and intangible
- SWOT analysis; assessment of internal and external issues

- Strategic planning process: (a) strategic plan, (b) business plan, and (c) budget allocation, monitoring, and controlling
- Plan implementation

Participative teams at the workplace is to employers' and workers' mutual benefit. The right balance between leadership and teamwork is clearly observed through employees' empowerment and subsequent transformation. Collective ideas and action plans need to be seriously considered by the management, and paths for implementation should be pursued. Authority figures need to establish trusting, consistent, and committed relations with their teams.

The changes implemented in a "participative" environment have a direct positive impact on teambuilding:

- Increased self-esteem, empowerment at work, opportunities for innovation and creativity, in a climate of acceptance, encouragement, and security.
- Opportunity to learn more at work, have more interesting tasks, and gain working experience.
- Even during difficult times in the market, when decision taking is far more important, employees may come up with new solutions to save the day.

In today's world of possibilities, craving for leaders and innovative thinking, business can never be "as usual." Nowadays, constant change requires fresh ways of seeing and doing things. Strategic decisions must be taken in a decisive manner, within a collective spirit. Great ideas should be implemented with no delays; otherwise, someone else will take them over.

In today's uncertain global market, leaders and managers strongly need their teams' support in order for the port or the company to survive and flourish.

Teambuilding entails robust connections among corporate personnel, which ideally are based on support, esteem, and mutual respect. Specific qualities and attributes necessary to establish powerful teams include the following:

- Recognizing all team members' contribution, appreciating each member's background and significance
- Understanding the members' motives and concerns
- Being in the position to overcome challenges
- Ability to resolve conflicts and communicate efficiently
- Maintaining top working standards at all times, including while under pressure, in unjust circumstances, when extended time and effort are needed, even during lack of appreciation
- The ability to reestablish damaged interactions
- Connecting in a dependable, trustworthy, and professional manner
- Retaining an optimistic professional frame of mind; not becoming despondent over setbacks, not becoming overwhelmed on smaller issues
- Establishing connections based on respect, retaining an environment of approval
- Deeply trusting the team, the leaders, and the company by retaining confidentiality and abstaining from spreading rumors

Teambuilding is a business strategy of doing things while operating interdependently within a group of professionals, as opposed to independent working methods.

While establishing the foundations of the team, it is necessary to identify the skills and potential, and motivate team members in a manner when all perform together, create together, and achieve together.

Teambuilding capabilities. When assembling a team, it is essential to assess the overall team potential. In accordance with LaFasto's theory, a competent and productive team consists of the following elements:

1. The team member, that is, every individual team member should be experienced, effective, and efficient.
2. Group connections: Suggestions, improvement, and methods associated with the assignments to be achieved should flow throughout the command line.
3. Problem-resolution management: An effective group should be able to resolve challenges and be centered on its objectives.
4. Effective group leadership is dependent upon leadership talents and skills. A competent leader has to be concentrated on the goals, encourage synergy and collaborative teams, stimulate group members, and supercharge performance through ongoing efforts.
5. Productive and satisfying team settings.

Team objectives. As opposed to solo playing career methods, teambuilding comprises of a more intricate environment, where joint efforts and joint responsibilities are required to bring the desired results and accountability.

The advantages of teambuilding include the following:

- Improved potential and adaptability in skills and abilities
- Increased productivity, value-adding participation
- Stimulating personal growth and development of talents
- Relying on group action to accomplish the set goals

A leader's capabilities will comprise of the following:

- Hiring and accumulating proficient team members
- Putting together straightforward and uplifting team aims
- Encouraging an atmosphere that stimulates growth
- Reliance and confidence; dependability
- Prioritizing goals
- Visibility in actions
- Gratification and appreciation
- Offer exterior assistance
- Promote an environment of expansion through ethics

REFERENCES

APM Terminals. 2013. New Nigerian Mega-Port. Available at http://www.apmterminals. com. Accessed August 18, 2013.

Burns-Kokkinaki, M. 2002. *Virtual Leadership*. London Metropolitan University.

Cheese, P. CNN. 2013. CNN money: What is so hard about corporate change? Available at http://management.fortune.cnn.com/2013/05/20/corporations-change-failure/. Accessed August 10, 2013.

IMO. 2011. STCW Manila Seafarer Training Amendments enter into force on January 1, 2012. Available at http://www.imo.org/MediaCentre/PressBriefings/Pages/67-STCW-EIF.aspx. Accessed August 15, 2013.

ITF. 2010. STCW: A guide for seafarers. International Transport Workers' Federation. Available at http://www.mptusa.com/pdf/STCW_guide_english.pdf.

Maersk. 2011. Maersk Hybrid Crane to Reduce the Carbon Footprint. Available at http://www.maersk.com/aboutus/news/pages/20110331-154630.aspx. Accessed August 21, 2013.

Maersk Drilling. 2013. Rig Fleet. Available at http://www.maerskdrilling.com/Documents/PDF/Mediacenter/Rig-Fleet-February2013.pdf. Accessed August 18, 2013.

Maersk Drilling. 2013. Fleet location. Available at http://www.maerskdrilling.com/drillingrigs/fleet/pages/fleet-location.aspx. Accessed August 18, 2013.

Maersk Fleet. 2013. Available at http://www.maerskfleet.com. Accessed August 18, 2013.

Maersk Line. 2013. Liner Ships. Available at http://www.maerskline.com and http://www.maersklinelimited.com. Accessed August 12, 2013.

Maersk Tankers. 2013. Available at http://www.maersktankers.com. Accessed August 18, 2013.

CHAPTER **10**

Port Authorities and Regulatory Framework

Zero Accidents, Zero Incidents, Zero Near Misses

Maritime Industry Motto

In order to retain its high standards of safety, security, environmental integrity, and so on, the maritime industry is heavily regulated for both port facilities and ships.

This chapter addresses the principal maritime regulations and provides a case study on the harmonization between global, national, and state regulations.

Chapter 11 further addresses the maritime regulatory framework and examines its potential harmonization with the offshore oil and gas (energy) industry.

A plethora of maritime laws, regulations, conventions, and so on, have been developed in order to safeguard the industry at a global, national and state (municipal) level.

a. *Global regulations*: At a global level, international organizations such as the International Maritime Organization (IMO), the International Organization for Standardization (ISO), the World Health Organization, and so on, have developed an efficient regulatory framework of a global radius that encompasses most nations.
b. *National regulations*: Each nation adopts rules and regulations that reflect the international standards to a certain extent, yet their priority is to promote the national needs and goals.

   Typically, advanced nations have the strictest regulations and the heaviest penalties, as their priority is to achieve quality and high standards of safety, security, environmental protection, and so on. Less advanced or certain rapidly advanced nations whose economies are based on cost efficiency typically select less stringent regulations, which will enable them to boost production and develop low-cost ports and trade zones.
c. *State or municipality regulations*: Additional regulations exist at a local level, depending on a nation's political regime and governance. For example, in the United States, or the European Union's Member States, *state regulations exist.*
d. China, on the other hand, is divided into municipalities that may have additional rules, on the basis of their local challenges and growth. Nevertheless, each municipality is subject to central government intervention. The National 5-Year Plan(s) determine each municipality's growth targets, opportunities for collaboration, and areas of specialization.

Not only are port authorities in charge of the safety of life, facilities, and assets within their ports, they are literally the areas where ships usually undergo inspections, surveys, and audits. Therefore, ports' decisions about strategy, structure, and regulatory frameworks will eventually determine their standards, safety culture, and overall performance.

A port's reputation is determined to a great extent by its track record of safety, performance, and regulatory compliance. Incidents that may harm a port's safety record may seriously affect its reputation and, consequently, its business growth (IHMA 2013).

Ports' Safety in the United States: The US Coast Guard has a statutory duty under the Ports and Waterways Safety Act of 1972 (PWSA), Title 33 USC §1221 to guarantee the safety and environmental integrity of US harbors and waterways. The PWSA assigns to the Coast Guard the duty to "...establish, operate and maintain vessel traffic services in ports and waterways subject to congestion." In addition, it allows the Coast Guard to request the carriage of electronic devices essential for engaging in the VTS system. The PWSA was founded in order to assure safety, good condition, and predictability on US harbors and waterways by employing essential waterways management methods. In 1996, the US Congress required the Coast Guard to review the VTS program and concentrate on user participation, compliance to minimum safety requirements, employing cost-effective techniques, implementing intelligent technology, and discovering public–private joint venture possibilities (NAVCEN 2013).

Ports' Safety in the United Kingdom: The Port Marine Safety Code (PMSC) ensures that all ports legally comply with a set of safety and efficiency rules and control marine operations within their authority. The PMSC and its "Guide to Good Practice" require the implementation of domestically established principles that ensure safe practices and optimum operational performance. Port authorities have both statutory and nonstatutory obligations. These responsibilities entail the obligation to preserve and assist in the port's safe utilization and a responsibility of caution from any damage or loss incurred as a result of negligence on behalf of the port authority. These responsibilities are harmonized with standard and particular capabilities that empower British port authorities to release these duties (UK Department for Transport 2012).

This enactment of a safety management system is structured on a sophisticated risk assessment methodology and aims to ensure risk identification and elimination to an endurable level that is as reduced as realistically feasible (Port of Tyne 2013).

The regulatory framework that port authorities and ships need to comply with is dealt with in this chapter, and it includes Safety (ISM), Security (ISPS), Occupational Safety and Health (OSHA in the United States, or OHSAS in the rest of the world), ISO 14001: Environmental Protection by the ISO, and the Vessel General Permit (VGP) by the US Environmental Protection Agency (VGP by EPA).

Among these regulations, ISM and ISPS are adopted by the IMO, and their compliance is ensured at a national level for the 170 member states that have ratified the multilateral treaty as of 2013. Some of these regulations were adopted by international organizations and are implemented at a global level (e.g., ISO 14001, OHSAS, HAZMAT), whereas others were adopted and implemented at a national level (e.g., OSHA, VGP, BWM in the United States).

Certification and verification are components that most of these regulations have in common. Both ashore (i.e., company) and shipboard verification and certification are issued by recognized organizations (ROs), which are appointed to carry out the regulatory certification on behalf of flag administrations. Some of the leading ROs are members of the International Association of Classification Societies (IACS, http://www.iacs.org.uk), namely, the American Bureau of Shipping (ABS 2005), Det Norske Veritas–Germanischer Lloyd Group (DNV-GL Group), Bureau Veritas (BV), Lloyd's Register (LR), Nippon Kaiji Kyokai (NK), China Classification Society (CCS), Korean Register of

Shipping (KR), Croatian Register of Shipping (CRS), Indian Register of Shipping (IRS), Polish Register of Shipping (PRS), Russian Maritime Register of Shipping (RS), and Registro Italiano Navale (RINA).

Over the past few years, the industry has initiated the harmonization of HSQEE (Health, Safety, Quality, Environment, and Energy) at a government and company level, where the ROs are again pivotal instruments of compliance and certification. Most of these regulations have identical patterns of compliance, certification, and a plan–do–check–act methodology, which enables their harmonization. This means that companies can follow a consistent process of compliance, audits, inspections, surveys, and incident investigation, and therefore save time and resources for governments, ports, and shipping companies. As an example, a modern-day auditor can now board a vessel and audit the regulatory compliance of Health, Safety, Quality, and Environment; use similar check-lists; follow a similar auditing process; and issue certificates of the same date, with the same expiration date, and so on. A few years ago, this process would have taken much more time, effort, and resources as it would require four different auditors to board the vessel at different times, most likely at four different ports. Other common elements in the above regulations include the industry's proactive stance through training, drills, commitment to lifelong learning, and the eagerness to learn from past mistakes. In fact, each year's maritime accidents, incidents, and near misses are being reviewed by the industry's conventions, governments and policy makers, and in-depth discussions take place in order to identify the root cause of the major accidents. Pertinent regulatory amendments take place, which may also require increased structural reliability (e.g., ice class, double hull/double tanks, etc.), improved ship designs (e.g., improved geometry of bulbous bows for wave-making resistance, etc.), and new technologies (e.g., energy conservation and alternative energy, ballast water treatment systems, etc.).

A significant characteristic of modern-day port management and shipping practices is the radical increase in automation and machinery use, while the number of employees both at port and ashore has been significantly reduced. As the man-to-machine ratio changes, it strongly suggests that fewer maritime personnel have to handle machinery breakdowns. Consequently, human error ratio tends to diminish, as the machinery failure ratio increases. This important element affects the regulatory compliance for all the following codes and rules, and it is prominent in the findings of incident investigation and root cause analysis (RCA), which will be discussed in this chapter.

10.1 ISM: INTERNATIONAL SAFETY MANAGEMENT

The International Safety Management (ISM) Code is an international standard for the safe operation of ships and for pollution prevention as adopted by the IMO Assembly Resolution A.741(18) of 1993 and amended by MSC.104(73), MSC.179(79), MSC.195(80), and MSC.273(85) (ISM, IMO 2010). The provisions of Chapter IX of the International Convention for the Safety of Life at Sea (SOLAS) and the ISM Code apply *to* more than 98% of the world's merchant shipping tonnage.

The ISM Code was developed pursuant to a number of very serious accidents that occurred during the late 1980s, that is, the *Herald of Free Enterprise* (1987) and the *Scandinavian Star* (1990). Some of the major maritime accidents were caused by human error, with contributing factors being management fault, lack of supervision, and lack of an impartial, knowledgeable, and reliable connection between the management and the seafarers.

The code is founded upon general rules and objectives due to the fact that that each company is a different entity, and no two companies are identical. The functional goal of the IMO is to produce an international set of standards for the safe management and operation of ships and for pollution prevention. This is accomplished by offering to the maritime industry a regulatory framework pertaining to the evolution of development, implementation, management, and operating practices.

The code recognizes the necessity for optimum management organization and dedication to safety and environmental protection throughout all levels of command, both ashore and onboard.

The code's objectives are as follows: (a) to ensure safety at sea, (b) to avert human injury or loss of life, (c) to eliminate environmental degradation, and (d) to prevent damage to property.

The company's objectives concerning safety management involve the following: (a) providing safe employment procedures during ship operations, (b) establishing safety measures against all identified hazards, and (c) enhancing employees' safety management skills both ashore and onboard ships.

The advantages of ISM concentrate on the emergency response and proactiveness associated with safety and environmental protection. Particularly, they address (a) safety record keeping, (b) sustainable efforts for improvement, (c) and belief on the part of policy makers, law enforcement officials, and the maritime industry in general that the company and its ships conform to requirements.

The ISM Code calls for every company to produce, implement, and sustain a safety management system consistent with the following requirements: (a) a safety and environmental protection plan; (b) guidelines and processes to assure safe operation of vessels, and environmental protection, in conformity with pertinent global and flag state laws; (c) outlined levels of authority and lines of communication involving ashore and shipboard employees; (d) processes for reporting incidents, nonconformities, and near misses in compliance with the code; (e) processes for proactive measures and emergency response scenarios; and (f) processes for internal and external audits and management evaluations.

Companies should obtain a Document of Compliance (DOC) or an Interim DOC once it has been confirmed that it conforms to the pertinent prerequisites of the ISM Code. An updated copy of the DOC ought to be kept onboard all ships managed or operated by the company.

A vessel ought to have a valid Safety Management Certificate (SMC) or Interim SMC once it has been confirmed that its shipboard management and its company function in compliance with the approved safety management system.

In November 2012, IMO's Maritime Safety Committee (MSC, 91st session) implemented a number of amendments to the International Convention for SOLAS, along with a new compulsory requirement for new ships to be designed to eliminate onboard noise and to safeguard crewmembers from noise; it also contemplated numerous other concerns, such as piracy and armed robbery against vessels.

10.2 ISPS: INTERNATIONAL SHIP AND PORT FACILITY SECURITY CODE

Within the maritime industry, there seems to be a fine line between safety and security. In fact, safety pertains to unintentional dangers such as natural threats (e.g., extreme weather), mechanical failure, or human error. Security, on the other hand, pertains to intentional threat and potential damage, mainly focused on piracy or terrorist attack. Well prior to the tragic events of September 11, the shipping industry was already committed in promoting safety and security and protection of human life, property, and the environment.

The International Ship and Port Facility Security (ISPS) Code is a wide-ranging set of procedures aiming to increase the security of port facilities and ships. It was developed pursuant to the identified threats to ships and port facilities during the 9/11 attacks in the United States. The ISPS Code is enforced through the International Convention for SOLAS, 1974, namely, SOLAS Chapter XI-2, "Special measures to enhance maritime security." The code features two parts, part A being mandatory and part B being recommendatory, that is, providing guidelines for implementation (IMO Globallast 2013). The ISPS Code is applicable to vessels on global voyages (i.e., passenger ships, cargo ships of 500 GT and greater, and mobile offshore drilling units) and the respective port facilities serving these vessels.

Fundamentally, the code ascertains that the security of ships and port facilities is a risk management action and in order to verify what security measures are suitable, a risk assessment must be applied in each specific situation. The intent behind the code is to offer a standardized, sustainable framework for risk assessment, empowering governments to balance out modifications in threat with the degree of vulnerability for ships and port facilities. This is achieved by means of identifying different security levels and respective security measures.

The key goals and objectives of the ISPS Code are as follows: (a) to identify security threats and implement security procedures; (b) to assign tasks and duties with regard to maritime security for government authorities, regional administrations, and ship and port sectors at a global and national level; (c) to gather and share security-relevant data; and (d) to offer security assessment guidelines, plans, and procedures.

The code does not stipulate particular measures that all vessels and port facilities should follow to assure the protection of the facility against security threats, owing to the several kinds and segments of these facilities. Rather, it defines a standardized, sustainable framework for risk assessment.

The shipboard and shipping company requirements pertain to company security officers, ship security officers, ship security plans, and ship security equipment.

The port security requirements include port facility security officers, port security plans, and port facility equipment. The compliance prerequisites for all involved in ISPS include security information being easily obtainable and accessible, entrance monitoring and controlling, and monitoring and controlling of movement of people and commodities (IMO Globallast 2013).

The ISPS Code has been enforced at a global level through the IMO. In the United States, a regulatory framework harmonizes the national security regulations with the global maritime security standards of the IMO, that is, SOLAS and the ISPS Code. In addition, the framework stipulates the provisions of the Maritime Transportation Security Act of 2002. Namely, the Code of Federal Regulations, that is, 33 CFR, Parts 101–107 comprise the national ship security regulations, together with procedures applicable to foreign ships in US territory waters.

10.3 OHSAS AND OSHA: OCCUPATIONAL SAFETY AND HEALTH ADMINISTRATION

The global maritime industry acknowledges the necessity of monitoring, controlling, and improving occupational health and safety performance by means of occupational health and safety management systems. Prior to the 1990s, the plethora of national and global standards, rules, and certifications pertaining to health and safety has brought adverse effects, as the industry needed a single consistent framework of compliance.

In the United States, the Occupational Safety and Health Administration (OSHA) is part of the United States Department of Labor and stipulated in the Public Law 91-596, December 29, 1970, with recent amendments. The US Congress developed OSHA through the Occupational Safety and Health Act of 1970, to ensure safe and healthful working conditions for employees, by establishing and implementing standards and by offering outreach, education, training, and assistance. The OSHA administrator is the Assistant Secretary of Labor for Occupational Safety and Health (OSHA 2013).

At a global level, the Occupational Health and Safety Advisory Services (OHSAS) Project Group released the OHSAS 18000 series in 1999. OHSAS is a British industry standard for occupational health and safety management systems, established to support British and global organizations' implementation of reliable occupational health and safety performance (OHSAS 2013).

10.4 VGP: VESSEL GENERAL PERMIT BY THE US ENVIRONMENTAL PROTECTION AGENCY

EPA presently handles vessel discharges with the 2008 VGP, which is in effect right up until December 19, 2013.

Since that date, discharges inadvertent to regular vessel operations are considered illegal except if approved by a National Pollutant Discharge Elimination System (NPDES) permit or excluded from NPDES permitting (US EPA 2013). VGP addresses 26 discharge forms, such as ballast water, antifoul hull coatings, deck runoff, bilge water, gray water, and so on.

In November 2011, the EPA recommended two Clean Water Act permits to control particular types of vessel discharges into US territory waters. The recommended permits would substitute individual VGPs issued in 2008 that will expire in December 2013. As suggested, the permits would be applicable to approximately 71,000 large domestic and foreign oceangoing vessels and possibly 138,000 smaller vessels. It is worth noting that different ship types (e.g., tankers, containers etc.) have different types of discharge, with different types of pollutants, oil, lubricants, hazardous materials, pathogens, and so on (Copeland 2013).

On March 28, 2013, EPA released the 2013 VGP effective as of December 19, 2013, to approve discharges incidental to the regular discharge of operations of commercial ships (VGP 2013). The 2008 VGP will be replaced by the latest VGP, officially referred to as NPDES 2013 VGP, including commercial vessels longer than 24.08 meters (i.e., 79 feet) but excluding military and recreational vessels.

The latest permit stipulates 27 particular discharge categories and will further recommend ameliorations to the performance of the permit procedure, in addition to describing discharge requirements.

In order to eliminate the risks of the release of nonindigenous species, the new permit features a more rigid discharge standard limit that will restrict the release of invasive aquatic species in ballast water. The permit includes specific environmental protection guidelines for the Great Lakes, a region that has experienced excessive damage from invasive species. It harmonizes federal standards with several Great Lakes states by requesting specific ships to implement extra safety measures to eliminate the likelihood of introducing new invasive species to US territory waters.

The 2013 final VGP additionally tries to decrease the administrative load for shipowners, managers, and operators, by reducing identical, overlapping reporting prerequisites, developing digital record-keeping options, and minimizing self-inspection requirements for ships that are out of service for prolonged periods.

10.5 ISO 14001: ENVIRONMENTAL MANAGEMENT SYSTEM

The ISO was established in the aftermath of World War II, to standardize machinery, construction, and industrial activities. Ever since, ISO has expanded and incorporates over 8000 distinctive standards, focusing in quality and environmental administration. The ISO 14000 series covers several elements of environmental compliance. ISO 14001 in particular is a voluntary industry standard common in the maritime industry. Companies are able to recognize and manage their environmental effects and therefore enhance their environmental efficiency. Other instruments from the ISO family include the ISO 14001:2004 and ISO 14004:2004 that specialize in environmental management systems (EMSs) (ISO 2013). ISO 14001 is the primary management system tool that stipulates the prerequisites for the development and compliance of an EMS. The key prerequisites for compliance are regulatory conformity, pollution prevention, and sustainable development of the EMS.

Ports and shipping companies can reap the benefits associated with ISO 14001 compliance, which include (a) sustainable environmental efficiency directed by the organization's leaders; (b) cost efficiency attained via resources control and elimination of waste in energy, water, and so on; (c) diminished risk of environmental pollution occurrences, leading to eliminated cleanup expenses or related penalties and fines; (d) proactive stance leading to a culture of preparedness; (e) enhanced brand name status, since the industry will recognize the port's or company's commitment; (f) enhanced earnings by means of eliminating expenditures; and (g) customer loyalty and satisfaction.

10.6 HAZMAT: HAZARDOUS MATERIALS; HAZWOPER: HAZARDOUS WASTE OPERATIONS AND EMERGENCY RESPONSE

At a global level, the committee of experts on the transport of dangerous goods of the United Nations Economic and Social Council establishes Model Regulations on the Transportation of Dangerous Goods. In the maritime industry, the IMO has produced the IMO Dangerous Goods Regulations for sea transport and the International Maritime Dangerous Goods Code as a section of the International Convention for SOLAS. Furthermore, the regulatory framework for HAZMAT is covered in IMO's International Convention for Standards of Training, Certification, and Watchkeeping, Regulation II/2, Sections A-II/2, B-V/b, and B-V/c. For rail transport, the Intergovernmental Organization for International Carriage by Rail has created the regulations pertaining to the International Carriage of Dangerous Goods by Rail as a component of the Convention for International Carriage by Rail.

The numerous accidents and disasters that have caused the loss of human life and severe environmental damage that related to the storage and transportation of hazardous materials (i.e., dangerous goods) verify the need for HAZMAT rules and laws. Among the different transportation modes by sea, land, and air, most accidents have occurred during road transportation. Unfortunately, even a small quantity of dangerous goods are capable of causing considerable damage. Hence, HAZMAT emergency response teams are exclusively trained professionals, capable of handling hazardous substances.

At a national level, government regulations are established to align with the UN dangerous goods regulatory framework. Hazardous materials in the United States and Canada, or dangerous goods in the rest of the world, are solids, gases, or liquids that may damage human health, animals and microorganisms, the environment, and property. In the United States, maritime transportation only represents 0.1% of the total number of HAZMAT accidents. Hazardous wastes are addressed in the Codes of Federal Regulations: 40 CFR 261.3 and 49 CFR 171.8. HAZMAT cargoes are typically demonstrated by diamond-shaped placards on the package, container, or storage area. The placards' colors represent the type of HAZMAT and frequently demonstrate the specified UN number indicating the commodity's unique chemical composition. Dangerous goods or HAZMAT belong to nine principal categories, that is, (1) explosives; (2) gases; (3) flammable liquids and combustible liquids; (4) flammable solid, spontaneously combustible, and dangerous when wet; (5) oxidizers and organic peroxides; (6) poison (toxic) and poison inhalation hazards; (7) radioactive; (8) corrosives; and (9) miscellaneous.

In the United States, regulations are stipulated in the Department of Transport (US DOT) Hazardous Materials Regulations (HMR); 29 CFR Part 1910, Occupational Safety and Health Standards, Subpart H, HAZMAT; 33 CFR Part 155, Oil and Hazardous Material Pollution; 40 CFR Parts 260–279 Hazardous Wastes; 46 CFR Part 98, Bulk Packages to allow for transfer of Hazardous Liquid Cargoes, 49 CFR Parts 100–185, the HAZMAT regulations (HMR).

The Hazardous Waste Operations and Emergency Response (HAZWOPER) relates to hazardous waste operations and emergency response performed in the United States under the OSHA. The OSHA standard 1910.120 of HAZWOPER stipulates the safety requirements to be fulfilled in the public and private sectors, as well as their subcontractors so as to carry out emergency response and clean-up operations. The HAZWOPER standard addresses five distinct aspects of operations, such as (a) clean-up operations mandatory by national, state, or local authorities pertaining to hazardous materials that are performed at uncontrolled waste sites; (b) corrective actions pertaining to clean-up operations at sites protected by the Resource Conservation and Recovery Act of 1976 as amended (42 U.S.C. 6901 et seq); (c) emergency response procedures for emissions of, or considerable risks of HAZMAT emissions, regardless of the hazard's area; (d) voluntary clean-up procedures at sites identified by national, state, or local authorities pertaining to hazardous materials that are performed at uncontrolled waste sites; and (e) operations concerning HAZMAT waste executed at storage area, processing, treatment, or disposal facilities as stipulated in 40 CFR Parts 264–265.

10.7 BWM: BALLAST WATER MANAGEMENT

The International Convention for the Control and Management of Ships Ballast Water and Sediments was implemented by IMO in 2004. Its standard responsibilities included (a) compliance with aim to avert, reduce, and eventually eliminate the exchange of dangerous aquatic microorganisms through the management and treatments for ships' ballast water; (b) control and management of ships' ballast water and sediments, aiming to minimize or eliminate the move of dangerous aquatic organisms; (c) ports' and terminals' sediment reception facilities, making certain that sufficient reception facilities are available for cleaning or repair of ballast tanks; and (d) investigation and surveillance of BW operations (IMO Globallast 2013).

A regulatory framework has been developed in order to tackle the marine invasive species issue connected with ships' ballast water treatment, specifically the standards of the IMO's Ballast Water Management Convention for the control and management of ships' ballast water and sediments for newly built and existing ships. The D1 standards involve ballast water exchange and specify the volume of water ballast, while the D2 standards relate to accredited ballast water treatment systems and define the quantities of invasive organisms residing in water after treatment. Both D1 and D2 standards have been in effect since 2012 with full compliance in 2016.

Following the critical environmental deterioration caused by invasive aquatic species, the IMO has created a set of rules and processes for management and control of vessels' ballast water, which function in addition to the International Ballast Water Management Convention. In 2012, the US EPA in partnership with the US Coast Guard and the US Department of Homeland Security founded the final rules on national requirements determining the allowable concentration of nonindigenous species in vessels' ballast water released in US waters. These rules initiate an authorization process for ballast water management systems and consequently amend the US Coast Guard's regulations for engineering equipment (US Army Corps of Engineers 2012).

Ballast water discharge generally includes a number of organic species including plants, animals, bacteria, and viruses. These species frequently include nonnative, harmful species that can trigger considerable environmental and economic destruction of native aquatic ecosystems. Ballast water discharges are considered to be the primary source of invasive species in US marine waters, thus influencing public health, environmental risks, and significant economic cost (Hazelwood 2004). The financial cost of US aquatic ecosystems exceeds $6 billion per year (Pimentel 1999). The National Invasive Species Act of 1996 was passed as a means of controlling the invasion of aquatic invasive species.

10.8 INCIDENT INVESTIGATION AND ROOT CAUSE ANALYSIS

As ship structures, sizes, and designs advance and crew sizes are reduced, the human factor/technology connection should be closely monitored, to ascertain safety, security, and environmental protection throughout ships' routine and emergency operations.

Port management and the shipping industry in general are heavily regulated and enjoy very high standards of safety and compliance compared to other global industries. The regulations and codes examined in this chapter have managed to tackle and seriously minimize the number of accidents, incidents, and near misses, resulting in a constantly improving record of safety, security, and compliance. At the same time, the industry requires to investigate each and every incident, as mandated by the global maritime laws (e.g., IMO's ISM Code) and also as required by most flag administrations and reputable companies in the industry.

Incident investigation is a procedure created to help the maritime industry become more efficient by investigating the circumstances involving an accident.

The primary goals of incident investigation include (a) prevention and averting future recurrence, (b) identification of potential threats or hazards, and (c) establishing appropriate measures, procedures, and regulations. The investigator needs to bear in mind that a close examination of the circumstances of an incident most frequently reveals that the true factors that have caused the damage are neither obvious nor straightforward.

Incident investigations need to focus on likely inadequacies and problematic areas pertaining to systems, processes, and equipment, mostly associated to the human factor. Once the RCA has revealed the underlying causes, corrective action must be implemented, aiming to improve overall functionality. Typically, the tools used within the industry include (a) information collection and analysis, (b) RCA, (c) suggestions for improvement, and (d) systematic filing and reporting of outcomes.

An RCA usually reveals a series of contributing elements that have triggered the accident. It is elementary to recognize the root causes and factors that have most likely contributed to the incident. The modern industry uses a plethora of tools for RCA, such as cause-and-effect diagrams, interrelation diagrams, brainstorming techniques, and so on.

Some of the cause mapping methods used for RCA include the following: (a) the "fishbone diagram," which may reveal the true cause by highlighting different functional segments, that is, the human factor, machinery, process, training, resources, and so on; (b) the "five whys" questions, which, if answered accurately, will drill down to the true cause; and (c) timeline analysis, which focuses on the chronological sequence of events. Once the true root cause has been identified and eliminated, the problem does not recur. This helps the industry protect human life, the environment, and the property, and prevent recurrence.

In port management and the shipping industry, RCA is a useful tool to identify problematic areas not exclusive to and not focused on the following: (a) change management and risk management; (b) occupational health and safety analysis; (c) production and business process; (d) failure analysis for manufacturing, maintenance, and engineering; and (e) quality control.

The lessons learned will help policy makers and port and marine managers develop a more proactive and efficient set of operational procedures.

The regulatory compliance standards of the maritime industry pertaining to safety, security, quality, and environmental protection are exemplary. And yet, disasters pertaining to safety, security, or environmental damage still occur. The occurrence of such serious accidents affirms the necessity to align risk assessment as a component of risk management and incident investigation. It is worth noting that risk management reflects the big picture, as it deals with risk identification, assessment, and prioritization.

Econometrics is the tool that will enhance a risk assessment and will enable the industry to prevent a crisis, on the basis of risk assessment models. The following econometric models can be used to assess risk in the maritime industry, taking into consideration the location, act of God (i.e., natural causes and disasters), security issues (i.e., deliberate attack from an adversary [terrorism or piracy]), human error, or events of unpredictable root cause.

i. Risk Assessment Formula for Asset Structural Integrity

The risk assessment formula for the structural integrity of assets applies to conducting a risk assessment of inanimate objects in the oil and gas and maritime industry. The assets to be evaluated in terms of risk and structural integrity include port facilities, terminals, ships, offshore platforms, oil refineries, pipelines, and so on.

$$R = \text{Lv} \times \text{SSEv} \times S \times H \times T \qquad (10.1)$$

where

R is the risk magnitude of an accident.

Lv is the asset's location; in particular, in acts of God (winds, hurricanes), it is the radius of maximum wind (RMW). The value (Lv) has the element of proactive evaluation of the risk possibilities based on the current location and the asset's exposure.

SSEv is the safety/security/environmental vulnerability of the particular area from a reactive perspective.

S is the asset's structural integrity (e.g., port, terminal, ship, etc.), considering internal structural factors such as fatigue, true structure strength, and external factors such as weather, temperature, maintenance, and age parameters.

H is the estimated hazardous material within the asset or in the vicinity of the asset. The risk formula will need to determine the hazard involved, that is, oil cargo stored at the port facilities, HAZMAT in containers, and so on.

T is the estimated exposure time to the risk. This is especially useful when the risk involved is prolonged (e.g., weather related) or in anticipation of outside help through repairs, maintenance, or a contingency plan. The time factor will increase the risk level, as it is analogous to the risk exposure (e.g., the longer the exposure time, the higher the risk).

ii. Risk Assessment and the Human Factor Formula

In the maritime industry, humans are also a key element of the risk assessment econometrics, having the capacity of responding to an emergency and modifying the accident's outcome, for example, by saving lives, preventing a terrorist attack on the facilities, following a contingency plan, isolating the hazard, or even evacuating the facility. Therefore, human coping aptitude (H) needs to be incorporated into a risk assessment, along with its particular vulnerabilities to different hazards. The human factor modifies the equation into

$$R = \frac{(\text{Lv} \times \text{SSEv} \times S \times P \times T)}{H} \qquad (10.2)$$

Again, R is the risk magnitude of an accident, Lv is the asset's location, SSEv is the safety/security/environmental vulnerability of the particular area from a reactive perspective, S is the asset's structural integrity, and P is the estimated number of pollutants within the asset or in the vicinity of the asset. The risk formula will need to determine the hazard involved (i.e., oil cargo at an oil terminal). T is the estimated exposure time to the risk.

H is the human factor empowered to influence the emergency outcome to a greater or lesser extent on the basis of factors such as the emergency response resources (outside communication, maintenance, technology) and also the asset's mobility (i.e., a hurricane at a port vs. a ship capable of deviation). In case of human error, the human factor will constitute a negative impact in the equation.

iii. Environmental Damage Assessment

In the subsequent equation, both anthropogenic and nonanthropogenic sources of damage are applied. The elements can be indexed by a finite number of real-valued parameters; hence, this is considered a parametric model.

A generalization of the multiplicative approach was defined as follows:

$$\text{Ed} = C\,(\text{NExp})^{\alpha} S^{\alpha 1} i S^{\alpha 1} i \ldots S_p^{\alpha 1} \qquad (10.3)$$

where Ed is the sum of all types of environmental damage to human life, flora and fauna, assets and natural resources caused by anthropogenic and nonanthropogenic pollution; C is a multiplicative constant; NExp denotes natural exposure: humans, animals, assets, nature, and resources in the vicinity of exposed areas, all multiplied by the frequency of hazardous occurrence; Si indicates socioeconomic variables; αi is the exponent of Si, which, for ratio purposes, may be negative.

10.9 INSPECTIONS, SURVEYS, AND AUDITS

Ports and ships are subject to inspections, surveys and audits.

a. *A marine surveyor* is a professional who performs surveys in order to evaluate, monitor, and document the condition and the products of ships, port facilities, cargoes, and so on. Surveyors may represent the government (i.e., Flag State), classification societies (i.e., the American Bureau of Shipping), private entities (i.e., prepurchase survey), and smaller craft surveyors.

   According to IMO, "all vessels should be surveyed and verified by officers of the flag State Administrations or Recognizing organizations (ROs)/surveyors nominated for the purpose in order to be issued certificates which establish their compliances with the requirements of IMO instruments, such as the Conventions of SOLAS, MARPOL, Load Lines, Tonnage," and so on.

   Surveys pertain to (a) hull and machinery, (b) forensic investigation, (c) condition and risk, (d) cargoes, (e) inland waterways navigation, (f) expert witness, and so on.

b. *Marine inspections* are typically carried out by a country's Port State Control/Coast Guard. Ship inspections aim to ensure a ship's overall seaworthiness, as well as structural and operational integrity. They focus on the following areas: (a) hull and machinery; (b) navigation equipment; (c) lifesaving and firefighting systems; (d) electrical systems, equipment, and wiring installation; (e) main/auxiliary engines and ships' propulsion; (f) boilers; (g) ship's security; and (h) pollution prevention.

c. Finally, *ship audits* pertain to the regulatory compliance in terms of safety (ISM), security (ISPS), quality (ISO), environmental protection (ISO 14001, VGP, BWT), and so on. Internal audits are carried out by the company's employees, which are certified auditors and typically work in the HSQE Department. External audits may be carried by the company's responsible organization, which typically is a classification society.

Audits, surveys, and inspections are useful tools that help proactiveness and preparedness. Their findings help identify areas for improvement and ensure a company's compliance and overall performance.

Ports are regions of great significance for the shipping industry. Global ports are not only responsible for compliance to all the aforementioned regulations; they also host most of the inspections, surveys, and audits onboard ships. To ensure the industry's integrity and sustain high overall standards, a series of policies, guidelines, and checks have been established for regulatory compliance pertaining to safety, security, environmental protection, and numerous other codes. Annual facts and figures presented by global and

TABLE 10.1 Example of Complementary Global, National, and State Regulations

Global Regulations	National Regulations (USA)
1. Security	
IMO → SOLAS → ISPS (International Ship and Port Facility Security Code)—Standards for ship and port security. All ships entering US waters must comply with parts A and B of the ISPS code.	MTSA Maritime Transportation Security Act of 2002 is the US national standard that mirrors ISPS and regulates US ports. It adds the requirement of Transportation Worker Identification Credential implementation.
	US ports must further comply with the regulations promulgated under MTSA 2002, specifically 33 CFR Parts 101 and 105. The USCG is the enforcing agency.
ISO 28000—International standard for security management. The standard provides a structure for a comprehensive security management program. The Security Management System (SMS) is a tool that may be used to help implement a port's required Facility Security Plan required under MTSA and 33 CFR Parts 101 and 105.	C-TPAT (Custom Trade Partnership against Terrorism) by US Customs and Border Protection (CBP) helps improve and manage supply chain security risks. CBP safeguards the homeland by eliminating the illegal trafficking of people and cargoes while assisting legitimate travel and trade.
	CSI—Container Security Initiative by CBP—targets high-risk containers coming into the United States.
	Safe Port Act of 2006 improves the US maritime and cargo security through layers of defense.
	Other—CBP regulations such as cruise terminal design standards; FIS (Foreign Inspection Station) Title VIII.

(continued)

TABLE 10.1 (Continued) Example of Complementary Global, National, and State Regulations

	Global Regulations	National Regulations (USA)
2. Safety	IMO → SOLAS CH: IX → ISM (International Safety Management Code) covers the minimum international standard for safe management and operation of ships and pollution prevention. SOLAS stands for International Convention for the Safety of Life at Sea. It is an international maritime safety treaty. The code pertains to ships and their shipping companies. Once in the United States, the US Coast Guard (USCG) enforces the code and makes sure vessels are in compliance. As far as safety goes, Port of Houston Authority (PHA) facilities are audited by the USCG. Some of the things the USCG looks for are as follows: fire extinguishers, emergency alarms, HAZMAT placement, records, certificates, and so on. Because some of the cargo comes across PHA docks, it is specifically regulated under 33 CFR 126 Port and Waterways Safety Act. Personal Safety and Social Responsibilities—Ports such as the PHA address these topics through training, exercises, and drills. Contingency plans are in place in case of an emergency. Constant communication and outreach to our tenants, vendors, stakeholders, and neighboring community ensure safety and accountability.	Ports and Waterways Safety Act (PWSA) of 1972. Ensuring the safety of ships entering and leaving US ports.

3. Occupational safety and health	OHSAS 18001 is a British standard pertaining to occupational health and safety management systems. It is a globally recognized occupational health and safety management systems standard.	OSHA—29 CFR Parts 1915—safety and health standards for Shipyard Employment, 1917—Marine Terminals, 1918—Safety and Health Regulations for Longshoring. For landlord ports, all parts pertain to the port's tenants/stevedores. Public ports (including public landlord ports) are "governmental entities" generally not subject to OSHA (unless they are in states that have federally approved OSHA programs that cover public employees), although their tenants and stevedores typically are.
4. Quality	ISO 9001—ISO Quality Management, which may serve as the precursor for environmental management systems and standards, that is, ISO 14000 certification.	
5. Environment	ISO 14000—Environmental management system. The PHA received its ISO 14000 Certification in 2002, becoming the first port authority to achieve that feat. The standard enhances the port's current system of environmental management and allows it to meet and exceed compliance with environmental laws and regulations. IMO's MARPOL—Ports must have a Certificate of Adequacy for the transfer/discharge of regulated waste. IMO's Ballast Water Management—IMO has initiated the global need to tackle the transfer of aquatic invasive species through shipping. See the MEPC resolution 50(31) of 1991 and the United Nations Conference on Environment and Development of 1992.	VGP—Enforced by the EPA and USCG for ships' ballast water treatment. OPA 90—Oil Pollution Act of 1990 USCG's Ballast Water Management 1990—Nonindigenous Aquatic Nuisance Prevention and Control Act, which aims to reduce or eliminate the spread of invasive species through the ballast water discharge. 1996—National Invasive Species Act pertaining to BW management practices.

(continued)

TABLE 10.1 (Continued) Example of Complementary Global, National, and State Regulations

	Global Regulations	National Regulations (USA)
6. Dangerous goods (global) and hazardous material (United States)	The United Nations Economic and Social Council issues the "*UN Recommendations on the Transport of Dangerous Goods*"	HAZMAT and HAZWOPER—All HAZMAT must be packaged, marked, classed, labeled, stowed, and segregated at facilities at ports such as the PHA. If an emergency were to occur, the PHA has shelter in place procedures that can be immediately activated. The PHA Fire Department is equipped and trained to deal with most HAZMAT situations. **Selected National and State Regulations** National (United States) Water Resources Development Act by the US Army Corps of Engineers pertains to permitting dredging. It drives the implementation of policy changes with respect to the Corps' water resource project and programs. State (Texas, USA) Chapter 68 of the Texas Water Code gives the PHA authorization to hold a directorship on the Houston Ship Channel Security District board. State (Texas, USA) Chapter 66 Houston Pilots Licensing and Regulation Act—Texas Transportation Code gives port commissioners the authority to serve as the Board of Pilot Commissioners.

Source: Compiled based on Port of Houston Authorities' national and state regulations for ports.

national maritime organizations reveal the continuous improvement and the industry's commitment to excellence. This positive outcome boosts ports' significant contribution to the national economy, national sovereignty, safety, and security.

10.10 GLOBAL AND NATIONAL REGULATORY COMPLIANCE FOR SHIPS

Ports and the maritime industry as a whole are subject to both *International* and *National Legislation and Regulations*. All internationally trading ships entering a port must comply with both global maritime laws and regulations and the national maritime regulations of the port they enter. Table 10.1 shows how ports need to comply with global, national, and state (regional, municipal, county, etc.) regulations. It is a small example, and non-exhaustive, but it shows some of the global, national, and state regulations, standards, and government initiatives with which ports need to comply or be familiar in the areas of security, safety, and environment. The Port of Houston, Texas (USA), is used as a case example.

REFERENCES

ABS. 2005. The ABS Guide for Ship Security(sec) Notation. Available at http://www.eagle. org. Accessed July 1, 2013.

Copeland, C. 2013. EPA's vessel general permit: Background and issues. Congressional Research Service, USA. 7-5700. Available at http://www.crs.gov; R42142. Accessed June 1, 2013.

Hazlewood, C. 2004. Statement of Catherine Hazlewood, The Ocean Conservancy, "Ballast Water Management: New International Standards and NISA Reauthorization," Hearing, House Transportation and Infrastructure Subcommittee on Water Resources and Environment, 108th Cong., 2nd sess., March 25, 2004.

International Harbour Master Association. 2013. The IHMA. Available at http://www. harbourmaster.org/ihma.php. Accessed August 6, 2013.

IMO Globallast. 2013. Available at http://globallast.imo.org. Accessed June 4, 2013.

ISM, IMO. 2010. Available at http://www.imo.org/OurWork/HumanElement/Safety Management/Pages/ISMCode.aspx. Accessed July 5, 2013.

ISO. 2013. ISO 14000 Environmental Management. Available at http://www.iso.org/iso/ home/standards/management-standards/iso14000.htm. Accessed June 2, 2013.

NAVCEN. 2013. The Navigation Center of Excellence. US Department of Homeland Security. Available at http://www.navcen.uscg.gov. Accessed August 6, 2013.

Occupational Health and Safety Advisory Services. 2013. Available at http://www.OHSAS. org. Accessed June 16, 2013.

Occupational Safety and Health Administration. 2013. Available at http://www.osha.gov. Accessed June 16, 2013.

Pimentel, D., Lach, L., Zuniga, R., and Morrison, D. 1999. Environmental and Economic Costs Associated with Non-indigenous Species in the United States, presented at AAAS Conference, Anaheim, CA, January 24, 1999. Available at http://www.news. cornell.edu/stories/1999/01/environmental-and-economic-costs-associated-non-indigenous-species. Accessed August 12, 2013.

UK Department of Transport. 2013. Port Freight Statistics, 2012. Final Figures. Available at https://www.gov.uk/government/publications/port-freight-statistics-2012-final-figures. Accessed August 6, 2013.

US Army Corps of Engineers (USACE). 2012. Ballast Water Management, Special Edition. Marine Design Center. Marine & Floating Plant Newsletter. Issue #16, May 2012. Available at http://www.nap.usace.army.mil/Portals/39/docs/MDC/MDC%20Newsletter_May%202012.pdf. Accessed August 12, 2013.

US Environmental Protection Agency. 2013. Available at http://www.epa.gov. Accessed June 4, 2013.

Vessel General Permit (VGP). 2013. Environmental Protection Agency, USA. Available at http://www.epa.gov/npdes/vessels. Accessed June 1, 2013.

CHAPTER 11

Ports as a Bridge to Maritime and Offshore Energy Activities

Engineers are extremely necessary for these purposes;
Wherefore it is requisite that, besides being Ingenious,
they should be Brave in proportion.

Isambard Kingdom Brunel (1806–1859)
British engineer and inventor of the first trans-Atlantic steamships
SS Great Western, SS Great Britain, SS Great Eastern, and so on

America's offshore oil exploration commenced in 1894 when Henry Williams drilled the first two wells in California beach, with the use of drilling and production piers. According to Santa Barbara County records, piers and drilled wells were designed in order to extend the Summerland oilfield offshore. This was the first offshore drilling wells from port-like pier constructions, signifying the strong connection between ports and offshore exploration (AOGHS 2013).

Since then, technological innovations have enabled the industry to expand its offshore explorations to further offshore and deeper waters, in order to meet the industry's insatiable demand for energy. The role of ports is becoming increasingly associated with the energy exploration, as their services and intervention are required for significant reasons. At present, 60% of the global oil and gas production is located in regions of increased sociopolitical turmoil. This, in combination with the lower-cost, lower-project-risk ratios, has forced the offshore industry to drill in remote offshore areas. Currently, approximately 30% of global oil and gas production stems from offshore regions, frequently of harsh environments and extreme weather conditions. Technological patents will increasingly enable the industry to drill in deeper waters, minimizing weather-related risks.

Again, as the most troubling questions for the offshore oil and gas industry relate to deliverability rather than availability, modern ports increasingly offer their infrastructure and superstructure to support the energy industry.

Offshore energy from other alternative sources is also increasing, benefiting from advances in technology and efficiency. In the first six months of 2013, Europe fully grid-connected 277 offshore wind turbines, with a combined capacity totaling over 1 GW. Overall, in 2013, 18 wind farms were under construction. Once completed, these wind farms will have a total capacity of 5111 MW. New offshore capacity installations during the first half of 2013 doubled compared to the same period the previous year and was just 121 MW less than the total of 2012 installations (EWEA 2013).

The interconnectivity between ports and the offshore energy industry is crucial to global ports and economies located in the vicinity of offshore drilling and production activities. According to the American Petroleum Institute, "for each direct job offshore, the industry supports three indirect and induced jobs onshore—including the cooks, suppliers and others servicing the industry" (API 2013).

On the other hand, ports also serve as a bridge that unites the maritime industry with offshore energy industry activities of all types. The purpose of this work is to highlight the opportunities and challenges entailed in a coalition among maritime and energy companies.

The first section of this chapter demonstrates the hybrid ports that either assimilate the design of offshore platforms or provide their unique services to the energy industry. The section further accentuates the immense opportunities for growth, by demonstrating what these innovative ports have achieved at a global level. The second section aims to draw the industry's attention as to the imminent necessity to harmonize the legal and regulatory framework between the maritime and energy industries for the sake of safety, security, environmental protection, quality, and economic prosperity. Once the regulatory harmonization has been achieved, the industry's potential for growth is limitless.

11.1 PORT OPERATIONS AND OFFSHORE DRILLING: WHEN PERFORMANCE EXCEEDS AMBITION

The significance of seaports can hardly be exaggerated: they facilitate global and national trade; they enable their clients to land their cargoes in their Free Trade Zones without the mediation of the customs officials; they serve as hub ports and distribute commodities throughout the supply chain(s). But most important, they boost global economy through their contribution in the energy sector, in three different levels:

1. Oil and gas transport from offshore platforms to port refineries to the port's nodes of logistics distribution networks. Energy products play a significant role in the global economy, trade, and industrial and domestic consumption. Most refineries are strategically located in the vicinity of seaports, and this signifies the importance of ports since the majority of oil and gas products are transported by sea. Namely, oil and gas are transferred from the land-based or offshore platforms to the refineries located in the vicinity of seaports. Crude oil is typically shipped through crude oil tankers or pipelines from the oil wells to refineries that are usually located in the vicinity of ports. The refined oil products consist of LPG, ethane, refinery gas, jet fuels, kerosene, aviation gasoline, motor gasoline, gas/diesel oil, fuel oil, petroleum coke, white spirit, lubricants, naphtha, bitumen, and paraffin waxes, along with other oil products. These are subsequently distributed throughout the logistics network, primarily through product tanker ships, and secondarily through pipelines and land transportation (rail, trucks).
2. Seaports' increasing investment in renewable energy, for example, offshore wind facilities.
   Renewable energy is an attractive, sustainable solution to nonrenewable fuels.

Offshore wind development in particular is an ideal way for modern ports to increase revenue and gain competitive advantage and differentiation. Ports may also entice alternative investment and business opportunities through establishing a configuration of offshore wind alliances at a state, national, or global level. It enjoys high rates of annual installed capacity led by China, the United States, and the European Union.

Ports increasingly demonstrate a keen fascination with offshore renewable energy for a plethora of reasons, including the following:
- Energy that is abundant and widely available while occupying minimum space.
- It lessens the local or national dependence on foreign energy.
- Being environmentally clean, it complies with the air emission regulations.
- Modern wind energy technology is closely associated with port technology and operations: ports are the ideal location not only for offshore farms (e.g., in ports' outer limits) but also for ashore wind farms in ports' land property.
- From a financial, commercial, and environmental standpoint, ports are the ideal consumers of renewable energy, which is highly compatible with maritime and port energy-generating technologies.
- A port's initial investment for infrastructure and wind power not only meets the regulatory standards to generate other forms of renewable sources but can also be utilized to generate other forms of renewable energy.

Typically, the energy development plans agreed among port authorities and renewable energy companies incorporate certain provisions, such as the following:
- Accessibility for maintenance and repairs; connectivity to electricity grids.
- Limitations as to the wind turbines' structural arrangement and location within the port, ensuring it will not constrain port or land traffic, and will not affect the port's electricity cables, port terminals, rail, or pipeline corridors.
- The siting of wind turbines must not restrict in any way the nautical access to and use of the ports.
- National energy administrations and spatial planners determine a noise zone ensuring that the proposed noise limitations are not surpassed.

3. Ports as launching platforms for the oil and gas offshore drilling activities.

As the offshore industry expands, the functionality of conventional seaports significantly grows, with positive financial rewards. A multitude of ports that facilitate in oil and gas offshore explorations are typically located in regions convenient for the transfer of rig personnel and supplies. Ports are the launching platforms where oil and gas logistics take place: not only do they collect the required equipment, spare parts, and provisions on behalf of supply vessels, but they also facilitate the regulatory checks in terms of safety, security, quality, environment, and so on.

The operational, technical, and regulatory structure of offshore units is quite similar to the port and maritime configuration. Hence, ports can easily facilitate the operational missions of offshore-related ships such as survey and seismic vessels, diving support and construction ships, exploration jack-up rigs, semisubmersible rigs, anchor handlers (AHTSs), and emergency rescue recovery vessels (ERRVs).

Moreover, ports offer their life-support services during offshore emergencies, that is, natural disasters or man-made accidents. In case of extreme weather or an emergency, ports serve as a safety platform that major oil companies use to evacuate rigs and platforms.

Mobile offshore drilling units (MODUs), platform support vessels (PSVs), and floating production installation ships (FPIs) typically switch out their crews anywhere from seven to 14 days (subject to geographical location) using helicopters—mostly for crew changes—and offshore support vessels (OSVs) for supplies. Crew boats are a cost-efficient alternative to helicopters (again subject to location).

The following is a case study on Harvey Gulf International Marine, a leading shipping company with specialty ships, that is, OSVs, multipurpose support vessels (MPSVs), and offshore supply and mooring line storage vessels.

HARVEY GULF INTERNATIONAL MARINE

In March 2011, Mr. Guidry commissioned the design of the very first US-flagged LNG-powered OSVs. The SV310DF 302 × 64 × 24.5 ft, FFV-1 vessels are dual-fuel designed to run primarily on LNG. The Wärtsilä LNGPac system provides for approximately 78,000 USG of fuel allowing for eight consecutive days' transit at continuous steaming with a cargo carrying capacity of 5520 MT DWT at load line and a transit speed of 13 knots. The vessels are powered by three Wärtsilä 6L34DF gensets delivering 7530 kW via the Wärtsilä Low Loss Concept electrical and automation system to two 2700 kW fixed pitch azimuthing stern thrusters and two 1280 kW fixed pitch transverse thrusters. The 42-person vessels are high block coefficient single chine hulls with 18,000 Bbls of liquid mud, 1545 Bbls of methanol, 10,250 cft of dry cement, 253,000 USG of fuel oil, and 621,000 USG of drill water. The vessels were designed by STX Marine, cutting-edge designer of OSVs.

In addition to being run by clean LNG, these vessels are designed in accordance with the most stringent environmental American Bureau of Shipping (ABS) notations "ENVIRO +, Green Passport."

Upon delivery, these vessels will exceed the most stringent EPA emissions requirements, thus improving best practices for safe marine and *offshore* operations.

By October 2011, the company's board had authorized and signed shipbuilding contracts for two of the very first US-flagged dual-fuel vessels for construction at Trinity Offshore LLC (Gulfport). Trinity Offshore, established in 2003, is a shipyard specializing in patrol ships, oil spill response ships, oil field support vessels, tug boats, inland and offshore barges, and luxury yachts.

By July 2013, the company had exercised options for four more vessels to bring the total to six, making Harvey Gulf the largest operator of gas-fueled OSVs in the world.

2012

Four Long-Term OSV Charters

In August 2012, the company was granted four long-term chartering contracts, for the *M/V Harvey Energy*, *Harvey Power*, *Harvey Liberty*, and *Harvey Champion*, thus becoming the first company to design, construct, and establish contracts for LNG-powered OSVs for US deepwater surveys, extraction, and recovery. These innovative LNG ships with state-of-the-art technology will significantly empower their customers to obtain drilling permits and comply with the latest environmental requirements.

New Drydock in Harvey's Terminal, Port Fourchon, Louisiana

In addition, the company launched a drydock with dimensions of 320′ × 120′ × 12′, with a lifting power of 9000 long tons, to be built in the company's terminal in Port Fourchon, Louisiana. The aim of this investment is to serve the company's fleet of 46 deepwater ships (i) by decreasing the repairs and maintenance expenditures and (ii) by reducing waiting times in drydocks. Moreover, the drydock will provide its services to other shipping companies.

Bee Mar Asset Purchase

In September 2012, the company reported the concluding of its fleet purchase agreement with Bee Mar, LLC, for 9 Offshore.

Supply vessels were purchased at a price of $243 million. This deal materialized the strategic plan of Harvey's CEO Shane Guidry in becoming the largest and greenest OSV owning/operators in the United States. Consequently, this move empowered the company to expand its operations in the US Gulf of Mexico, Alaska, Mozambique, and Israel.

Company's Fleet Expansion and Investment of $540,000,000

In May 2013, Harvey concluded three purchase contracts for $540,000,000, which brought up the company's capital expenses to $1,700,000,000 since August 2008.

a. Eastern Shipbuilding Contract

The first contract was established among Harvey Gulf International Marine and Eastern Shipbuilding Group for the building of two STX MPSV340H 327′ × 73′ × 29.5′ heavy lift construction ships, named *M/V Harvey Sub-Sea* and *M/V Harvey Blue-Sea*.

Additionally, Harvey Gulf has an option vessel, the *M/V Harvey Intervention*, with 250 mt modular handling tower for well intervention activities (see Figure 11.1).

b. TY Offshore, Gulfport, Mississippi

The second contract was concluded among Harvey Gulf International Marine and TY Offshore for the building, ownership, and operation of the sixth dual-fuel offshore vessel. The shipbuilding expenditure for these ships exceeded 20% of the cost of conventional OSV ships. Nevertheless, this addition has enabled Harvey Gulf to become the largest owner/operator of clean burning LNG OSVs globally.

c. Gulf Offshore Logistics of Lafayette, Louisiana

The third agreement entailed the finalizing of an asset acquisition with Gulf Offshore Logistics for 11 dynamically positioned Class 2 offshore supply and fast supply vessels. This purchase, particularly the addition of the fast supply vessels, has brought increased diversity to Harvey Gulf.

The First LNG Marine Bunkering Facility in the United States

In June 2013, the company's CEO, Shane Guidry, announced Harvey Gulf's new plans to build and operate the first LNG marine bunkering facility in the United

FIGURE 11.1   Heavy lift construction ships (from left to right), *M/V Harvey Blue-Sea*, *M/V Harvey Intervention*, and *M/V Harvey Sub-Sea*. (Courtesy of Harvey Gulf International Marine, LLC.)

States, situated at its fleet facility in Port Fourchon, Louisiana. This project will be a principal acquisition to the rapidly expanding national LNG supply infrastructure, accommodating vital operations of the oil and gas industry's OSV fleet operating on LNG. The facility's engineering, procurement, and construction development will be undertaken by "CH•IV International of Houston, Texas" EPC contractors.

The facility will include two sites, each having 270,000 gallons of LNG storage capacity, and each being capable of transferring 500 gallons of LNG per minute. The tanks will be stainless steel Type "C" pressure vessels with vacuum insulation and carbon steel exteriors. Apart from the facilities' principal function of accommodating the oil and gas industry, they will additionally support over-the-road vehicles that burn LNG.

The bunkering facility will be a significant acquisition to the developing LNG supply infrastructure in the United States, sustaining vital operations of the oil and gas industry's OSV fleet running on LNG. The company's CEO, Shane Guidry, proclaimed that "to date, Harvey Gulf is the only company in North America that has committed $400M USD to build, own and operate LNG-powered OSVs as well as two LNG fueling docks."

INTERVIEWS

Mr. Shane J. Guidry, Harvey's Chairman and CEO

Mr. Chad Verret, Senior Vice President, Alaska and LNG Operations

Mr. Mike Carroll, Senior Vice President New Construction and Chief Naval Architect

REFERENCES

Harvey Gulf International Marine, LLC, 2013 (http://www.harveygulf.com) (accessed August 15, 2013).

Harvey Gulf Press Release LNG Contract 10.7.11.

Harvey Gulf Press Release: June 10, 2013. Shane Guildry, CEO of Harvey Gulf, Building First LNG Bunkering Facilities in America.

Harvey Gulf Press Release: May 2, 2013. Harvey Gulf acquires and orders vessels totaling $540,000,000.00.

Other Case Studies: Oil and Gas Ports

Examples of global ports that are specialized in serving oil and gas projects include the following:

i. Açu Superport, Brazil (Porto do Açu)

Brazil's "Highway to China" or the "Rotterdam of the Tropics"

Brazil and in particular Rio de Janeiro are among the world's top offshore havens, with a magnitude of energy reserves and offshore drilling prospects.

The development of a mega-port in the region is strongly associated with the energy sector yet expands to the container and dry bulk sector.

Açu Superport (Figure 11.2), one of the world's largest ports, is the ambitious concept of and is designed by Eike Batista, Brazil's wealthiest man (Darlington 2010). With an initial investment of $2.7 billion, this state-of-the-art venture has four key purposes (Paschoa 2013):

a. The initial investment objective of Mr. Batista was for the port to become a major export corridor for his energy and mining corporations under his conglomerate EBX Group, consisting of five major trade companies: (i) OGX

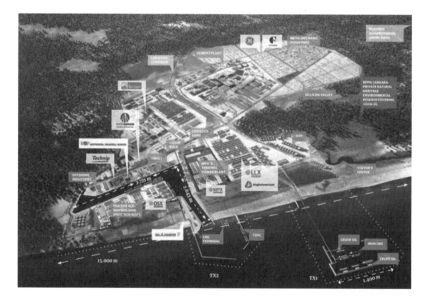

FIGURE 11.2   Açu Superport illustration. (Courtesy of LLX.)

for oil and gas, (ii) MPX for energy, (iii) OSX for offshore services and equipment, (iv) LLX for logistics, and (v) MMX for mining (EBX Group Brazil 2013).

Açu's commercial perspectives were broadened owing to Brazil's increasing leadership role in the global oil and gas offshore production. Mr. Batista's commercial conquests captured the attention of the Brazilian government and major investors such as BP's and EBX partnership that formed "Marine Fuels X," Brazil's oil giant Petrobras, offshore industries, thermal power plants, LNG and crude oil terminals, and so on.

Hence, the port's new vision expanded into the following:

a. Facilitating the trade agreements between Brazil and China.
b. Attracting global investors by significantly improving Brazil's logistics network. Brazil is determined to address the bottlenecks and infrastructure challenges that have previously restricted its power to attract global investors. Açu's mega-trade corridor of 10 berths will now eliminate ships' turnaround time and cargo distribution, which in the past could take approximately 1–2 months.
c. Achieving lower operation costs through the dispersion of commercial activities: the combination of wet and dry cargoes will ensure the port's sustainability throughout different market cycles and commodities' demand fluctuations.

ii. The Port of Pembroke, United Kingdom

Pembroke Port (Figure 11.3) is a distinctive energy port, as in addition to the freight and combined passenger vessels, it manages oil-rig supply ships as well. The port is positioned in a location well suited for marine renewable system testing and manufacturing and handles onshore wind turbines. Moreover, the port handles project equipment and heavy lifts for the oil and gas industry (Marine Energy Pembrokeshire 2013).

FIGURE 11.3   The Port of Pembroke. (Courtesy of Marine Energy Pembrokeshire 2013.)

Pembroke Port and Tidal Energy's Innovative "Tidal Energy Generation" Project

An innovative project on tidal energy has been initiated under a partnership between the port authorities and Tidal Energy Ltd. of Cardiff. The Pembroke Port was selected as a suitable location for the operation of the DeltaStream "Tidal Stream Turbine Trials" (Figure 11.4). The agreement also includes the unit's storage, installation, implementation, and maintenance of the "tidal generation unit" (Barry 2013; Port of Pembroke 2013; Tidal Energy 2013).

iii. The Port of Onne, Nigeria

The port of Onne, in Nigeria, apart from providing efficient and timely delivery of equipment, supplies, and provisions offshore, is also the world's greatest Oil and Gas Free Zone. In addition, Onne facilitates the transport of spare parts for oil field machinery from the port's free zone all in the Gulf of Guinea, via Gabon, Cameroun, Congo, and Ivory Coast, all duty free. The port of Onne has been acknowledged as a primary facilitator for the oil and gas industry, in terms of oil production, handling, and distribution. In addition, it meets the oil and gas industries' needs through established services such as fabrication yards, pipe coating, marine suppliers, and so on (Nigerian Ports Authority 2013).

iv. Kakinada Seaports, India

Sembmarine Kakinada Ltd. (SKL) is a joint venture company between Kakinada Seaports and Sembawang Shipyard that manages and operates a marine and offshore facility catering to offshore drilling units and merchant vessels trading or operating in Indian waters (Sembcorp 2009). SKL is strategically located in the East Coast of India, one of the world's key oil and gas exploration areas near India's major ports of Vishakhapatnam and Chennai.

Ports' Offshore Expansion: Assimilating the Offshore Platform Structures

Offshore ports are an attractive alternative to dredging operations: through an offshore structure, ports are capable of expanding and overcoming their land space and draft limitations. Offshore ports typically serve larger ships, such as crude oil carriers and new-generation containers. This increasing trend not only helps seaports generate increased revenue but also boosts local economy, generates jobs, and promotes national growth.

FIGURE 11.4   Tidal energy turbines. (Courtesy of Tidal Energy, Cardiff 2013.)

FIGURE 11.5   Louisiana Offshore Oil Port (LOOP). (Courtesy of LOOP LLC.)

i. Louisiana Offshore Oil Port (LOOP), United States

Since its inception in 1981, the Louisiana Offshore Oil Port (LOOP) (Figure 11.5) has been a distinctive link in America's energy infrastructure. It is a deepwater seaport located in open waters, 18 nautical miles off the Gulf of Mexico, in the vicinity of Port Fourchon. Because of its deep draft, LOOP is the only US port able to accommodate the largest tankers, up to 700,000 DWT, that is, ultra large crude carriers and very large crude carriers, in addition to the vast amounts of oil produced within the Gulf of Mexico. The port distributes large amounts of crude oil to allocated refineries and provides short-term storage space services (LOOP 2013). Through its underground pipelines and its aboveground tank farm, LOOP handles over 1.2 million barrels of global crude oil daily, plus 325,000 barrels from offshore platforms situated in the US Gulf (COQA 2013). LOOP obtains crude oil from two of the US Gulf's largest oil fields (Mars and Thunder Horse), extracted by some of the largest offshore floating platforms globally.

ii. Port of Venice, Italy—Offshore Terminal

The Port of Venice keeps abreast with the new developments in the maritime industry. Its new Offshore Terminal Off-Port-Limits, with a natural draft of 20 meters (65.62 feet), will be able to host the world's new-generation ultra-large vessels and thus generate increased revenue for Venice. The innovative port design (made in Italy) offers an offshore alternative to other ports' dredging operations (Port of Venice 2013). This strategic endeavor will offer Italy a competitive advantage within the Mediterranean Sea's historical passages:

a. From the Far East and the Middle East via the Suez Canal

b. From the Black Sea via Bosphorus to the Mediterranean

c. Accommodating the "Mediterranean Oil and Gas Fever" pursuant to the vast reserves discovered in Israel, Greece, Cyprus, and so on

CASE STUDY

PORT FOURCHON (LOUISIANA, USA): THE GULF'S ENERGY CONNECTION

Port Fourchon (Louisiana, USA) was designed as a multipurpose port facility specialized as a shore platform for offshore oil support provider. On top of its enormous

FIGURE 11.6   Port Fourchon. (Courtesy of Port Fourchon.)

local hydrocarbon importance, Port Fourchon (Figure 11.6) serves as a land base for LOOP, which manages 20% of the nation's global and local oil, and 50% of US refining capacity. LOOP is the only US deepwater port able to offload the world's largest tanker ships. Its strategic geographical location by the US Gulf presents a logistical benefit for drilling rigs' maintenance and repairs, whereas a favorable legal system additionally advances the port's capacity to entice the drilling rig service and refurbishment industry segment.

FACTS AND FIGURES

Port Fourchon is experiencing unparalleled growth as an immediate result of oil and gas activity in the Gulf of Mexico:

- Port Fourchon by itself presently services 90% of deepwater structures in the Gulf of Mexico, a region that produces over 54% of overall US crude oil and 52% of total US natural gas.
- Approximately 95% of its tonnage handled at the port pertains to oil and gas.
- All equipment and supplies required to support the oil and gas industry are processed as cargo. Roughly 30% of the entire tonnage moves to and from the port by inland barge prior to being sent to or from an offshore supply vessel.
- Port Fourchon organizes the transportation of around 15,000 people to offshore locations each month.
- Seventy percent of the port's tonnage moves to and from the port by rail or truck prior to being moved to or from an offshore supply vessel or helicopter.

- Oil production is anticipated to exceed 1.6 million barrels of oil per day.
- Roughly 270 large supply vessels navigate through the port's channels on a daily basis.
- More than 1.5 million barrels of crude oil daily are shipped via pipelines via the port.

Market forecasts expect that the majority of all drilling rig activities off the Louisiana, Mississippi, and Alabama shorelines will take place in the Port Fourchon service area, with a commercial life of at least 40 more years. The port's rising market segments consist of (a) deepwater oil and gas, (b) oil rig repairs, and (c) maritime activities and logistics.

These facts and figures demonstrate that Port Fourchon is the most prominent deepwater and offshore support base in the US Gulf (Hornbeck Offshore 2013; Port Fourchon 2013; Ports of Louisiana 2013).

11.2 REGULATORY HARMONIZATION AMONG PORTS, THE OFFSHORE INDUSTRY, AND OSVs

Unlike a conventional scientific textbook, this section aims to assume a policy-making stance and suggest certain common elements between the maritime and energy industries, in particular safety, security, environmental protection, HAZMAT (dangerous goods), and quality.

Offshore units are an innovative segment of hybrid products that align the shipping and the oil and gas industry: a crossbreed of ships based on the "derived demand" concept will generate a crossbreed of ports. Figure 11.7 verifies the connectivity between the maritime industry and oil and gas drilling, by demonstrating a variety of rig types by Maersk Drilling (Maersk 2013).

A serious legal and regulatory issue that must be addressed is the fact that this moneymaking "crossbreed," despite its utility, is globally treated in a contradictory or inconsistent manner: according to the facts and figures available from numerous inspections carried out at a global level, in the eyes of the law, these floating constructions are sometimes regarded as vessels and, at other times, as fixed offshore structures. Table 11.1

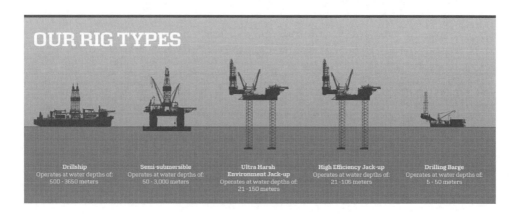

FIGURE 11.7   Maersk rig types.

TABLE 11.1   Regulatory Compatibility among Fixed and Mobile Units

	Fixed Units		Mobile Units	
Regulations	Floating Offshore Facilities	Ports	Commercial Ships	Offshore Drilling Units
Safety Natural disaster Act of God	X	X	X	X
Marine pollution	X	X	X	X
Pollution prevention	X	X	X	X
Environmental liability	X	X	X	X
Security Man-made disaster	X	X	X	X
Structural integrity		X	X	X
Offshore drilling	X			

Source: Compiled by M.G. Burns based on regulatory data from EUROVOC Descriptor, European Union (2013), IMO (2013), and USCG (2013) (http://eur-lex.europa.eu/LexUriServ/LexUriServ.do?uri=CELEX:52011PC0688:EN:NOT).

shows the regulatory compatibility among (a) fixed units, that is, ports and floating offshore facilities, and (b) mobile units, that is, standard commercial ships and offshore drilling units.

11.2.1 OSVs: The Link between Ports and Offshore Exploration Activities

In order to better understand the role of the ports and regulatory harmonization in the offshore industry, the offshore industry's most prevalent designs are analyzed:

1. FPSOs: Floating Production Storage and Offloading Units
   Consistent to the "time is money" principle, the industry frequently converts commercial tanker ships of different sizes into FPSOs. This fact demonstrates the close connection and common concepts between offshore activities and the shipping industry.
   The first FPSO design was invented in 1977 by Shell Oil Company and was built in Spain. Currently, over 200 FPSOs serve in the offshore industry. The FPSO design, building, and tanker ship conversion operations are undertaken in specialized yards globally.
   Since 1980, a variety of different mooring configurations have been developed. For example, FPSOs can be designed with an additional turret mooring system in their bows (an innovation of SBM Offshore company), enabling them to weathervane around the turret, hence smoothly float under any course of the waves, currents, and winds (SBM Offshore 2013).
   FPSOs are especially constructed to support the offshore industry through (a) the handling of hydrocarbons, (b) the use of its deck for the storage of offshore spares, equipment, and so on, and (c) the storage of crude oil in their tanks and the subsequent unloading onto tanker ships: cargo transfer to commercial tankers takes place at regular intervals, depending on the FPSO's oil storage capacity

and the tanker's cargo carrying capacity. Occasionally, oil is transported via pipeline units connected to the shore.

For all these reasons, the FPSO concept is a significant component to the offshore sector.

Other structures similar to the FPSO concept include *FPIs* (floating production installation ships) and the most recent innovation, *FLNGs* (LNG floating production facilities).

2. FSOs: Floating Storage and Offloading Units

This is a floating structure similar to FPSOs, yet lacking the processing facilities. This tank design is utilized by the offshore oil and gas industry with the intention of receiving oil or gas cargoes produced from offshore structures in its vicinity and storing them temporarily until they can be loaded onboard tanker ships or moved through pipeline systems.

3. OSVs: Offshore Support Vessels

This wide category encompasses various highly specialized boat categories. Broadly speaking, OSVs move and store machinery and supplies and transfer on-signing and off-signing personnel between offshore platforms, seaports, and so on. The ships are usually equipped with dynamic positioning systems. OSVs are classified as per the services they offer, their main categories being the following:

a. PSVs: Platform Supply Vessels

Platform supply vessels (PSVs) are designed to support offshore platforms throughout the initial stages of drilling and production and subsequent stages of offshore development and servicing. During different phases of oil and gas exploration, PSVs are able to supply the offshore facilities with necessary machinery, drilling equipment, spare parts, fuel, lubricants and fluids, water, and so on. The ships may be equipped with firefighting or oil recovery systems.

b. Anchor Handling and Towing Vessels (AHTs)

AHTs may still serve as supply vessels, in particular the *Anchor Handling, Towing and Supply Vessels* (AHTSs), yet their distinctive design enables them to
  i. Tow or tug oil rigs
  ii. Position them in designated locations of precise latitude and longitude
  iii. Secure them through anchors
  iv. Participate in emergency response and incident prevention operations, as these ships are powerful enough

Their features include the following:

a. Amplified towing power or pulling force, which is achieved by their powerful engines combined with towing winches

b. Anchor handling facilities, which include quick anchor release and having an open stern to allow the decking of anchors

Other categories whose services are indicated by their names include pipe-laying (PL), crew boats (CBs), standby and rescue vessels (SRVs), emergency rescue recovery vessels (ERRVs), multipurpose service vessels (MPSVs), etc.

Regulatory harmonization may therefore be achieved through the following process:

a. Distinguishing the principal structures, platforms, and ship designs of the offshore industry into
  • *Mobile structures* (e.g., MODUs, OSVs, etc.).

This fleet is subject to maritime regulations and serves as a bridge among (i) seaports and the regional economy, (ii) the energy industry, and (iii) the maritime industry.

- *Fixed structures* (e.g., floating offshore facilities, including floating production storage and offloading units [FPSOs], floating storage and offloading units [FSOs], floating production units [FPUs], platforms, etc.). These structures are subject to an offshore regulatory framework.

b. Distinguishing the regulations and rules into *global* and *national*
c. Establishing a comprehensive matrix table, where the industry's prevailing regulations are classified into
    i. Mobile—Maritime, for ships and ports (also see Chapter 10)
    ii. Mobile—Offshore Maritime, that is, for MODUs
    iii. Fixed, that is, for production platforms

Once the classification has been made, an attempt to bridge these industries will be made, and a regulatory framework that encompasses these three classifications will be proposed, as per item (c).

Table 11.2 classifies the principal global regulations for mobile and fixed offshore units and thus reveals the harmonization potential among the maritime and offshore oil and gas industries.

11.2.2 Offshore Regulations in the United States

The Bureau of Ocean Energy Management, Regulation, and Enforcement

The Bureau of Ocean Energy Management (BOEM) and the Bureau of Safety and Environmental Enforcement (BSEE) have replaced the Bureau of Ocean Energy Management, Regulation, and Enforcement (BOEMRE), previously the Minerals Management Service (MMS).

BOEM is in charge of controlling the development of US offshore resources from an environmental and economic standpoint. Its capabilities comprise offshore leasing, resource assessment, analysis, and management of oil and gas exploration and development plans. In addition, it handles renewable energy development, National Environmental Policy Act analysis, and environmental research (BOEM 2013).

BSEE is in charge of safety and environmental administration of offshore oil and gas operations, which comprises permitting and inspections of offshore oil and gas operations. Its capabilities pertain to the safety and environmental regulations from a development and enforcement standpoint; inspections; permitting offshore exploration, development, and production; offshore regulatory programs; oil spill response; and recently formed training and environmental compliance programs (BSEE 2013).

The Minerals Management Service

The MMS is a bureau in the US Department of the Interior in charge of handling the country's oil, gas, and various other mineral resources on the outer continental shelf (OCS). Under its authority as stipulated in the Outer Continental Shelf Lands Act and other government bodies, the MMS oversees functions such as exploration, drilling, production, development, storage, well servicing, pipeline shipping, workover activities, and so on.

TABLE 11.2 Harmonization Potential for Maritime, Oil, and Gas Regulations: Global Regulations for Mobile and Fixed Offshore Units

Regulations	Ships	Mobile Units — OSVs (Offshore Support Vessels) MODUs (Mobile Drilling Units)	Fixed Units — Production Platforms
Security			
ISPS: International Ship and Port Facility Security Code	X	X (ISPS Code, Part A, 3.1).	N/A Applicable regulation
Safety			
COLREG 1972: Convention on the International Regulations for Preventing Collisions at Sea	X	X	72 COLREGS and Inland Rules Act will hinder the vessel's ability to maneuver within proximity of *offshore platforms.* Federal Register, Volume 75, Issue 2 (Tuesday, January 5, 2010) *http://www.gpo.gov/fdsys/pkg/FR-2010-01-05/html/E9-31226.htm* X
INMARSAT 1976: Convention on the International Maritime Satellite Organization	X	X	X
ISM: International Safety Management Code (or International Management Code for the Safe Operation of Ships and for Pollution Prevention);	X	IMO's Maritime Safety Committee adopted, by resolution MSC.266(84), the Code of Safety for Special Purpose Ships, 2008 (*2008 SPS Code*). Special Purpose Ship Safety Certificate Special Personnel	
LL Protocol 1966; 1988: International Convention on Load Lines (LL).	X	X	N/A
SAR 1979: International Convention on Maritime Search and Rescue	X	X	X

(continued)

SOLAS Convention 1974; Protocols 1978; 1988: The International Convention for the Safety of Life at Sea (SOLAS), and its pertinent certification	X	X
STCW 1978: International Convention on Standards of Training, Certification and Watchkeeping for Seafarers (STCW), 1978	X	N/A
SALVAGE 1989: International Convention on Salvage	X	N/A
SUA 1988: Convention for the Suppression of Unlawful Acts Against the Safety of Maritime Navigation	X	N/A
SUA 1988 Protocol for the Suppression of Unlawful Acts Against the Safety of Fixed Platforms Located on the Continental Shelf, 1988	N/A	X
Other safety certifications — SLC: Safety Construction Certificate	X	
SLE: Safety Equipment Certificate	X	
SLR: Safety Radio Certificate	X	Code of Safe Practice for the Carriage of Cargoes and Persons by Offshore Supply … and Persons by Offshore Supply Vessels (OSV CODE)
CLC Convention 1969; Protocols 1976; 1992: International Convention on Civil Liability for Oil Pollution Damage	X	X

TABLE 11.2 (Continued) Harmonization Potential for Maritime, Oil, and Gas Regulations: Global Regulations for Mobile and Fixed Offshore Units

Regulations	Mobile Units	Fixed Units
FUND Convention 1972; Protocols 1976; 1992; 2000; 2003: International Convention on the Establishment of an International Fund for Compensation for Oil Pollution Damage	X	X
Intervention Convention 1969 and Protocol 1973: International Convention Relating to Intervention on the High Seas in Cases of Oil Pollution Casualties	X	X
LC 1972; 1996: Convention on the Prevention of Marine Pollution by Dumping of Wastes and Other Matter	X	X
MARPOL 73/78: International Convention for the Prevention of Pollution from Ships, 1973, as modified by the Protocol of 1978 relating thereto	X	N/A
· Annex IV (Sewage), optional	X	X
· Annex V (Garbage), optional	X	X
· MARPOL IOPP: International Oil Pollution Prevention Certificate	X	X
· MARPOL IAPP: International Air Pollution Prevention Certificate	X	X
OPRC 1990: International Convention on Oil Pollution Preparedness, Response and Cooperation	X	X

Construction	SOLAS REGULATIONS: 1. Safety regulations for different ship types 2. The new SOLAS regulation II-1/3-10. Goal-based construction standards for new ships—IMO *http://www.imo.org/blast/mainframe.asp?topic_id = 1017*		OSV Guidelines, 2007: Guidelines for the design and construction of offshore supply vessels MODU Code Code for the construction and equipment of mobile offshore drilling units	
Claims Liabilities	LLMC 1976 Convention and 1996 Protocol on Limitation of Liability for Maritime Claims (LLMC), 1976	X		
	HNS 1996: International Convention on Liability and Compensation for Damage in Connection with the Carriage of Hazardous and Noxious Substances by Sea			
	HNS 2000: Protocol on Preparedness, Response and Cooperation to Pollution Incidents by Hazardous and Noxious Substances		(LHNS Guidelines) Guidelines for the Transport and Handling of Limited Amounts of Hazardous and Noxious Liquid Substances in Bulk on Offshore Support Vessels (Resolution A.673(16))	X

Source: Courtesy of M.G. Burns, based on data from (IMO, 2013; and USCG, 2013. Available at http://eur-lex.europa.eu/LexUriServ/LexUriServ.do?uri=CELEX:52011PC0688:EN:NOT; ABS, 2013. Available at http://www.eagle.org [retrieved August 7, 2013]; MEPC, 2006. Marine Environment Protection Committee. Resolution MEPC.158(55). Adopted on October 13, 2006. Available at http://www.marinesafetyforum.org/upload-files/notices/amm-11.06-marpol-draft-resolution-673%2816%29.pdf [retrieved June 26, 2013]; and IMO, ISPS Code, 2013. Available at http://www.imo.org/blast/blastDataHelper.asp?data_id=22047&filename=266%2884%29.pdf [retrieved August 21, 2013]).

TABLE 11.3   Bridging Maritime, Oil, and Gas through US Regulations and Training

1	**Risk Management** 33 CFR PART 103: (3) Local public safety, crisis management and emergency response agencies; 6CFR27 Chemical Facility Anti-Terrorism Standards; 40 CFR 68: Chemical accident prevention provisions; 40 CFR 68 Subpart B: Hazard Assessment; 40 CFR 68 Subpart E: Emergency response; 40 CFR Subpart G: Risk Assessment Plan; 40 CFR 68.10(b)(1) *risk management* program; 40 CFR §68.190(b) environmental risk management plan, EPA; 12 CFR 225.175—*risk management* record keeping; 17CFR240.15c3-4-Internal risk management control systems.
2	**The Sea Survival Training** SWET (Shallow Water Egress Training)/HUET (Helicopter Underwater Egress Training). CFR Reference: API RP T-7, T-4, T-1, and USCG Title 33; Offshore Orientation (API RP 75 and API RP T-1); Confined Space Entrant/Attendant (OSHA 29 CFR 1910.146); Fall Protection (OSHA 29 CFR 1926.500); Hazmat (HM 126) (DOT 49 CFR Part 171); Hazardous Communications (OSHA 29 CFR 1910.1200); Respiratory Protection (OSHA 29 CFR 1910.134).
3	**OIM/Barge Supervisor and Ballast Control Operator (BCO)** 46 CFR 10 Licensing of marine personnel; 46 CFR 10.474—License for *ballast control operator*; 46 CFR 10.920—Subjects for MODU licenses. Regulations; 46 CFR 11.920. Examinations and Licensing; 46 CFR 11.474—Officer endorsements as *ballast control operator*; 46 CFR 15: Shipping; manning of barge supervisor or *ballast control operator*.
4	**Oil, Gas, and Sulfur Operations: Well Cap** 33 CFR: Navigation and Navigable Waters; Chapter I: Coast Guard, Department of Homeland Security; Subchapter O: Pollution; Part 151: Vessels Carrying Oil, Noxious Liquid Substances, Garbage, Municipal Or Commercial Waste, and Ballast Water, Subpart A: Implementation of MARPOL 73/78 and the Protocol on Environmental Protection to the Antarctic Treaty as it Pertains to Pollution from Ships: Oil Pollution; 29 CFR 151.10—Control of Oil Discharges; OSHA Industry Standard for Personal Protective Equipment (PPE), 29 CFR 1910.132(A), For the Failure to Provide and Use Flame-Resistant Clothing (FRC) in Oil and Gas Well Drilling, Servicing, and Production-Related Operations; CFR 30: Mineral Resources; 30 Chapter II: Minerals Management Service, Department of the Interior; Subchapter B: Offshore; Part 250: Oil and Gas and Sulfur Operations in the Outer Continental Shelf; Subpart P: Sulfur Operations; 31 250.1608—Well Casing and Cementing.
5	**ISPS according to IMO; 33** *CFR* **103: TITLE 33** Navigation and Navigable Waters, (3) Local public safety, crisis *management* and emergency response agencies; (a) The Area Maritime Security (AMS) Committee. 33CFR104: Maritime Security Vessels; 33CFR105: Maritime Security: Facilities; NVIC 11-02, "Security Guidelines for Facilities"; SOLAS XI-2; Marine Transportation Security Act of 2002 (MTSA) § 70102: Port Security; MTSA National Maritime Transportation Security Plan; § 70103: Mandates Security Plans; MTSA—Facility and Vessel Response Plans required—May be included in the Vessel/Facility Security Plan (Sec. 70103) § 70104: Transportation Security Incident; § 70105: Transportation Security Cards; MTSA § 70108: Foreign Port Assessments; MTSA § 70111: Enhanced Foreign Crew Identification; MTSA §70112: Established National Maritime Security Advisory Committee; Established Port Security Committees; MTSA § 70113: Maritime Intelligence Acceptable to USCG, Developed under ILO Standards.

(continued)

TABLE 11.3 (Continued) Bridging Maritime, Oil, and Gas through US Regulations and Training

6 **BOSIET (Basic Offshore Safety Induction and Emergency Training) (40 hours)**
This training meets the UK OPITO standards in line with the Offshore Petroleum Industry Training Organization.
These standards could be aligned with the US Regulatory framework as follows:
(1) Maritime and Offshore Regulatory Compliance all of the following;
(2) Safety Management; (1, 2) 29 CFR 1910 OSHA *Safety Management* of Highly Hazardous Chemicals; Explosives and Blasting Agents; 33 CFR 96.340—Navigation *Safety Management* Certificate; 40 CFR 68—Protection of Environment, EPA, Chemical Accident Prevention Provisions.
(3) Hazard Identification; (3) Carriage of dangerous goods competency, Table A-II/2 of the STCW Code, US DOT HMR; 29 CFR 1910; 33 CFR 155 Oil of HAZMAT Pollution; 40 CFR Hazardous Wastes; 49 CFR 100-185 HAZMAT Regulations.
(4) Safe Working Practice; (4) USCG STCW Section Ch. VI; Table VI/I; Section A-VI/1; A-VI/2; A-VI/3; A-Vi/4; 46 CFR 10.205—Validity of a merchant mariner credential; 46 CFR 11.202; Basic safety training or instruction.
(5) Incidents Control;
(6) Emergency Response; 33 CFR PART 103: (3) Local public safety, crisis *management* and emergency response agencies; 6CFR27 Chemical Facility Anti-Terrorism Standards; 40 CFR 68: Chemical Accident Prevention Provisions; 40 CFR 68 Subpart B: Hazard Assessment; 40 CFR 68 Subpart E: Emergency Response; 40 CFR Subpart G: Risk Assessment Plan;
40 CFR 68.10(b)(1) *risk management* program; 40 CFR §68.190(b) environmental risk management plan, EPA; 12 CFR 225.175—*risk management* record keeping; 17CFR240.15c3-4-Internal risk management control systems.
(7) Helicopter emergency, CFR Reference: API RP T-7, T-4, T-1, and USCG Title 33; Offshore Orientation (API RP 75 and API RP T-1); Confined Space Entrant/Attendant (OSHA 29 CFR 1910.146); Fall Protection (OSHA 29 CFR 1926.500); Hazmat (HM 126) (DOT 49 CFR Part 171); Hazardous Communications (OSHA 29 CFR 1910.1200); *Respiratory Protection (OSHA 29 CFR 1910.134).*

7 **DP (Dynamic Positioning)**
This training meets the Nautical Institute, UK standards, which could be aligned with the US Regulatory framework through the following US Codes of Federal Regulations:
33 CFR 143.207; 33 CFR 156.120; 46 CFR 4.05;
ARPA 33 CFR 164.38—*Automatic radar plotting aids (ARPA)*; 32 CFR PART 767: National Defense.
ECDIS 15 CFR Subpart A Certification, type-approved display system, such as an Electronic Chart Display and Information System (*ECDIS*);
33 CFR 83—Inland Navigation Rules codification. 33 CFR 164—Navigation Equipment; Electronic Chart Display and Information System (*ECDIS*); 33 CFR 164.33 nautical charts. 46 CFR 11.480 Radar Observer Recertification.
GMDSS 47 CFR Subpart W—Global Maritime Distress and Safety System (*GMDSS*); Telecommunication. 47 CFR 80 Subpart W. *GMDSS* equipment under FCC Regulation; 46 CFR 12.25-45; *GMDSS* At-Sea Maintainer; Code of Federal Regulations—Part 12: Certification of Seamen.

Source: Courtesy of M.G. Burns.

MMS makes sure that the processes from oil drilling to production, storage, and distribution are conducted safely and in line with the environmental regulations, based on the concepts of resource conservation. MMS furthermore provides right-of-use and support in the facilities development and maintenance, as well as rights-of-way for subsea umbilical and power cable systems, subsea pipelines, and other critical apparatus (MMS/USCG 2004, 2008).

The US Coast Guard (USCG)

The USCG, US Department of Homeland Security, is a significant safety regulator related to the US OCS. Both MMS and the USCG have mutual authority and duty to evaluate and endorse the structural design of nonship shaped floating platforms. Two memoranda of agreement have been drafted (MMS/USCG 2004, 2008) to determine their division of accountabilities for mobile offshore drilling units (MODU systems and subsystems), as well as fixed and floating offshore facilities. Their duties extend to root cause analysis and accident/incident investigations, civil penalty charges, oil spill planning, emergency preparedness, and emergency response (USCG 2013).

Table 11.3 recommends selected training courses to meet the regulatory requirements of the offshore industry in the United States and bridge its activities with the maritime industry.

REFERENCES

ABS. 2013. Available at http://www.eagle.org. Retrieved August 7, 2013.

AOGHS. 2013. American Oil and Gas Historical Society. Available at http://aoghs.org/offshore-2/offshore-oil-history/. Accessed August 1, 2013.

API. 2013. American Petroleum Institute. The State of American Energy. Report 2013. Available at http://www.api.org/~/media/Files/Policy/SOAE-2013/SOAE-Report-2013.pdf. Accessed August 12, 2013.

Barry, S., WalesOnline. 2013. Pembroke Port chosen as base for tidal stream turbine trials. WalesOnline, June 17, 2013. Available at http://www.walesonline.co.uk/business/business-news/tidal-energy-use—pembroke-4342444. Accessed August 8, 2013.

Bureau of Ocean Energy Management (BOEM). 2013. Available at http://www.boem.gov.

Bureau of Safety and Environmental Enforcement (BSEE). 2013. Available at http://www.bsee.gov.

Coast Guard (USCG). 2013. US Department of Homeland Security. Available at http://www.uscg.mil.

COQA. 2013. Crude Oil Quality Association. Available at http://www.coqa-inc.org/Simoncelli%20on%20LOOP.pdf. Accessed August 5, 2013.

Darlington, S. 2010. CNN. Brazil's richest man builds huge port. Available at http://www.cnn.com/2010/WORLD/americas/10/21/brazil.port/index.html. Accessed August 3, 2013.

EBX Group Brazil. 2013. EBX Group, Available at http://www.ebx.com.br. Accessed August 8, 2013.

EWEA. 2013. Available at http://www.ewea.org/fileadmin/files/library/publications/statistics/EWEA_OffshoreStats_July2013.pdf. Accessed August 6, 2013.

Hornbeck Offshore. 2013. Available at http://www.hornbeckoffshore.com. Accessed August 8, 2013.

IMO. 2013. Available at http://www.imo.org. Retrieved August 21, 2013.

IMO, ISPS Code. 2013. Available at http://www.imo.org/blast/blastDataHelper.asp?data_id=22047&filename=266%2884%29.pdf. Retrieved August 21, 2013.

LOOP Port. The Louisiana Offshore Oil Port (LOOP). Available at http://www.loopllc.com/Home. Accessed August 6, 2013.

Maersk. 2013. Available at http://www.maersk.com. Accessed August 21, 2013.

Marine Energy Pembrokeshire. 2013. Pembroke Port facilities. Available at http://www.marineenergypembrokeshire.co.uk/about/port-facilities/pembroke-port. Accessed August 6, 2013.

MEPC. 2006. Marine environment protection committee, MEPC. RESOLUTION MEPC.158(55). Adopted on October 13, 2006. Available at http://www.marinesafetyforum.org/upload-files//notices/amm-11.06-marpol-draft-resolution-673%2816%29.pdf. Retrieved June 26, 2013.

MMS/USCG 2004. MMS/USCG MOA: OCS-01 on mobile offshore drilling units, effective September 30, 2004. Available at http://www.mms.gov/MOU/PDFs/MOA-MMSUSCGOCS01-30September2004.pdf.

MMS/USCG 2008. MMS/USCG MOA: OCS-04 on floating offshore facilities, effective February 28, 2008. Available at http://www.uscg.mil/hq/cg5/cg522/cg5222/docs/mou/FLOATING_OFFSHORE_FACILITIES.pdf.

Nigerian Ports Authority. 2013. Port of Onne, Nigeria, 2013. Available at http://www.nigerianports.org/AboutUsOnnePort.aspx. Accessed August 4, 2013.

OGP Standards, Committee. 2010. Available at http://www.ogp.org.uk/pubs/426.pdf.

Paschoa, C., MarineLink. 2013. Açu Superport: A modern Port concept for Brazil. Available at http://www.marinelink.com/news/superport-concept-modern354879.aspx. Accessed August 5, 2013.

Port Fourchon. Available at http://www.portfourchon.com/explore.cfm/aboutus/portfacts/. Accessed August 3, 2013.

Port of Pembroke, Wales, UK. 2013. Available at http://pembroke.ports-guides.com/. Accessed August 12, 2013.

Port of Venice, Italy. 2013. Available at https://www.port.venice.it/en/the-offshore-terminal.html. Accessed August 4, 2013.

Ports of Louisiana. 2013. Available at http://portsoflouisiana.org/. Accessed August 2, 2013.

SBM Offshore. 2013. Turret Mooring Systems. Available at http://www.sbmoffshore.com/what-we-do/our-products/turret-mooring-systems/. Accessed August 2, 2013.

Sembcorp. 2009. Sembcorp Marine's Sembawang Shipyard and Kakinada Seaports sign joint venture to establish a marine and offshore facility in India's East Coast. Available at http://www.maritime-executive.com. Accessed August 4, 2013.

Tidal Energy Ltd, UK. 2013. Available at http://www.tidalenergyltd.com. Accessed August 6, 2013.

CHAPTER 12

The Future of Ports

To make sense of new and strange phenomena, one must be prepared
to play with ideas. And I use the word "play" advisedly:
Dignified people, without a whimsical streak,
almost never offer fresh insights in economics
or anywhere else.

Paul Krugman
Winner of the Nobel Prize in Economics

It is one of the main propositions of this book, that growth in the maritime industry is only achieved through an assertive stance, with the understanding that great opportunities and novel ideas entail above-average risk levels. In a highly competitive global market, low risk equals to business stagnation and lack of growth. The second principal contention is the necessity to keep abreast of the global market with a wholehearted commitment to go beyond the surface of maritime economics and market analysis, especially because the demand for ports and ships is a derived demand; that is, it is not a demand "per se," but is based upon the demand for the cargoes they handle and distribute—from extraction to production.

Hence, a political and socioeconomic perspective is necessary to foresee the future trends of port ownership, management, and operations, which encompasses a global view of politics, macro- and microeconomics, trade agreements, national treaties, allies, and competitors. From a commodity-centered viewpoint, an examination of the major global commodities, currencies, and so on, is necessary.

12.1 PORT DEVELOPMENT STRATEGY: ELEMENTS OF LONG-TERM STRATEGIC PLANNING

12.1.1 Strategic Port Planning

Strategic planning is to a port, what chart plotting is to a ship's master: while a ship's master will plot a course on a nautical chart taking into consideration the bearing or direction of the ship's forward course, distance, time, speed, and so on, the port manager will strategically plot the port's future direction and positioning in the global market, whereas appraising elements such as the following:

- The structural integrity of the port's corporate pyramid
- The element of time and growth rate
- The distance between the port and its global clients, or between the port's current position and its future goals

The contemporary business world is one of increasing chaos, complexity, and uncertainty, generated by accelerating change and more intense global competition. These are now permanent features of the global maritime landscape, necessitating rapid response and adaptation to ever-changing and unpredictable circumstances. Consequently, port managers and industry leaders are continually dealing with the issue of how to redesign their organizations to ensure that they remain competitive. While planning ahead, the element of time is crucial. Ports, just like other service-provider industries and manufacturers, typically produce multiple development plans, with different time frames, most usually:

- Quarterly plans
- Annual plans
- Three-year plans
- Five-year plans (initiated by China)
- Ten-/twenty-/thirty-year plans

Japanese Corporations seem to be the leaders in the Multiple-Centuries plans:

- A 300-Year Plan by Softbank Corp., led by CEO Masayoshi Son, Japan's second richest man (McCombs and Alpeyev 2012)
- A 500-Year Plan by Mitsubishi, the Japanese Conglomerate

Strategic port planning is an ongoing decision-making procedure that port managers should follow:

a. Develop a port's Vision and Mission Statement that will serve to (i) reestablish its market position, (ii) motivate its employees, and (iii) demonstrate to the market a climate of power, direction toward growth, and achievements.
b. Appraise the port's current market positioning and examine opportunities for future expansion.
c. Evaluate their current resources and estimate future resources, both tangible (i.e., assets, funds, and income obtained) and intangible (i.e., market reputation, niche, innovation, talent, well-established partnerships with valuable clients, alliance with logistics networks), and so on.
d. Set strategic goals, that is, the big picture: setting the standards for the port's direction in the foreseeable future, as well as medium and long term.
e. Determine the tactical path, that is, the way of achieving the strategic goals. This is achieved via gathering the means needed to accomplish the port's future strategy. This is achieved based on the port's current status and resources, as well as future opportunities and threats. Table 12.1 reveals the basic components of strategic port planning.

12.1.2 Tactical Port Planning

Tactical port planning is a set of comprehensive tools or processes that will zoom into the port's corporate structure and different departments, with the purpose of deciding upon the action needed and the resource allocation arrangement for the port to meet its strategic goals within a given period.

TABLE 12.1 The Components of Strategic Port Planning

1	Port Vision and Mission—Current versus Future
2	Market Positioning

- Current market positioning: people, place, product, price, promotion
- Future market positioning: people, place, product, price, promotion
- Establishing the port's competitive edge and market niche

3   Corporate Structure—Current versus Future

4   Resource Management and Planning

- Innovation and talent utilization
- Resource utilization—factors of production
- Financing and investment equity
- Budget allocation, monitoring and controlling
- Cash flow management

5   Competitor Profile Analysis

- Target markets, establishing market segments
- Customers' profile, demographics, geographic analysis
- Customers financial analysis, purchasing power, and SWOT analysis
- General: market demand/supply equilibrium
- Specific: customers' needs versus port supply

6   Customer Profile Analysis:

- Target markets, establishing market segments
- Customers' profile, demographics, geographic analysis
- Customers financial analysis, purchasing power, and SWOT analysis
- General: market demand/supply equilibrium
- Specific: customers' needs versus port supply

7   Strategic Market Analysis (Current): Global–Regional–National–State Level

- Economic analysis
- Commodities market analysis
- Shipping and transport market analysis

8   Strategic Market Forecasting (Future Projections): Global–Regional–National–State Level

- Economic analysis
- Commodities market analysis
- Shipping and transport market analysis

9   Marketing Plan: Public Relations, Marketing and Customer Service

12.1.3 Port Planning and the Factors of Production

More specifically, port planning focuses on the input/output ratio, that is, examines the optimum arrangement for the port's factors of production to generate the desired output:

- *Entrepreneurship*: The port's capital on leadership, strategy, innovation, and talent.
- *Labor*: This is an extension of entrepreneurship: successful ports do not have a stagnant workforce: they recruit labor with the potential of working their way up to the entrepreneurial level. This is achieved with sufficient training and

hands-on experience and is made possible through the recruitment of employees with inherent strategic, leadership, and teamwork qualities.

- *Land*: that is, the port's strategic geographic position, infrastructure, and accessibility within the logistics networks and the markets: the land in terms of commercial asset value and layout efficiency, the land's ability to meet the market's demand, land expansion through landfill operations, and so on.
- *Capital*: that is, the port's revenue, profitability, return on capital, cash flow available; potential for subsidies or investment obtained from the private or public sector, and so on; also, the port's superstructure, as an extension of investment.

The element of time is of paramount importance in the factors of production, as in the real business world, resource availability is nonlinear but seems to fluctuate subject to supply and demand, seasonality, political, or other factors.

12.2 FORECASTING THE MARKET

Market forecasting is a fundamental component of strategic planning and market analysis, as it uses past and current market trends, facts, and figures. Its aim is to project the future trends of a specific market segment or the overall status of a specific industry, for a given period.

The accuracy of forecasting is directly related to the quality and accuracy of data input.

12.2.1 Port Management and Forecasting Areas

Port managers can develop market projections in numerous areas, market segments, and for different reasons. Because of the integration of markets, objectives, facts, and figures, the forecasting areas may frequently overlap, or use similar metrics. Furthermore, from a supply/demand perspective, it is worth noting that the demand for ports is a derived demand based on the demand for ships, which in turn are dependent upon the demand for cargoes, and so on. A concise selection of forecasting areas is found in Figure 12.1.

An indicative list of the principal forecasting areas includes the following:

a. *Global economy forecasting*, which focuses on the value-for-money aspects. Hence, it is the value of the commodities rather than the volume of goods transported. The value of the major global commodities, that is, oil and gas products, precious metals, wheat, coffee, and other food products, as well as metals, minerals, and construction-related products, is illustrated in Table 12.2 and Figure 12.2 (energy commodities) and in Table 12.3 and Figure 12.3 (other liquid and dry commodities).

   The projected values of factors of production are estimated, as well as currency fluctuations and rates of exchange, funding and subsidy opportunities, inflation versus deflation, and labor cost. Also, socioeconomic factors are examined such as purchase power, unemployment, and so on.

b. *Global market forecasting*, with a focus on the commercial aspects such as the volume of goods transported, commodity supply and demand variables, trade routes, trade agreements, trade barriers, and so on. An evaluation of this sector

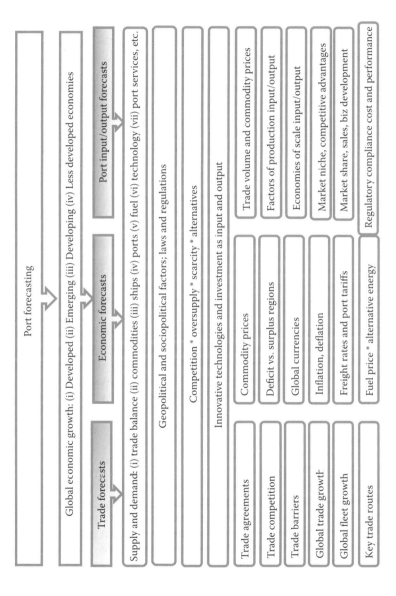

FIGURE 12.1   Elements of forecasting: an overview of the shipping industry. (Courtesy of M.G. Burns.)

TABLE 12.2   Commodity Types: Energy (1980–2012)

Commodity Types	Energy			
Trade	Liquid Bulk Trade			Dry Bulk Trade
Ship Types	Oil and Gas Carriers			Bulk Carriers
Year	Crude Oil ($/BBL)	Gas ($/mmbtu)	LNG ($/mmbtu)	Coal ($/mt)
1980	36.87	2.9	5.7	40.14
1981	35.48	3.29	6.03	53.62
1982	32.65	3.46	6.05	54.77
1983	29.66	3.32	5.55	38.19
1984	28.56	3.21	5.24	31.91
1985	27.18	3.08	5.23	32.91
1986	14.35	2.79	4.1	33.91
1987	18.15	2.13	3.35	34.91
1988	14.72	2.02	3.34	35.91
1989	17.84	1.89	3.28	36.91
1990	22.88	2.26	3.64	37.91
1991	19.37	2.3	3.99	38.91
1992	19.02	2.17	3.6	39.91
1993	16.84	2.4	3.51	40.91
1994	15.89	2.18	3.18	41.91
1995	17.18	2.23	3.45	42.91
1996	20.42	2.79	3.67	43.91
1997	19.17	2.61	3.91	44.91
1998	13.06	2.25	3.02	45.91
1999	18.07	2.2	3.14	46.91
2000	28.23	4.08	4.71	47.91
2001	24.35	4.01	4.63	34.2
2002	24.93	3.2	4.28	26.62
2003	28.9	4.7	4.73	30.04
2004	37.73	5.09	5.13	56.19
2005	53.39	7.62	5.99	48.3
2006	64.29	7.6	7.08	50.66
2007	71.12	7.77	7.68	64.05
2008	96.99	11.13	12.53	123.36
2009	61.76	6.33	8.94	65.31
2010	79.04	6.34	10.85	89.52
2011	104.01	7.26	14.66	116.41
2012	105.01	7.11	16.55	91.09

Source: World Trade Organization (WTO) Reports 1980–2012, Available at http://www.wto.org; International Monetary Fund (IMF) Primary Commodity Prices 1980–2012. Accessed at http://www.imf.org/external/np/res/commod/index.aspx; Spatafora, N. and Irina T. 2009. Commodity terms of trade: The History of booms and busts, IMF Working Paper 09205, September. 2009. Available at http://www.wto.org/english/res_e/publications_e/wtr10_forum_e/wtr10_13july10_e.htm.

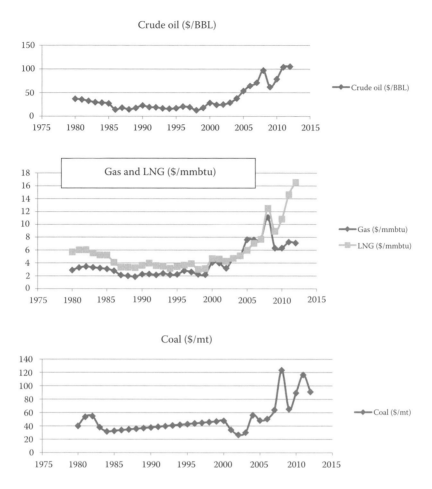

FIGURE 12.2 Energy commodities. (Courtesy of World Trade Organization [WTO] Reports 1980–2012, Available at http://www.wto.org; International Monetary Fund [IMF] Primary Commodity Prices 1980–2012. Accessed at http://www.imf.org/external/ np/res/commod/index.aspx; Spatafora, N. and Irina T. 2009. Commodity terms of trade: The History of booms and busts, IMF Working Paper 09205, September 2009. Available at http://www.wto.org/english/res_e/publications_e/wtr10_forum_e/wtr10_13july10 e.htm.)

will reveal for example the national production factors and countries with trade balance or trade imbalance. A focus on individual commodities may reveal their trade route patterns from raw materials extraction to their transportation and outsourcing for further processing, up to their final distribution to consumers.

A key characteristic of the post-2008 global market is that the emerging economies (BRIC, e.g., Brazil, Russia, India, China) seem to have a more rapid growth rate than the developed economies.

GDP is the total of gross value added by all national production plus taxation and less the subsidies not incorporated in the products' value. The GDP estimation excludes depreciation write-offs of manufactured goods or for natural resources exhaustion or deterioration (http://data.worldbank.org/indicator/ NY.GDP.MKTP.KD.ZG).

TABLE 12.3　Commodity Types: Liquid and Dry (1980–2012)

	Liquid Bulk	Dry Commodities—Foods		Dry Commodities—Nonfood						
	Vegetable Oils ($/mt)	Foods—Dry Bulk (cents/kg)	Foods—Dry Bulk ($/mt)	Nonfood ($/mt)	Nonfood ($/cubic meter)	Nonfood (cents/kg)	Metals and Minerals ($/mt)	Metals and Minerals (cents/kg)	Precious Metals ($/troy oz)	Iron Ore, CFR Spot ($/dry mt)
Year	Oils: Groundnut, Palm, Coconut, Palm kernel, Soybean	Barley, Meat—Chicken, Meat—Sheep, Cocoa, Coffee, Tea, Shrimps, Sugar	Copra, Groundnuts, Soybeans, Soybean meal, Fish Meal, Rice, Maize, Sorghum, Wheat, Bananas, Oranges	Tobacco, Woodpulp	Logs, Sawn-wood	Plywood, Rubber, Cotton	Phosphate rock, DAP, TSP, Urea, Potassium Chloride, Nickel, Aluminum, Copper	Lead, Tin, Zinc	Platinum, Gold, Silver	Iron Ore
1980	678.47	242.86	377.98	1406.20	309.85	580.93	1404.08	614.73	435.92	28.09
1981	672.55	211.52	374.79	1432.30	238.92	579.67	1207.82	524.38	305.92	28.09
1982	485.42	229.98	291.64	1525.94	220.59	593.81	985.44	470.54	237.34	32.50
1983	617.28	248.49	337.64	1541.85	208.87	598.27	1030.96	472.59	285.75	29.00
1984	906.02	257.81	325.01	1647.80	228.01	635.67	996.71	453.27	242.14	26.15
1985	641.92	222.78	264.71	1514.07	227.44	554.36	984.39	423.78	205.51	26.56
1986	366.33	239.94	248.44	1574.73	229.85	610.36	856.84	244.05	279.39	26.26
1987	404.71	220.65	260.87	1679.21	287.33	684.96	1089.54	268.68	336.40	25.30
1988	513.94	228.25	336.07	1603.78	291.50	659.46	2440.95	298.31	324.64	24.30
1989	518.38	221.12	319.56	1998.64	316.54	787.68	2332.38	362.22	298.86	27.83
1990	509.31	212.11	337.85	2103.35	396.68	810.61	1715.77	280.31	286.63	32.50
1991	530.17	202.00	335.60	2048.43	353.35	818.14	1548.92	242.32	247.35	34.76

1992	502.48	195.36	302.94	2001.25	382.67	812.92	1383.82	262.74	235.79	33.10
1993	511.88	211.72	297.33	1559.62	486.13	749.60	1101.41	217.65	246.03	29.09
1994	693.56	238.65	301.67	1597.06	486.28	762.63	1336.82	233.64	264.85	26.47
1995	728.46	245.75	338.89	1748.45	482.80	783.14	1709.43	262.53	271.23	28.38
1996	651.85	234.69	367.07	1814.75	447.54	775.29	1503.16	265.48	263.34	30.00
1997	685.93	272.04	389.63	2044.02	437.78	825.68	1430.86	252.90	243.64	30.15
1998	710.20	256.35	375.13	1922.28	364.83	743.97	1032.56	236.45	224.00	31.00
1999	616.43	230.72	316.36	1774.48	378.09	657.79	1188.69	232.75	220.31	27.59
2000	451.17	227.77	291.37	1820.25	387.28	675.44	1569.95	233.93	276.21	28.79
2001	389.22	220.65	315.61	1760.96	344.60	635.88	1187.05	194.88	268.30	30.03
2002	473.68	195.92	323.95	1598.39	353.15	601.43	1277.47	176.40	284.70	29.31
2003	633.28	207.23	348.97	1585.88	392.32	624.98	1682.35	207.92	353.37	31.95
2004	711.43	214.22	394.46	1690.50	424.21	670.22	2394.13	348.23	420.41	37.90
2005	654.28	220.66	428.28	1712.56	439.00	687.54	2648.61	324.58	449.56	65.00
2006	647.03	227.09	450.22	1833.90	482.53	753.25	4306.25	444.86	585.56	69.33
2007	964.22	242.38	542.24	2041.01	553.86	842.87	6042.29	678.64	671.68	122.99
2008	1338.28	275.92	702.08	2204.45	666.64	938.42	4236.79	749.18	820.40	155.99
2009	828.11	264.05	578.80	2424.90	565.78	931.54	2881.42	564.94	730.34	79.98
2010	1123.41	297.46	625.05	2600.01	591.92	1021.71	4142.85	823.83	951.52	145.86
2011	1558.25	343.43	742.95	2692.35	660.13	1104.36	4542.28	1021.61	1107.97	167.75
2012	1376.47	311.14	749.47	2532.60	611.89	1009.80	3698.15	838.03	1083.83	128.50

Source: World Trade Organization (WTO) Reports 1980–2012, Available at http://www.wto.org; International Monetary Fund (IMF) Primary Commodity Prices 1980–2012. Accessed at http://www.imf.org/external/np/res/commod/index.aspx; Spatafora, N. and Irina T. 2009. Commodity terms of trade: The History of booms and busts, IMF Working Paper 09205, September 2009. Available at http://www.wto.org/english/res_e/publications_e/wtr10_forum_e/wtr10_13july10_e.htm.

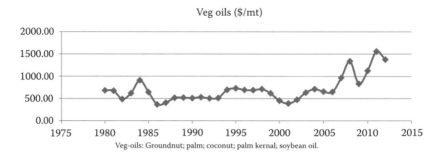

Veg-oils: Groundnut; palm; coconut; palm kernal; soybean oil.

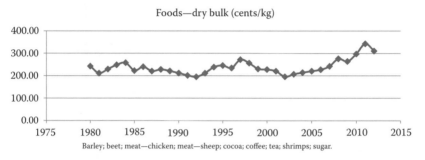

Barley; beet; meat—chicken; meat—sheep; cocoa; coffee; tea; shrimps; sugar.

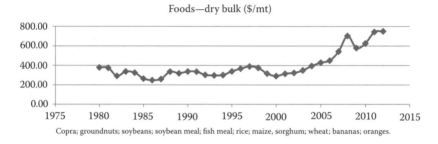

Copra; groundnuts; soybeans; soybean meal; fish meal; rice; maize, sorghum; wheat; bananas; oranges.

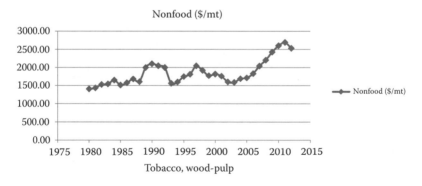

Tobacco, wood-pulp

FIGURE 12.3 Commodity types: liquid and dry (1980–2012). (Courtesy of World Trade Organization [WTO] Reports 1980–2012, Available at http://www.wto.org; International Monetary Fund [IMF] Primary Commodity Prices 1980–2012. Accessed at http://www.imf.org/external/np/res/commod/index.aspx; Spatafora, N. and Irina T. 2009. Commodity terms of trade: The History of booms and busts, IMF Working Paper 09205, September. 2009. Available at http://www.wto.org/english/res_e/publications_e/wtr10_forum_e/wtr10_13july10_e.htm.)

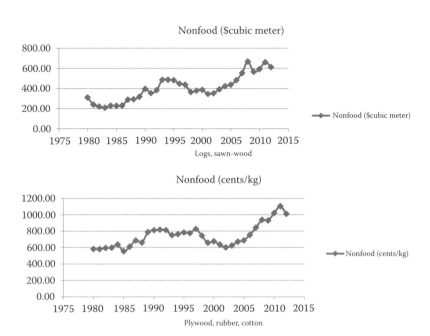

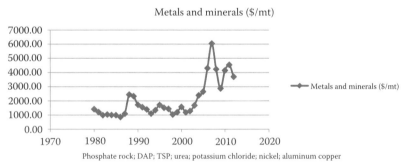

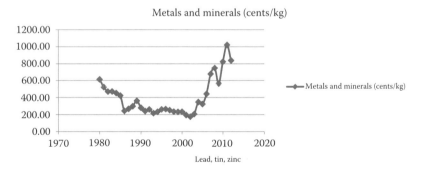

FIGURE 12.3 (Continued) Commodity types: liquid and dry (1980–2012). (Courtesy of World Trade Organization [WTO] Reports 1980–2012, Available at http://www.wto.org; International Monetary Fund [IMF] Primary Commodity Prices 1980–2012. Accessed at http://www.imf.org/external/np/res/commod/index.aspx; Spatafora, N. and Irina T. 2009. Commodity terms of trade: The History of booms and busts, IMF Working Paper 09205, September. 2009. Available at http://www.wto.org/english/res_e/publications_e/wtr10_forum_e/wtr10_13july10_e.htm.)

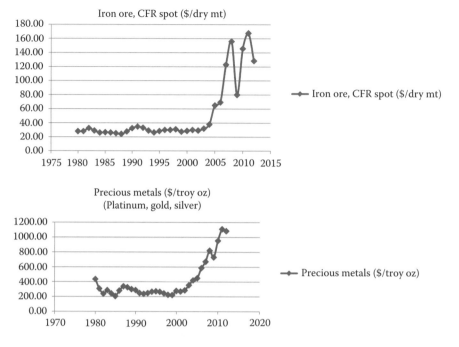

FIGURE 12.3   (Continued) Commodity types: liquid and dry (1980–2012). (Courtesy of World Trade Organization [WTO] Reports 1980–2012, Available at http://www.wto.org; International Monetary Fund [IMF] Primary Commodity Prices 1980–2012. Accessed at http://www.imf.org/external/np/res/commod/index.aspx; Spatafora, N. and Irina T. 2009. Commodity terms of trade: The History of booms and busts, IMF Working Paper 09205, September. 2009. Available at http://www.wto.org/english/res_e/publications_e/wtr10_forum_e/wtr10_13july10_e.htm.)

    c. *Supply networks forecasts*, which seek to verify whether proper commodities are found at the proper geographical markets within the proper time. These estimations enable trade and transport companies to satisfy market demand and also adopt a "just in time," "lean and agile" system.

    d. *Regulatory forecasts and incident and emergency forecasts.* These exceed the HSQE forecasts and expand to numerous significant emergency-prone areas and accidents, as well as the areas of occupational health, safety, security, quality, the environment, and so on.

       • The safety element includes "act of God," that is, extreme weather phenomena, asset's structural integrity, equipment failure, and so on

       • Security includes sea piracy and terrorism

    e. *Port traffic and utilization forecasts*, including port traffic, cargo volumes, berth occupancy, land usage, port infrastructure, and superstructure utilization.

    f. *Port performance/terminal and berth performance/departmental performance.*

    g. *Geopolitical/sociopolitical forecasts*, which may include the sociopolitical climate, social unrest, strikes, political instability, turmoil, and any such disruptions that may affect the port's continuity of business, as well as input and output in a specific geographical area. Critical events such as the ones that occurred in Egypt, Syria, and Libya, to name a few, apart from the human and social concerns raised, are also associated with production and transportation disruptions, which directly affect ports, ships, industries, and economies.

The closure of the Suez Canal, for example, has previously occurred five times, one of which lasted for 8 years. Similar events raise the need for larger ships (e.g., ultra large crude carriers and very large crude carriers). Also, when such events affect oil- and gas-producing countries, the price of energy goes up. This is good news for the energy and tanker markets, but an additional financial burden for the rest of the industry. More shipowners will turn to dual-energy consumption, that is, fuel and liquefied natural gas (LNG). Hence, more LNG port facilities will be available to accommodate the increased demand, and this will be the trend for numerous years ahead.

h. *Innovation and technology forecasts*, which may encompass the potential need to invest in new technology, the possibility of existing technology becoming obsolete, productivity fluctuations owing to technology, and so on. Port innovation is typically the result of the following:
  i. The need to grow through differentiation
  ii. The need to handle a larger volume of cargoes
  iii. The need to comply with environmental, safety, security, or other regulations and laws (e.g., ballast water management; fuel emissions and low sulfur limits, etc.)
i. *Competition forecasts*, which encompass business intelligence aspects related to the following:
  • *Substitute ports*, potential loss or gain of business owing to the performance of substitute ports
  • *Alternative transportation modes*, that is, rail, trucks, pipelines, airplanes, and so on
j. *Business development, sales and market share forecasts*, which focus on the port's business growth and are based on all the other forecasting elements explained in this section.

These forecasting areas may be distinguished into measurable, postmeasurable, and nonmeasurable:

A. *Measurable forecasting values*, that is, in terms of money, time, volume and size, input, and so on. For example, commodities value and volume, currency value and rates of exchange, port traffic, national GDP, national trade balance, products consumption, production, imports, and exports, to name but a few. Typically, these measurements reflect the present condition and are incorporated into econometric formulas, many of which are included in this book.
B. *Postmeasurable forecasting values*, which include factors that can only be measured upon the completion of a certain event. For example:
  • Political turmoil
  • Strikes
  • Acts of God and extreme weather phenomena
  • Emergency events such as environmental pollution, safety, or security threats
    Typically, these events cannot be measured in the present, that is, as the emergency is still ongoing, but can be assessed in the aftermath, that is, when losses have been verified (loss of life, injury, asset and cargo damage, commercial losses, etc.).

C. *Nonmeasurable forecasting values*, which reflect factors such as changes in the political regime, port strategy, workers' absenteeism and its effects upon input, and so on. Also, when numerous changes occur simultaneously, it may be difficult to measure the individual contribution within the cause and effect or input and output metrics.

Since these events are unforeseen, it is only through a complex thinking pattern or the infamous "market feeling" that entrepreneurs may be capable of observing the imminent market changes. In fact, this "sixth sense" is a hot topic in modern marketing and entrepreneurial schools.

The 2008 Global Crisis is a radical example of how an unexpected event of considerable impact may not be traced by the industry's "measurable" methods of forecasting. As can be seen by the 2007 and 2008 market reports available on the Internet, the vast majority of economists, entrepreneurs, and forecasters considered 2008 to be a promising year in terms of prosperity and global growth. And yet, there were a few entrepreneurs whose strategic actions reveal that they were prepared for this global collapse. Some of these attributed their success to the special "market feeling."

Many entrepreneurs and pioneers take pride of their "sixth sense" or the special skills to accurately forecast unforeseen circumstances and assess nonmeasurable values, among them, Steve Jobs (CEO of Apple), Albert Einstein (Nobel Prize in Physics), Isaac Asimov, Alexis Carell (Nobel Prize in Medicine), and other executives. In fact, it is this ability that can ensure corporate longevity and safeguard the port or the company from sudden and unanticipated changes of political, monetary, or other nature.

12.2.2 The Risk Element in Forecasting

Far better it is to dare mighty things, to win glorious triumphs,
even though checkered by failure, than to take rank with those poor spirits
who neither enjoy much nor suffer much, because they live in the gray twilight
that knows neither victory nor defeat.

Theodore Roosevelt
26th President of the United States

The risk element in forecasting is multidimensional, and yet it pertains to the ports' strategies:

- First, it is the commercial and marketing risk: inaccurate forecasting will mislead the port's marketing and business development efforts by approaching the wrong clients in the wrong market segments.
- Second, it is the investment and financial risk pertaining to budget allocation; market diversification; market entry or exit; commercial, financial, and operational priorities; and so on. Table 12.4 clearly shows a port's strategic choices and the associated risk levels.

TABLE 12.4 Port Strategy and Risk Levels

		Resources	
		Existing resources	**Investment in new resources**
Markets	Existing markets	Increasing market share	Service and capacity proliferation
		Low risk	Medium risk
	Entry in new markets	Business development	Market diversification
		Medium risk	High risk

12.2.3 Forecasting Methods and Tools

In addition to the forecasting areas, another dimension to be examined is the wide variety of forecasting tools and methods, many of which are used to develop different computer software programs currently available in the industry.

The principal forecasting methods include the following:

a. *Quantitative versus qualitative forecasts*

While quantitative forecasting methods are based upon accurate, measurable and proven quantitative data, and information being available, qualitative methods are based on market experience.

b. *Naïve forecasting method*

This is a simple and cost-effective method based on which all other, more refined and advanced forecasting methods are developed. It assumes that any future projection is equal to the historical average.

c. *Time-series methods*

These methods typically employ historical facts and figures within an econometric model in order to assess prospective trends within different times into the future. Some popular time-series methods include autoregressive moving average model variations, weighted moving average, exponential smoothing, growth curve, and linear and extrapolation forecasting techniques.

In terms of domain or focal point, time-series forecasting is classified into the following:

i. *Time domain analysis*, where the variable is typically calculated in terms of time. This analysis is useful in examining among others port activity, commercial patterns, and market fluctuations, in terms of time. Line charts are frequently used to demonstrate these trends.

ii. *Frequency domain analysis*, where the variable is usually estimated in terms of frequency, for example, spectral analysis.

Other forecasting methods include the following:

d. *Causal forecasting*, whose focal point is to identify the underlying cause that will trigger changes in the economy, trade patterns, labor, and so on. Its techniques are usually variations of regression and auto-regression analytical models.

e. *Software, IT forecasting.*

f. *Judgmental forecasting methods*, based on probabilistic models, statistical analysis, alternative scenario analysis, the Delphi method of experts, and so on.

As previously seen in Figure 12.3, there are numerous critical areas of port management where forecasting is requested, and numerous forecasting econometric formulas have already been covered in this book. The wide variety of tools and methods available over time have produced an endless combination of econometric formulas. Among these, the gravity model is examined in this section.

The Gravity Model of Trade

Variations of the gravity model are among the most prevailing forecasting formulas that are being widely used in the maritime industry. The gravity model is based on Newton's Law of Universal Gravitation (1665), based on Johannes Kepler's "Third Law" (1609). Its application in global trade was initiated by Jan Tinbergen (1962), with the following possibilities:

- Demonstrating commercial and financial prospects among two or more seaports
- Revealing friction and incoherence within a transport segment, sea-trade route, or among two ports
- Suggesting which factors of commercial attraction are of major or minor significance

The trade gravity model can be used to evaluate the trade potential among two countries or two ports:

$$F_{ij} = G \frac{M_i M_j}{D_{ij}} \tag{12.1}$$

where F_{ij} is trade flow among two countries; M is the economic mass of each country, estimated by national GDP over a given year; G is a constant; and D is the distance among two countries.

This equation demonstrates that a nation's GDP, that is, its economic mass, will enable the financially strongest nation to achieve a trade surplus, that is, positive balance of trade, where its exports exceed its imports, whereas the weakest economy will suffer from a trade deficit, that is, a negative balance of trade. The stronger economy typically exerts increased purchasing and bargaining power over the other nation. Hence, it can import cheaper goods and force its products to be exported at favorable prices and larger quantities.

The gravitational-trade force among the two nations is directly proportional to the nations' GDP and inversely proportional to the distance among the two nations.

12.3 LEADING THE WAY

Finish the Race Without Ever Shrinking Back
Πέρας ἐπιτέλει μὴ ἀποδειλιῶν

Delphic Epigram, Ancient Greece

In the era of globalization, ports have assumed a powerful role in the supply chain management and logistics networks: they operate in their own space, which are *geographical* by their physical position; *strategic* by their political, economic, and natural resources; and *business oriented* by their profitable commercial activities and service to the community.

Ports have the power to determine the outcome of groundbreaking trends and events, such as the following:

1. *The "Triple-E" mentality* for *energy efficiency*, *cost efficiency*, and *environmental integrity* will be the industry's guide for prosperity.
2. The *rising cost of petroleum energy* and amplified environmental regulations such as for low sulfur, "single source of emissions," ballast water systems, and so on, will increase the need for green technologies.
3. An increased demand for ship designs with dual-fuel burning, that is, both LNG and IFO, will spur an increased shipbuilding activity and more ports with LNG bunkering facilities.
4. Heavy port investment aims to reshuffle the global market share and define new trade routes. Intense port competition will lead to the distinction of the distinguished "mega-ports" and numerous "feeder ports." The risk of overinvestment is present.
5. The global production shifting to emerging economies. Asian overpopulation will seek for balance between a labor-active population, unemployment rates, and other demographic challenges.
6. The new energy-production regions such as the United States, Latin America, Africa, the Mediterranean basin, Baltic Sea and North Atlantic, Australia, and so on, will initiate a series of intercontinental trade agreements and changes in labor wages in search of cost-effectiveness and competitive advantages.
7. New trade routes and a stronger economic bond will be developed within the American continent, America, and the oil-rich African nations. New trade agreements will be established between China and Europe, as well as Asia and the Middle East.
8. A global currency crisis and the possibility of an experimental monetary system to break out in 2014/2015, with the possibility of the bursting of global market bubbles.
9. The expansion of Panama Canal by 2015 will facilitate NAFTA and other transcontinental trade and transport. Other developments in the continent's infrastructure such as the Guatemala Corridor and the Nicaragua Canal are expected.
10. The United States establishing its leadership as a global oil and gas exporter in 2016.
11. The opening of the Arctic Circle by 2016, and the empowerment of the North Atlantic and Far Eastern ports. New Trade Agreements will now encompass this new trade route.
12. Africa's economic and trade role will be heightened with extremely positive economic prospects. In particular, (a) the oil-rich Gulf of Guinea in West Africa (Nigeria, Ghana, Ivory Coast) will be developed as a global energy center, and (b) Tanzania's mega-port will promote regional growth.

"Survival of the fittest" is a theory formulated by Darwin (1869) and Spencer (1864) two centuries ago, to express the struggle for survival, in which only the ones that are best adapted to existing conditions are able to survive and expand. This book was designed to serve as a sourcebook to maritime, logistics, energy, and commodity professionals who wish to survive, expand, and remain competitive in the global marketplace.

REFERENCES

Darwin, C. 1869. *On the Origin of Species by Means of Natural Selection: Or the Preservation of Favoured Races in the Struggle for Life.* D. Appleton and Company, 1st American Edition 1860. 432 p.

IMF. 2013. http://www.imf.org/external/pubs/ft/weo/2013/update/02/. Retrieved in August 27, 2013.

International Monetary Fund (IMF) Primary Commodity Prices 1980–2012. Accessed at http://www.imf.org/external/np/res/commod/index.aspx.

McCombs Dave and Alpeyev Pavel, Bloomberg. 2012. Softbank Founder has 300-year plan in pursuit of sprint Nextel. http://www.bloomberg.com/news/2012-10-11/softbank-founder-has-300-year-plan-in-pursuit-of-sprint-nextel.html. Retrieved in January 12, 2013.

Spatafora, N. and Irina T. 2009. Commodity terms of trade: The History of booms and busts, IMF Working Paper 09205, September. 2009. Available at http://www.wto.org/english/res_e/publications_e/wtr10_forum_e/wtr10_13july10_e.htm.

Spencer, H. 1864. *Principles of Biology.* Williams and Norgate Publishers, London. 365 p.

Tinbergen, J. 1962. Shaping the World Economy: Suggestions for an International Economic Policy. New York: Twentieth Century Fund. The first use of a gravity model to analyze international trade flows.

World Bank. 2013. http://data.worldbank.org/indicator/NY.GDP.MKTP.KD.ZG. Retrieved in August 3, 2013.

World Trade Organization (WTO) Reports 1980–2012. Available at http://www.wto.org.

Index